Body size in mammalian paleobiology

Body size in mammalian paleobiology

Estimation and biological implications

Edited by

JOHN DAMUTH
University of California, Santa Barbara

BRUCE J. MACFADDEN
Florida Museum of Natural History

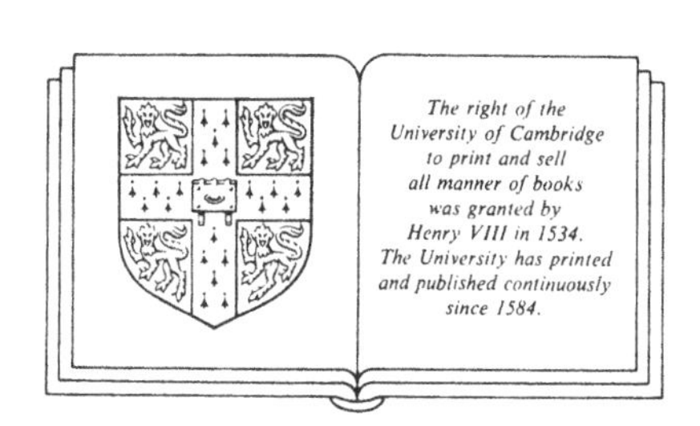

CAMBRIDGE UNIVERSITY PRESS
Cambridge
New York Port Chester Melbourne Sydney

CAMBRIDGE UNIVERSITY PRESS
Cambridge, New York, Melbourne, Madrid, Cape Town, Singapore, São Paulo

Cambridge University Press
The Edinburgh Building, Cambridge CB2 2RU, UK

Published in the United States of America by Cambridge University Press, New York

www.cambridge.org
Information on this title: www.cambridge.org/9780521360999

First published 1990
This digitally printed first paperback version 2005

A catalogue record for this publication is available from the British Library

Library of Congress Cataloguing in Publication data

Body size in mammalian paleobiology : estimation and biological
implications / edited by John Damuth, Bruce J. MacFadden.
p. cm.
ISBN 0-521-36099-4
1. Mammals – Size – Congresses. 2. Adaptation (Biology) –
Congresses. I. Damuth, John Douglas, 1952– . II. MacFadden,
Bruce J.
QL739.B63 1990
599 – dc20 87–71308
CIP

ISBN-13 978-0-521-36099-9 hardback
ISBN-10 0-521-36099-4 hardback

ISBN-13 978-0-521-01933-0 paperback
ISBN-10 0-521-01933-8 paperback

The excerpt on page 49, from *The Time Machine,* by H. G. Wells,
appears with the permission of A. P. Watt Ltd. on behalf of the
Literary Executors of the Estate of H. G. Wells.

Contents

Preface

Recent years have seen an increased interest among neontologists in the importance of body size as a major influence on an animal's adaptation to its environment and its place in the community. Along with this, there has been a surge of research by paleobiologists that uses estimates of body size for fossil species, particularly Cenozoic mammals.

This book is the result of a workshop held in Gainesville, Florida, at the University of Florida in April 1987. The purpose of the workshop was to bring together people with mutual interests in using body size as a tool in studies of mammalian evolution and paleobiology. We felt that much complementary research was being done by different investigators and that it was time to try to assess the relative reliability and comparability of body mass estimation techniques for fossil mammals, and to do so in the context of some of the applications of body mass estimates. We restricted our attention to the mammals for two reasons: Mammalian body size and related topics are currently highly active areas of rescarch in both neontology and paleontology; and body mass estimation in other important fossil vertebrate groups, which have no (or only a few) extant members, necessarily requires different kinds of approaches both to estimation and to assessment of the outcome. In addition to review chapters, we sought primary research results, because we wanted this book to serve both as a practical introduction to the subject and as an indicator of where the field is heading.

John Damuth is grateful to the Smithsonian Institution for supporting him as a Visiting Scientist under the auspices of the Evolution of Terrestrial Ecosystems Program, Department of Paleobiology, 1986–8.

We thank the Florida Museum of Natural History at the University of Florida for allowing us to host the Body Size Workshop there in April 1987.

We also thank Peter-John Leone of Cambridge for his encouragement during the development of this book.

Santa Barbara, California — *John Damuth*
Gainesville, Florida — *Bruce J. MacFadden*

Contributors

John Damuth
Department of Biological Sciences
University of California
Santa Barbara, California

John F. Eisenberg
Florida Museum of Natural History
University of Florida
Gainesville, Florida

Mikael Fortelius
Department of Geology
University of Helsinki
Finland

Theodore I. Grand
Department of Zoological Research
National Zoological Park
Washington, DC

Richard C. Hulbert, Jr.
Florida Museum of Natural History
University of Florida
Gainesville, Florida

Christine M. Janis
Division of Biology and Medicine
Brown University
Providence, Rhode Island

William L. Jungers
Department of Anatomical Sciences
School of Medicine
SUNY at Stony Brook
Stony Brook, New York

Bruce J. MacFadden
Florida Museum of Natural History
University of Florida
Gainesville, Florida

Brian K. McNab
Department of Zoology
University of Florida
Gainesville, Florida

Virginia C. Maiorana
Department of Ecology and Evolution
University of Chicago
Chicago, Illinois

Robert A. Martin
Department of Biology
Berry College
Mt. Berry, Georgia

V. Louise Roth
Zoology Department
Duke University
Durham, North Carolina

Christopher Ruff
Department of Cell Biology and Anatomy
Johns Hopkins University School of Medicine
Baltimore, Maryland

Kathleen M. Scott
Department of Biological Sciences
Rutgers University
Piscataway, New Jersey

Blaire Van Valkenburgh
Department of Biology
University of California
Los Angeles, California

Body size in mammalian paleobiology

1

Introduction: body size and its estimation

JOHN DAMUTH AND BRUCE J. MACFADDEN

Body size in biology and paleobiology

In recent years there has been growing interest in the biological implications of body size in animals. Body mass is correlated with a host of metabolic and physiological variables; with ecologically relevant characteristics such as life history traits, diet, population density, population growth rate, home range size, and behavioral adaptations; and with larger-scale patterns in community structure and biogeography (Brown & Maurer 1989; Calder 1984; Damuth 1981, 1987; Eisenberg 1981, this volume; Emmons, Gautier-Hion, & Dubost 1983; Fleming 1973; Jarman 1974; McMahon & Bonner 1983; McNab, this volume; Peters 1983; Schmidt-Nielsen 1984).

At the same time, there has been increasing use of body size in a wide range of applications in vertebrate paleobiology. Studies of functional morphology often employ measurements that must be related to body size before functional interpretations can be applied to fossil species (e.g., Emerson 1985; Gantt 1986; Kay 1975; Thomason 1985). Inferences concerning the metabolism and energetics of fossil vertebrates often depend critically on body mass estimates of fossil species, and body size has figured prominently in evolutionary explanations of the evolution of vertebrate metabolic physiology (McNab 1987; Thomas & Olson 1980; Tracy, Turner, & Huey 1986; Turner & Tracy 1986). The evolution of body proportions and scaling relationships themselves, and the resulting functional implications, form an important area of paleobiological research (Fortelius 1985, this volume; Gingerich & Smith

In *Body Size in Mammalian Paleobiology: Estimation and Biological Implications,* John Damuth and Bruce J. MacFadden, eds.

1985; Gould 1974, 1975; Jerison 1973; Kay 1975; Martin & Harvey 1985; Radinsky 1978; Scott 1985).

Body size plays a major role in studies of mammalian paleoecology. In fact, based upon modern knowledge, a mammal's body size may be the most useful *single* predictor of that species' adaptations. In addition to allowing reconstruction of trophic role and habitat preference, relationships between body size and ecological factors help us to make inferences about many more complex ecological characters of fossil mammals. For example, interpretations of life history characters and social behavior have played a part in explanations of the evolution of horns in ungulates (Janis 1982) and explanations of megafaunal susceptibility to extinction (Kiltie 1984; McDonald 1984). Faunal and community structure can be characterized in part by body size range and distribution, and has been used in studies of community evolution and energetics, and in the interpretation of climate and ancient vegetation (Andrews & Nesbit Evans 1979; Andrews, Lord, & Nesbit-Evans 1979; Collinson & Hooker 1987; Farlow 1976, 1987; Fleagle 1978; Janis 1982, 1984; Legendre 1986; Van Couvering 1980; Van Valkenburgh 1985, 1988).

The mammalian fossil record shows many examples of body size change within lineages, including the sometimes spectacular cases of gigantism and dwarfism on islands. Explanation of the island trends is still a challenge (Geist 1987; Hooijer 1967; Marshall & Corruccini 1979; Maiorana, this volume; Martin, this volume; Roth, this volume; Sondaar 1977, 1987; Thaler 1973).

Because body size is a character exhibited by every species, it potentially has an important role to play in studies of the tempo and mode of evolution (e.g., Gingerich 1976; Gould & Eldredge 1977; MacFadden 1987). Cope's rule – that size tends to increase within lineages or taxa – is one of the most widely discussed general macroevolutionary patterns (though until recently there were few studies that used actual body-mass estimates in quantitative studies of the phenomenon) (Gould 1988; MacFadden 1987; Newell, 1949; Rensch, 1959; Stanley 1973). At least the Pleistocene and Cretaceous mass extinction events have been notably size-selective among terrestrial vertebrates, affecting the larger species more strongly (Martin & Klein 1984; Padian & Clemens 1985). It has been suggested that frequent "megafaunal" extinctions or periods of size reduction may be characteristic of fossil vertebrate faunas, and may happen more locally and on a different time scale than the extinctions of the proposed long-term (i.e., 26 million years [myr]) global cycle

(Bakker 1977; Prothero 1985; Raup & Sepkoski 1984; Webb 1984). The presumed different extinction probabilities (and other ecological differences) exhibited by large and small vertebrates have implications for explaining the sorting of species in mammalian clades, but as yet this has not received much attention (but see Martin, this volume; Stanley 1973).

Finally, it is clear that many taphonomic processes operate upon vertebrate remains in a size-selective or size-dependent way (Behrensmeyer & Hill 1980; Behrensmeyer, Western & Dechant Boaz 1979; Voorhies 1969). Knowledge of body size can thus be a useful tool in the interpretation of mammalian fossil assemblages (Badgley 1986; Damuth 1982; Wolfe 1975).

Rationale and organization of this book

It may seem from the preceding sketch that body mass, having already been so widely used, must not be a problematic character to infer for fossil species. Certainly, size would appear to be one of the most straightforward characters that a fossil could exhibit. But the picture of satisfying accuracy and general compatibility that this widespread use suggests is misleading. We cannot measure body mass directly for fossil species, and must derive estimates from skeletal remains that are usually fragmentary and incomplete. Some taxa are represented by better fossil material than are others, and some have no living representatives. Some groups, such as the primates, have received far more detailed attention and have been analyzed by more sophisticated techniques than have others (e.g., Jungers 1985, this volume; Ruff, this volume). Estimates for species of different taxa typically derive from different regressions or other estimator techniques, and are based on different body parts. Studies vary considerably in the degree of precision they demand of body mass estimates. Some require only that relative masses be correct, some require only species averages, and some require accurate estimates for particular fossil specimens.

Workers have tended to develop and employ estimation techniques suited primarily to the study at hand, and intended to yield the level of accuracy that a particular study requires. Often, a skeletal element well represented among the fossil species is measured in a group of extant relatives and regressed upon body mass; a high correlation coefficient is taken as a sign of success and the regression equation is used to estimate fossil body masses. Too little attention has been paid in the past to the choice of the reference sample of modern species, and to the range of probable error in the resulting estimates. As is shown by some of the contributions in this volume, regressions involving characters commonly used to estimate body masses of fossil species do not always produce what would be satisfactory results, even when reapplied to the extant species upon which the regression is based.

We sought several different kinds of chapters for this book. Some contributors work primarily with the modern fauna, investigating the significance of size. Others are here primarily concerned with the techniques of estimating mass in fossil mammals accurately. We decided to include this range of topics, so that there would be in one volume a compendium of techniques and basic practical information, and a source of ideas (including caveats) and signposts to the literature about the various applications of body size in paleobiological studies. The tables of Chapter 16 (Appendix) combine in one place, for the reader's convenience, the regression equations discussed in the preceding chapters. We have also included here additional regression equations provided by some authors, which are based on characters that may not be as reliable as those that the authors have judged to yield the "best" estimates of body mass for a particular group. Nevertheless, they may be of some value in cases where the "best" characters are not preserved for the fossils in question. We warn the reader not to use the equations of Chapter 16 uncritically, and also not to bypass the reading of the relevant chapter.

General conclusions

The general theme that emerged at the Gainesville workshop, and that runs through book, is that *body mass estimation and functional morphological interpretation are not separable.* The reason is that, in using data from modern species to derive estimation equations or other means of estimation, one must choose a set of modern species that exhibit a similar relationship between body parts and body mass. This requires identification of broad functional/morphological groupings, which may or (at least as often) may not fall within traditional taxonomic lines (Damuth; Fortelius; Janis; Van Valkenburgh, this volume). As Grand (this volume) reminds us, "body mass" is a composite character, whose components differ in animals pursuing different modes of life. The different ways the same anatomical element may be related to overall mass in different species can to some extent be predicted from functional considerations.

An analogous statement can be made for interpretations: Dividing the modern fauna into functional groups improves predictive power of body mass for both physiological and ecological variables (Eisenberg; McNab, this volume). In particular, diet is related to much of the variation found in both kinds of variables.

Practical conclusions and caveats

A number of general practical conclusions, recommendations, and cautions concerning body mass estimation in fossil mammals can be abstracted from the conference proceedings and the papers published here:

1. Estimates based upon certain limb measurements, if available, appear to be substantially more reliable than those based on cranial or dental measurements. However, much more comparative work needs to be done on the scaling of postcranial elements in modern forms before we can use limb measurements for most groups. Proximal limb elements are more reliable correlates of body mass than are distal elements in ungulates and primates (Ruff; Scott, this volume). The problem with using fossil limb elements, however, is that they first must be taxonomically assignable.
2. Because of their identifiability and preservability, teeth will continue to be of importance in body mass estimation. However, dental measures alone, even when restricted to functionally similar groups, may not be accurate enough for all purposes. Percent standard errors below 30 are rare in the dental regressions reported here.
3. In dealing with taxa with high ontogenetic dental variation (e.g., high-crowned horses), the worker must carefully consider which wear stage(s) should be used to predict body mass. Such variation differs among taxa and among different dental measures and among different teeth within the same species.
4. For ungulates, tooth length measurements are generally better predictors than widths or areas, as the latter vary more with diet.
5. Techniques using more than one morphological variable (e.g., multiple regression) can increase accuracy of predictions when feasible and appropriate. Particularly, combinations of dental and postcranial measures, including body length, can result in relatively accurate estimates for well-represented species (percent standard error < 30).

6. As in any other statistical prediction procedure, it is unwise to extrapolate regression predictions beyond the range of the available data for modern forms. There is no guarantee that relationships holding within this range will be applicable to smaller or larger forms (e.g., MacFadden & Hulbert, this volume). In such cases use of a more general regression (such as "all ungulate" or "all mammal" regressions), with the resulting loss of precision, may be preferable. This consideration adds particular uncertainty to our estimates of the very largest fossil species, which may lie beyond the size range of any living mammal; our sample of living megafauna is also highly restricted and composed of species exhibiting only a few body forms.
7. The use of general regression equations, even those with small standard errors, does not guarantee accuracy for every species. In particular, care should be taken to recognize fossil species that may be aberrant in one or more characters that for most species might yield good estimates (e.g., the enlarged third molars of suids, or the enlarged second molars of amynodonts).
8. Estimates remain only estimates. There are certainly numerous unrecognized sources of error for fossil species. The statistical errors reported here for regressions on living forms are underestimates of the actual inaccuracy in estimates for extinct species. To the extent that fossil species deviate from the average of the modern population used, inaccuracy will increase.

Future research directions

Some directions for future research in mammalian body-mass estimation are also evident. The need for broader investigation of the scaling of postcrania has already been mentioned. Craniodental and postcranial scaling relationships analogous to those presented here are unknown for many extant mammalian groups. These include most rodent taxa, insectivores, small marsupials, and various "oddball" groups such as the edentates. We need further refinement of functional groups among all of the mammals, as this could dramatically increase the accuracy of our predictions. Finally, there are extinct groups, such as the North American Merycoidodontidae (oreodonts), that exhibit what appear to be unique body proportions and craniodental relationships, but for which we have complete material for some members. Using the well-represented species, reliable body-mass estimates may be obtainable, and from these species correction factors could be derived for use on the more fragmentary material representing the other members of the group.

In summary, we are just beginning to develop the empirical base and the analytical tools necessary for reliable reconstruction of the body masses of fossil species. Further refinement of body mass estimation techniques will make possible the reconstruction of a wide range of biological aspects of fossil species and faunas, and an exciting new dimension of time can be added to ecological and evolutionary studies that heretofore have had to rely primarily on patterns observed in the extant biota.

References

Andrews, P., & Nesbit Evans, E. 1979. The environment of *Ramapithecus* in Africa. *Paleobiology* 5:22–30.

Andrews, P., Lord, J. M., & Nesbit-Evans, E. M. 1979. Patterns of ecological diversity in fossil and modern mammalian faunas. *Biol. J. Linn. Soc.* *11*:177–205.

Bakker, R. T. 1977. Tetrapod mass extinctions – a model of the regulation of speciation rates and immigration by cycles of topographic diversity. In *Patterns of Evolution, As Illustrated by the Fossil Record,* ed. A. Hallam, pp. 439–468. Amsterdam: Elsevier.

Badgley, C. 1986. Taphonomy of mammalian fossil remains from Siwalik rocks of Pakistan. *Paleobiology 12*:119–142.

Behrensmeyer, A.K., & Hill, A. P. (eds.). 1980. *Fossils in the Making: Vertebrate Taphonomy and Paleoecology.* Chicago: Univ. of Chicago Press.

Behrensmeyer, A. K., Western, D., & Dechant Boaz, D. E. 1979. New perspectives in vertebrate paleoecology from a Recent bone assemblage. *Paleobiology* 5:12–21.

Brown, J. H., & Maurer, B. A. 1989. Macroecology: The division of food and space among species on continents. *Science 243*:1145–1150.

Calder, W. A., III 1984. *Size, Function, and Life History.* Cambridge, Mass.: Harvard University Press.

Collinson, M. E., & Hooker, J. J. 1987. Vegetational and mammalian faunal changes in the Early Tertiary of southern England. In *The Origins of Angiosperms and Their Biological Consequences,* ed. E. M. Friis, W. G. Chaloner, & P. R. Crane, pp. 259–304. Cambridge: Cambridge University Press.

Damuth, J. 1981. Population density and body size in mammals. *Nature 290*:699–700.

Damuth, J. 1982. Analysis of the preservation of community structure in assemblages of fossil mammals. *Paleobiology 8*:434–446.

Damuth, J. 1987. Interspecific allometry of population density in mammals and other animals: the independence of body mass and population energy-use. *Biol. J. Linn. Soc. 31*:193–246.

Eisenberg, J. F. 1981. *The Mammalian Radiations.* Chicago: University of Chicago Press.

Emerson, S. B. 1985. Jumping and leaping. In *Functional Vertebrate Morphology,* ed. M. Hildebrand, D. M. Bramble, K. F. Liem, & D. B. Wake, pp. 58–59. Cambridge, Mass.: Belknap Press.

Emmons, L. H., Gautier-Hion, A., & Dubost, G. 1983. Community structure of the frugivorous-folivorous forest mammals of Gabon. *J. Zool., Lond. 199*:209–222.

Farlow, J. O. 1976. A consideration of the trophic dynamics of a Late Cretaceous large-dinosaur community. *Ecology 57*:841–857.

Farlow, J. O. 1987. Speculations about the diet and digestive physiology of herbivorous dinosaurs. *Paleobiology 13*:60–72.

Fleagle, J. G. 1978. Size distributions in living and fossil primate faunas. *Paleobiology 4*:67–76.

Fleming, T. H. 1973. Numbers of mammal species in North and Central American forest communities. *Ecology 54*:555–563.

Fortelius, M. 1985. Ungulate cheek teeth: developmental, functional, and evolutionary interrelations. *Acta Zool. Fennica 180*:1–76.

Gantt, D. G. 1986. Enamel thickness and ultrastructure in hominoids: with reference to form, function, and phylogeny. In *Comparative Primate Biology,* Vol. 1: *Systematics, Evolution, and Anatomy,* ed. D. R. Swindler & J. Erwin, pp. 453–475. New York: Alan R. Liss.

Geist, V. 1987. On speciation in Ice Age mammals, with special reference to cervids and caprids. *Can. J. Zool. 65*:1067–1084.

Gingerich, P. D. 1976. Paleontology and phylogeny: patterns of evolution at the species level in Early Tertiary mammals. *Am. J. Sci. 276*:1–28.

Gingerich, P. D., & Smith, B. H. 1985. Allometric scaling in the dentition of primates and insectivores. In *Size and Scaling in Primate Biology,* ed. W. L. Jungers, pp. 257–272. New York: Plenum Press.

Gould, S. J. 1974. The origin and function of "bizarre" structures: antler size and skull size in the "Irish elk," *Megaloceros giganteus. Evolution 28*: 191–220.

Gould, S. J. 1975. On the scaling of tooth size in mammals. *Am. Zool. 15*: 351–362.

Gould, S. J. 1988. Trends as change in variance: a new slant on progress and directionality in evolution. *J. Paleontol. 62*:319–329.

Gould, S. J., & Eldredge, N. 1977. Punctuated equilibria: The tempo and mode of evolution reconsidered. *Paleobiology 3*:115–151.

Hooijer, D. A. 1967. Indo-australian insular elephants. *Genetica 38*: 143–162.

Janis, C. 1982. Evolution of horns in ungulates: ecology and paleoecology. *Biol. Rev. 57*:261–318.

Janis, C. 1984. The significance of fossil ungulate communities as indicators of vegetation structure and climate. In *Fossils and Climate,* ed. P. J. Brenchley, pp. 85–104. New York: John Wiley.

Jarman, P. J. 1974. The social organization of antelopes in relation to their ecology. *Behaviour 58*:215–267.

Jerison, H. J. 1973. *Evolution of the Brain and Intelligence.* New York: Academic Press.

Jungers, W. L. (ed.). 1985. *Size and Scaling in Primate Biology.* New York: Plenum Press.

Kay, R. F. 1975. The functional adaptations of primate molar teeth. *Am. J. Phys. Anthropol. 43*:195–216.

Kiltie, R. A. 1984. Seasonality, gestation time, and large mammal extinctions. In *Quaternary Extinctions: A Prehistoric Revolution,* ed. P. S. Martin & R. G. Klein, pp. 299–314. Tucson: University of Arizona Press.

Legendre, S. 1986. Analysis of mammalian communities from the Late Eocene and Oligocene of southern France. *Palaeovertebrata 16*:191–212.

MacFadden, B. J. 1987. Fossil horses from "Eohippus" (*Hyracotherium*) to *Equus:* scaling, Cope's law, and the evolution of body size. *Paleobiology 12*:355–369.

McDonald, J. N. 1984. The reordered North American selection regime and Late Quaternary megafaunal extinctions. In *Quaternary Extinctions: A Prehistoric Revolution,* ed. P. S. Martin & R. G. Klein, pp. 404–439. Tucson: University of Arizona Press.

McMahon, T. A., & Bonner, J. T. 1983. *On Size and Life.* Scientific American Library. New York: W. H. Freeman.

McNab, B. K. 1978. The evolution of endothermy in the phylogeny of mammals. *Am. Nat. 112*:1–21.

Marshall, L. G., & Corruccini, R. S. 1979. Variability, evolutionary rates, and allometry in dwarfing lineages. *Paleobiology 4*:101–119.

Martin, R. D., & Harvey, P. H. 1985. Brain size allometry: ontogeny and phylogeny. In *Size and Scaling in Primate Biology,* ed. W. L. Jungers, pp. 147–173. New York: Plenum Press.

Martin, P. S., & Klein, R. G. 1984. *Quaternary Extinctions: A Prehistoric Revolution.* Tucson: University of Arizona Press.

Newell, N. D. 1949. Phyletic size increase – an important trend illustrated by fossil invertebrates. *Evolution 3*:103–124.

Padian, K., & Clemens, W. A. 1985. Terrestrial vertebrate diversity: episodes and insights. In *Phanerozoic Diversity Patterns: Profiles in Macroevolution,* ed. J. W. Valentine, pp. 41–96. Princeton: Princeton University Press.

Peters, R. H.. 1983. *The Ecological Implications of Body Size.* Cambridge: Cambridge University Press.

Prothero, D. R. 1985. North American mammalian diversity and Eocene–Oligocene extinctions. *Paleobiology 11*:389–405.

Radinsky, L. 1978. Evolution of brain size in carnivores and ungulates. *Am. Nat. 112*:815–831.

Raup, D. M., & Sepkoski, J. J., Jr. 1984. Periodicity of extinctions in the geologic past. *Proc. Natl. Acad. Sci. U.S.A. 81*:801–805.

Rensch, B. 1959. *Evolution Above the Species Level.* New York: Columbia University Press.

Schmidt-Nielsen, K. 1984. *Scaling: Why Is Animal Size So Important?* Cambridge: Cambridge University Press.

Scott, K. M. 1985. Allometric trends and locomotor adaptations in the Bovidae. *Bull. Am. Mus. Nat. Hist. 179*:197–288.

Sondaar, P. Y. 1977. Insularity and its effect on mammal evolution. In *Major*

Patterns of Vertebrate Evolution, ed. M. K. Hecht, P. C. Goody, & B. M. Hecht, pp. 671–707. New York: Plenum Press.

Sondaar, P. 1987. Pleistocene man and extinction of island endemics. *Mém. Soc. Géol. France [N.S.] 150*:159–165.

Stanley, S. M. 1973. An explanation for Cope's rule. *Evolution 27*:1–26.

Thaler, L. 1973. Nanisme et gigantisme insulaires. *La Recherche 37*:741–750.

Thomas, R. D. K., & Olson, E. C. 1980. *A Cold Look at the Warm-Blooded Dinosaurs.* Boulder, Colo.: Westview Press.

Thomason, J. J. 1985. Estimation of locomotory forces and stresses in the limb bones of Recent and extinct equids. *Paleobiology 11*:209–220.

Tracy, C. R., Turner, J. S., & Huey, R. B. 1986. A biophysical analysis of possible thermoregulatory adaptations in sailed pelycosaurs. In *The Ecology and Biology of Mammal-like Reptiles,* ed. N. H. Hotton, III, P. D. MacLean, J. J. Roth, & E. C. Roth, pp. 195–206. Washington, D.C.: Smithsonian Institution Press.

Turner, J. S., & Tracy, C. R. 1986. Body size, homeothermy and the control of heat exchange in mammal-like reptiles. In *The Ecology and Biology of Mammal-like Reptiles,* ed. N. H. Hotton, III, P. D. MacLean, J. J. Roth, & E. C. Roth, pp. 185–194. Washington, D.C.: Smithsonian Institution Press.

Van Couvering, J. A. H. 1980. Community evolution in East Africa during the Late Cenozoic. In *Fossils in the Making: Vertebrate Taphonomy and Paleoecology,* ed. A. K. Behrensmeyer & A. P. Hill, pp. 272–298. Chicago: University of Chicago Press.

Van Valkenburgh, B. 1985. Locomotor diversity within past and present guilds of large predatory mammals. *Paleobiology 11*:406–428.

Van Valkenburgh, B. 1988. Trophic diversity in past and present guilds of large predatory mammals. *Paleobiology 14*:155–173.

Voorhies, M. R. 1969. Taphonomy and population dynamics of an Early Pliocene vertebrate fauna, Knox County, Nebraska. *Univ. Wyoming, Contrib. Geol. Special Paper,* No. 1, 69 pp.

Webb, S. D. 1984. Ten million years of mammal extinctions in North America. In *Quaternary Extinctions: A Prehistoric Revolution,* ed. P. S. Martin & R. G. Klein, pp. 189–210. Tucson: University of Arizona Press.

Wolfe, R. G. 1975. Sampling and sample size in ecological analyses of fossil mammals. *Paleobiology 1*:195–204.

2

The physiological significance of body size

BRIAN K. MCNAB

All organisms are defined in terms of a distinctive combination of characters, including those associated with anatomy, physiology, behavior, and reproduction. Fundamental to the definition of a species is its body size, food habits, and thermal biology, which interact to form the basis for derived characters like its distribution. If such patterns can be specified in well-known living species, they can be applied to species about which our knowledge is unavoidably fragmentary – namely, to rare and extinct species. The view advocated here is that studies on the biology of living mammals have generated a series of size-related relationships for physiology and reproduction that can be applied (with care) to *ecologically similar* fossil species to suggest ecologically relevant aspects of their physiology and reproduction. As an example, Lillegraven, Thompson, McNab, and Patton (1987) use size-related aspects of physiology in a reconstruction of the origin of eutherian mammals.

The principal impediment to the application of these relationships is that they invariably are fitted to idiosyncratic sets of data derived from living species, and therefore contain much residual variation. More often than not, this variation is "swept under the rug" in the unfortunate belief that it represents an "error" term. In fact, a detailed analysis generally shows that much of this residual variation contains biologically important variation (see, for example, McNab 1988a). This variation is ignored at the peril of imprecision in predicting the characteristics of living and extinct species. Consequently, *the more that is known about*

In *Body Size in Mammalian Paleobiology: Estimation and Biological Implications,* John Damuth and Bruce J. MacFadden, eds.

the biology of fossil mammals, the more that can be inferred about the size-related aspects of their physiology, behavior, and reproduction.

Most size-related patterns seen in living species are complex; they are most robust when derived from a phylogenetically, morphologically, and ecologically uniform array of species. Some limitations in the ability to analyze size relationships, then, are potentially present if the pattern is derived from a biased sample of the living fauna, or if the fauna is a biased sample of those capable of existing. This difficulty is at the heart of the controversy as to whether dinosaurs were, or were not, endothermic (see Thomas & Olson 1980). Our judgments on this question are directly or indirectly based on an analysis of the energetics of mammals, which may not provide an adequate basis for judging the physiology of dinosaurs: No mammals are known to be true inertial homoiotherms, which may have been the case in many large dinosaurs (McNab & Auffenberg 1976; Reid 1987).

The plan of this article is (1) to summarize the information derived from extant species of mammals on size-related aspects of physiology and reproduction, and (2) to caution against the blind use of these relations, being mindful of the many factors other than size that contribute to physiology and reproduction.

The influence of body size on the physiology and reproduction of mammals

Many studies since the 1960s and 1970s have demonstrated the quantitative correlation of physiological functions with body size, which is usually defined in terms of body mass, expressed in either grams or kilograms. These studies have been summarized by Peters (1983), Calder (1984), and Schmidt-Nielsen (1984).

Because the interest here is in the application of various physiological functions to fossil mammals, I shall summarize those physiological functions that are most likely to have a consequence for the ecology, distribution, and evolution of mammals, the subjects of greatest interest to paleontologists. These functions will be grouped into the headings of (1) metabolism, (2) temperature regulation, (3) locomotion, (4) ingestion, (5) respiration, (6) water balance, (7) reproduction and growth, and (8) production. Some appropriate equations for these relations are summarized in Table 2.1. All equations for these relations have a power form, in part following the original suggestion of Huxley (1932) and in part owing to its mathematical simplicity; other mathematical models,

Table 2.1. *Selected scaling functions for physiological variables in mammals*

System and factor	Equations[a]	Reference
Metabolism		
Basal rate		
All mammals	M (J/h) = 86.9 $m^{0.69}$	Hayssen & Lacy 1985
All mammals	M (J/h) = 69.3 $m^{0.71}$	McNab 1988a
Marsupials	M (J/h) = 50.0 $m^{0.74}$	McNab 1988a
Eutherians	M (J/h) = 70.9 $m^{0.72}$	McNab 1988a
Field expenditures		
Marsupials	M (J/h) = 491.8 $m^{0.58}$	Nagy 1987
Eutherians	M (J/h) = 140.0 $m^{0.81}$	Nagy 1987
Temperature regulation		
Minimal basal rate for continuous endothermy	M (J/h) = 312.0 $m^{0.33}$	McNab 1983
Locomotion		
All terrestrial animals	v_{max} (km/hr) = 2.70 $m^{0.38}$	Bonner 1965
	M (J/h) = 0.026 $m^{0.68} \cdot v$ + 0.048 $m^{0.70}$	Taylor et al. 1982
Ingestion		
Carnivores	M (J/h) = 330.3 $m^{0.70}$	Farlow 1976
Herbivores	M (J/h) = 264.6 $m^{0.73}$	Farlow 1976
Herbivores	M (J/h) = 247.9 $m^{0.73}$	Nagy 1987
Respiration		
Frequency	f (1/min) = 322.1 $m^{-0.26}$	Stahl 1967
Water balance		
Water intake	Wi (ml/h) = 0.010 $m^{0.88}$	Adolph 1943
Evaporative water loss	EWL (ml/h) = 0.0043 $m^{0.86}$	Altman & Dittmer 1968
Urine:plasma concentration ratio	C_u/C_p = 24.7 $m^{-0.10}$	Calder & Braun 1983
Reproduction		
Gestation period	GP (d) = 12.0 $m^{0.24}$	Millar 1981
Neonatal mass	NM (g) = 0.045 $m^{0.89}$	Millar 1981
Litter size	l = 24.1 $m^{-0.30}$	Millar 1981
Growth		
Growth rate	GR (g/d) = 0.04 $m^{0.69}$	Millar 1977
Growth rate	GR (g/d) = 0.04 $m^{0.72}$	Case 1978
Growth constant	K (1/d) = 0.126 $m^{0.30}$	Zullinger et al. 1984
Production		
Population	P/Biomass (J/g/yr) = 24180 $m^{-0.27}$	Farlow 1976
Homoiotherms	r_{max} (1/d) = 0.040 $m^{-0.27}$	Fenchel 1974
Organisms	r_{max} (1/d) = 0.025 $m^{-0.26}$	Blueweiss et al. 1978

[a]Body mass (m) is in grams; velocity (v) is in km/h.

namely polynomials, are difficult to interpret biologically, even though they may be more "accurate" in describing the data. These relations are discussed here in relevant detail; some functions are included because they have an obvious application to paleontology, whereas others are included because of their importance to physiological ecology.

Metabolism

The function having the broadest implications probably is the association of rate of energy expenditure with body size. This relation was first described for basal rate of metabolism by Kleiber (1932), and by many others since then. The difficulty with any description of rate of metabolism as a function of body size in endotherms is that it complexly depends on ambient temperature, so that several scaling relationships can be described, each with its own implications and mathematical characteristics. In theory, the basal rate is the lowest rate of metabolism in an endotherm that is compatible with temperature regulation (for an exception, see McNab 1988b), and therefore represents a minimal expenditure of energy. The advantage of using the basal rate is that it is comparable in all species (McNab 1989), not that it directly estimates field energy expenditures. The most recent estimates of the scaling relation for basal rate in mammals have been made by Hayssen and Lacy (1985) and McNab (1988a). The difficulty with these scaling relations, and all others, is that they incorporate the influence of various factors other than body mass on basal rate (see last section), even though they are not explicitly acknowledged in the equations.

Most early descriptions of the scaling of basal rate dwelled on the power of the metabolism function. The power for total basal rate was usually thought to fall between 0.67 and 0.75. McMahon (1973) tried to derive 0.75 from physical principles, and Heusner (1982) argued that the "true" *intra*specific power is 0.67, *inter*specific powers (presumably) having no particular biological significance. Comparative biologists, however, are particularly interested in *inter*specific comparisons. Hayssen and Lacy (1985) found that the total basal rate was proportional to mass raised to the 0.69 power in a sample of 297 species, and McNab (1988a) found that it was proportional to mass raised to the 0.71 power in a sample of 321 species. A detailed analysis of this relation suggests that its power does not simply reflect the influence of body mass (McNab 1988a). As is to be expected, these scaling relations reflect the sample of species used to construct the relationship, so that caution should prevail when one concludes that a particular relation represents the "true" relation between two variables, here basal rate and body mass, independent of sample composition.

Although the basis of the relationship between basal rate and body mass is unclear, its consequences are obvious: Large mammals have higher total rates of metabolism than small species. The increase in rate associated with an increase in mass is not proportional to mass, but to mass raised to a power less than 1.0. Using the power 0.71, a doubling of mass leads to a 64% increase in basal rate, and a halving of mass leads to a 39% decrease in total basal rate. So, contrary to a widely held belief, even among some physiologists, an increase in body size requires an *increase* in the rate of energy expenditure, and in the amount of food and energy available in the environment. In other words, *an increase in mass does not reduce energy expenditure.* A mammal can respond to a shortage in available energy by a *reduction* in mass or by a reduction in rate of metabolism at a fixed mass.

More is probably known about the residual variation in the basal rate–mass function than for any other scaled relationship. At a given mass, basal rate varies by a factor of 3:1 to 10:1 (McNab 1988a). Much of this variation is related to food habits. Mammals that specialize on grass and on vertebrates tend to have basal rates that are two or more times those of species of the same body mass that specialize on fruits, the leaves of woody plants, and invertebrates (McNab 1986). The correlation of basal rate with food habits is accentuated at masses greater than a kilogram, presumably because energy and nutrient demand, as noted, increase with mass. So, as mass increases, the differential liabilities associated with the various foods, such as seasonal availability, digestibility, and toxicity, are increasingly encountered by the consuming mammal.

Additional factors that contribute to the residual variation around the basal metabolism–body mass curve are activity level (and muscle mass), temperature regulation, and climate. For example, fruit and leaf eaters have especially low basal rates if these mammals have arboreal habits (McNab 1986). This association occurs because arboreal species, at least at masses greater than a kilogram, are often characterized by small muscle masses. Basal rate is proportional to the fraction of total mass that is muscle (McNab 1978b). Small mammals that go into daily or seasonal torpor have lower basal rates than similarly sized species that do not enter torpor (McNab 1983, 1988a). Species living in cold climates tend to have higher basal rates than those living in warm climates, but this correlation may partially reflect the distribution of foods with respect to climate.

The principal difficulty with using a basal-rate function to estimate rates of energy expenditure in living or fossil mammals is that most mammals spend little time in a standard state. Basal rates may be a convenient index to mammalian energy expenditures, but they do not represent those expenditures directly. Recent improvements in the ability to measure directly energy expenditures in the field, especially by means of doubly labeled water, permitted the development of scaling relations for field expenditures in mammals (see Nagy [1987] for a summary). These relations differ from basal-rate functions in both intercept and power, indicating that eutherian field expenditures are approximately double those of basal rate. The difficulty with scaling field estimates is that they incorporate the multiplicity of factors operating at the time the measurements were made, including season, activity level, reproduction, etc. This makes field measurements very difficult to interpret, as their scaling functions surely do not reflect the influence of body mass alone. At present, the best analysis of energy expenditure derives from combining field measurements with laboratory measurements of rate of metabolism at ecologically relevant temperatures (McNab 1989).

Temperature regulation

Although temperature regulation is often considered in qualitative terms (i.e., a particular species regulates body temperature or it does not), McNab (1983) demonstrated that a unique relationship exists between the basal rate of metabolism and body mass in those species characterized by continuous endothermic temperature regulation. In this relation, total basal rate is proportional to mass raised (approximately) to the 0.33 power. All mammals having basal rates that fall below this relation enter daily torpor, at least upon occasion. One consequence of this relation is that all small marsupials enter torpor in association with their low rates of metabolism. From the viewpoint of paleontology, this relation may be difficult to apply, unless one can establish that some factor sets basal rate in a species so low that it falls below this scaling relation and, therefore, that the species is likely to enter daily torpor.

Locomotion

All mammals move from one place to another and search for food and mates. Some have specialized behaviors, such as digging shelters and long-distance migration. Each of these activities requires an expenditure of energy. Extensive measurements in mammals have demonstrated that the cost of terrestrial locomotion increases with body mass and the velocity of movement (see Taylor, Heglund, & Maloiy [1982] for a summary). Body mass in this relation is raised (approximately) to the 0.69 power. Because large mammals can run faster than small mammals (Bonner 1965), they may save time by moving a distance rapidly, but they do not conserve energy by doing so.

Ingestion

The intake of food can be measured directly, or estimated from the energy and nutrient requirements of species, when the assimilation efficiency is known. Farlow (1976), summarizing food intake for carnivorous and herbivorous mammals, showed that total intake is proportional to mass raised to the 0.70 and 0.73 powers, respectively. These equations are similar to those proposed by Nagy (1987) for field expenditures. They are approximately four times the basal rate; that is, assimilation efficiency is about 50%.

Respiration

Relatively little use has been made in paleontology of the correlations of respiration with body mass. One use was in explaining the evolution of the secondary palate in the phylogenies of the mammals and the bauriamorph therapsids. The development of the secondary palate was associated with a marked decrease in body mass, a shift from ectothermy to endothermy, and a consequent increase in respiratory frequency (McNab 1978a). Respiratory frequency is proportional to mass raised to the -0.26 power (Stahl 1967).

Water balance

Total water intake and evaporative water loss are proportional to mass raised to powers that range from 0.86 to 0.88, which means that large species require more water than small species and that small species need more water on a mass-specific basis. The capacity of mammals to produce a concentrated urine relative to the plasma, in contrast, scales proportionally to mass raised to the −0.10 power (Calder & Braun 1983). This function means that small species are more parsimonious with water than are large species, which may be necessary because of the greater rate of water turnover associated with a small mass.

Reproduction and growth

These processes are essential for survival of every species, and both vary with body mass. Reproduction bears a complex relationship to mass. Like other time functions (Lindstedt & Calder 1981), gestation period and the time from birth to sexual maturity increase with mass raised approximately to the 0.25 power, whereas neonate mass varies with maternal mass raised to the 0.89 power, and litter size is proportional to mass raised to the −0.30 power. Total growth rates (in grams per day, g/day) are proportional to mass raised to the 0.69 power (Millar 1977), or to the 0.72 power (Case 1978). Growth, measured in terms of a Gompertzian exponential-growth rate constant, is proportional to mass raised to the −0.30 power (Zullinger, Ricklefs, Redford, & Mace, 1984), which if converted to growth in grams per day, has a 0.70 power, a value similar to those described by Millar and Case.

Production

Net population production is proportional to population metabolism (Humphreys 1979; Lavigne 1982; McNeil & Lawton 1970), which means that production should scale proportionally to mass raised to the 0.75 power, which is the case (Farlow 1976; Lavigne 1982). As a consequence, production per unit biomass should be proportional to mass raised to the −0.25 (= $m^{0.75}/m^{1.00}$) power. In fact, Fenchel (1974) and Blueweiss et al. (1978) showed that the maximal growth constant of populations, r_{max}, which is the ultimate measure of effective production by a population, is estimated to be proportional to −0.27 and −0.26, respectively. That the concept of production can be applied to fossil faunas was demonstrated by Farlow (1976), who examined living mammalian communities to evaluate the possibility that large herbivorous dinosaurs had levels of production representative of endotherms.

Scaling basal rate in mammals: a cautionary tale

All mass functions, as noted, are calculated from data collected on a sample of living species. The powers and intercepts of these relations therefore vary as the species composition of the sample varies (McNab 1988a), a condition that has led to differing interpretations. For example, various authors (Glazier 1985; Hennemann 1983; McNab 1980; Schmitz & Lavigne 1984) have argued that much of the residual variation in the reproduction and growth of mammals is related to ecologically relevant variations in (basal) rate of metabolism. Others (Hayssen 1984; Robinson & Redford 1986) have denied this correlation. The correctness or error in this suggestion, however, is not as relevant here as is the realization that all scaling relationships are approximations at best and incorporate many factors other than body size. So these relations should be viewed as approximations, not as physical or biological "laws." The complexity of scaling relations can best be demonstrated in the basal rate of metabolism in mammals.

The factors setting basal rate in mammals have been under dispute recently, and this dispute has significance for the interpretation of scaling functions generally. McNab (1980, 1986) argued that food habits independent of taxonomic affiliation are an important factor influencing the basal rate of mammals, whereas Hayssen and Lacy (1985) and Elgar and Harvey (1987) maintained that most of the residual variation is related to taxonomic affiliation.

The comparative contributions of body mass, food habits, activity level, climate, and taxonomic affiliation to basal rate are difficult to separate because these factors are often correlated with one another. Some of these complexities have been explored by McNab (1988a). For example, if all 321 species of mammals for which reasonable estimates of basal rate exist are pooled, a curve similar to the Kleiber curve is obtained (Table 2.2). If, however, the curve is restricted to the 272 eutherians, a parallel, but slightly higher curve is obtained. These curves are similar because 85% of the species belonging to the inclusive curve are eutherians. In contrast, a curve consisting only of the 46 marsupials available is significantly lower than the eutherian curve. Many biologists, including MacMillen and Nelson (1969) and Dawson and Hulbert (1970), have concluded that marsupials have lower basal rates than eutherians. But that conclusion is more complicated than appears at first glance. All small marsupials go into torpor, but most small eutherians do not. Marsupials that do not enter torpor have a scaling relation for basal

Table 2.2. *Basal rate of metabolism of mammals*

Condition	Species (n)	Equations[a]
All mammals	321	$M(\mathrm{J/h}) = 69.3m^{0.713}$
Marsupials		
All	46	$M(\mathrm{J/h}) = 50.0m^{0.737}$
No torpor	32	$M(\mathrm{J/h}) = 67.0m^{0.698}$
Eutherians		
All	272	$M(\mathrm{J/h}) = 70.9m^{0.716}$
No torpor	212	$M(\mathrm{J/h}) = 76.7m^{0.711}$
No invertebrate-eating specialists	219	$M(\mathrm{J/h}) = 65.1m^{0.741}$

[a]Body mass (m) is in grams. All equations from McNab (1988a).

rate similar to that found in eutherians that do not enter torpor (Table 2.2). Thus, some of the difference in basal rate found between eutherians and marsupials is related to the differential occurrence of torpor in these two groups: The scaling relationship of basal rate to mass includes a correlation of torpor with body mass and a correlation of torpor with basal rate. Furthermore, if one eliminates all invertebrate-eating specialists from the eutherian curve, the remaining eutherians have a lower, but steeper, curve than do all eutherians. This change occurs because a disproportionate number of small, invertebrate-eating mammals at small size were excluded from the curve. That is, invertebrate-eating is related to body size and to basal rate. I conclude that all factors that are related to body mass and to basal rate are incorporated into the basal-rate–body-mass function, unless steps are taken explicitly to prevent the inclusion of these factors, which is unlikely to occur as long as the data are grouped by taxa. This conclusion applies to all scaling relations.

Summary

Paleontologists can greatly profit in the analysis of the biology of extinct faunas by using the extensive set of correlations that have been shown in living animals between various aspects of physiology and body mass. These so-called "scaling" functions limit the range of possible physiological configurations found in organisms. This range is further reduced if additional knowledge about the organisms is available, including food habits, activity level, climate, and phylogeny. The more information that is known about an extinct species, the more one is able to specify its physiology and behavior. Nevertheless, knowledge of body size alone is sufficient to reduce significantly the range of possible combinations of physiology and behavior found in a species.

Of course, the observations of paleontologists have potential value for physiological ecologists. Thus, observations on the body size and bone structure of dinosaurs have made major contributions to the debate on the physiology of dinosaurs (Reid 1987; Thomas & Olson 1980). Detailed descriptions of the reduction in body mass and of the concomitant evolution of the secondary palate in cynodont and bauriamorph therapsids have contributed to our understanding of the evolution of endothermy in the early phylogeny of mammals (McNab 1978a). These observations demonstrate that some physiological combinations used in the past are not used today, a conclusion that makes a significant contribution to our understanding of the potentiality for physiological adaptation to the environment. The morphological observations made by paleontologists that will contribute further to our understanding of the evolution of physiological function are difficult to predict. Without doubt, however, they will include the concept of body size.

References

Adolf, E. F. 1943. *Physiological Regulations*. Lancaster, Penn.: Cattell Press.

Altman, P. L., & Dittmer, D. S. 1968. *Metabolism*. Bethesda, Md.: Federation of American Societies of Experimental Biology.

Blueweiss, L., Fox, H., Kudzma, V., Nakashima, D., Peters, R., & Sams, S. 1978. Relationships between body size and some life history parameters. *Oecologia 37*:257–272.

Bonner, J. T. 1965. *Size and Cycle: An Essay on the Structure of Biology*. Princeton, N.J.: Princeton University Press.

Calder, W. A. 1984. *Size, Function and Life History*. Cambridge, Mass.: Harvard University Press.

Calder, W. A., & Braun, E. J. 1983. Scaling of osmotic regulation in mammals and birds. *Am. J. Physiol. 244*:R601–R606.

Case, T. J. 1978. On the evolution and adaptive significance of postnatal growth rates in the terrestrial vertebrates. *Q. Rev. Biol. 53*:243–282.

Dawson, T. J., & Hulbert, A. J. 1970. Standard metabolism, body temperature, and surface areas of Australian marsupials. *Am. J. Physiol. 218*:1233–1238.

Elgar, M. A., & Harvey, P. H. 1987. Basal metabolic rates in mammals: allometry, phylogeny and ecology. *Funct. Ecol. 1*:25–36.

Farlow, J. O. 1976. A consideration of the trophic dynamics of a Late Cretaceous large-dinosaur community (Oldman formation). *Ecology 57*:841–857.
Fenchel, T. 1974. Intrinsic rate of natural increase: the relationship with body size. *Oecologia 14*:317–326.
Glazier, D. S. 1985. Relationship between metabolic rate and energy expenditure for lactation in *Peromyscus. Comp. Biochem. Physiol. 80A*:587–590.
Hayssen, V. 1984. Basal metabolic rate and the intrinsic rate of increase: an empirical and theoretical reexamination. *Oecologia 64*:419–421.
Hayssen, V., & Lacy, R. C. 1985. Basal metabolic rates in mammals: taxonomic differences in the allometry of BMR and body mass. *Comp. Biochem. Physiol. 81A*:741–754.
Hennemann, W. W. 1983. Relationship among body mass, metabolic rate and the intrinsic rate of natural increase in mammals. *Oecologia 56*:104–108.
Heusner, A. A. 1982. Energy metabolism and body size. I. Is the 0.75 mass exponent of Kleiber's equation an artifact? *Respir. Physiol. 48*:1–12.
Humphreys, W. F. 1979. Production and respiration in animal populations. *J. Anim. Ecol. 48*:427–453.
Huxley, J. S. 1932. *Problems of Relative Growth*. London: Methuen.
Kleiber, M. 1932. Body size and metabolism. *Hilgardia 6*:315–353.
Lavigne, D. M. 1982. Similarity in energy budgets of animal populations. *J. Anim. Ecol. 51*:195–206.
Lillegraven, J. A., Thompson, S. D., McNab, B. K., & Patton, J. L. 1987. The origin of eutherian mammals. *Biol. J. Linn. Soc. 32*:281–336.
Lindstedt, S. L., & Calder, W. A. 1981. Body size, physiological time and longevity of homeothermic animals. *Q. Rev. Biol. 56*:1–16.
MacMillen, R. E., & Nelson, J. E. 1969. Bioenergetics and body size in dasyurid marsupials. *Am. J. Physiol. 217*:1246–1251.
McMahon, T. 1973. Size and shape in biology. *Science 179*:1201–1204.
McNab, B. K. 1978a. The evolution of endothermy in the phylogeny of mammals. *Am. Nat. 112*:1–21.
McNab, B. K. 1978b. Energetics of arboreal folivores: physiological problems and ecological consequences of feeding on an ubiquitous food supply. In *The Ecology of Arboreal Folivores*, ed. G. G. Montgomery, pp. 153–162. Washington, D.C.: Smithsonaian Institution Press.
McNab, B. K. 1980. Food habits, energetics, and the population biology of mammals. *Am. Nat. 116*:106–124.
McNab, B. K. 1983. Energetics, body size, and the limits to endothermy. *J. Zool., Lond. 199*:1–29.
McNab, B. K. 1986. The influence of food habits on the energetics of eutherian mammals. *Ecol. Monogr. 56*:1–19.
McNab, B. K. 1988a. Complications inherent in scaling basal rate of metabolism in mammals. *Q. Rev. Biol. 63*:25–54.
McNab, B. K. 1988b. Energy conservation in a tree-kangaroo (*Dendrolagus matschiei*) and the red panda (*Ailurus fulgens*). *Physiol. Zool. 61*:280–292.
McNab, B. K. 1989. Laboratory and field studies of the energy expenditure of endotherms: a comparison. *Trends Ecol. Evol. 4*:111–112.
McNab, B. K., & Auffenberg, W. 1976. The effect of large body size on the

temperature regulation of the Komodo dragon, *Varanus komodoensis*. *Comp. Biochem. Physiol.* *65A*:345–350.

McNeil, S., & Lawton, J. H. 1970. Annual production and respiration in animal populations. *Nature* *225*:472–474.

Millar, J. S. 1977. Adaptive features of mammalian reproduction. *Evolution* *31*:370–386.

Millar, J. S. 1981. Post partum reproductive characteristics of eutherian mammals. *Evolution* *35*:1149–1163.

Nagy, K. A. 1987. Field metabolic rate and food requirement scaling in mammals and birds. *Ecol. Monogr.* *57*:111–128.

Peters, R. H. 1983. *The Ecological Implications of Body Size.* Cambridge: Cambridge University Press.

Reid, R. E. H. 1987. Bone and dinosaurian "endothermy." *Modern Geol.* *11*:133–154.

Robinson, J. G., & Redford, K. H. 1986. Intrinsic rate of natural increase in Neotropical forest mammals: relationship to phylogeny and diet. *Oecologia* *68*:516–520.

Schmidt-Nielsen, K. 1984. *Scaling: Why Is Animal Size so Important?* Cambridge: Cambridge University Press.

Schmitz, O. J., & Lavigne, D. M. 1984. Intrinsic rate of increase, body size, and specific metabolic rate in marine mammals. *Oecologia* *62*:305–309.

Stahl, W. R. 1967. Scaling of respiratory variables in mammals. *J. Appl. Physiol.* *22*:453–460.

Taylor, C. R., Heglund, N. C., & Maloiy, G. M. O. 1982. Energetics and mechanics of terrestrial locomotion. I. Metabolic energy consumption as a function of speed and body size in birds and mammals. *J. Exp. Biol.* *97*:1–21.

Thomas, R. D. K., & Olson, E. C. 1980. *A Cold Look at the Warm-Blooded Dinosaurs.* AAAS, Selected Symp. No. 28.

Zullinger, E. M., Ricklefs, R. E., Redford, K. H., & Mace, G. M. 1984. Fitting sigmoidal equations to mammalian growth curves. *J. Mammal.* *65*:607–636.

3

The behavioral/ecological significance of body size in the Mammalia

JOHN F. EISENBERG

In 1984, Calder published the comprehensive review *Size, Function, and Life History,* which attempts to document the overwhelming importance of body size in determining the form and function of a species (individual) within an ecosystem. I recommend this book to anyone interested in problems of biological scaling.

Given a mammal (endothermic and sometimes homoiothermic), Brian McNab (this volume) has eloquently summarized the physiological consequences of becoming large or remaining small. There is an implied directionality in that statement which derives from the fact that I am addressing paleontologists. Most researchers will concede that in the evolutionary history of mammals, the primitive condition was a small body size; larger body size was often a derived character within any of the separate lineages. As McNab has pointed out, when various physiological and behavioral values are plotted against body size in a double common-log plot, it is possible to perform a regression analysis and specify the form of the curve in terms of an allometric equation. He has also pointed out, and I wish to emphasize, that the best-fit regression line often has a great many scattered points around it, and an explanation of the residual values can be extremely enlightening to a biologist. In fact, most of the scatter can be accounted for by considering either phylogenetic affinity or trophic adaptation. This will become evident as I proceed with my discussion.

In *Body Size in Mammalian Paleobiology: Estimation and Biological Implications,* John Damuth and Bruce J. MacFadden, eds.

Mammalian body size and niche

Some years ago, I began to analyze the size of mammals and study the distribution of size classes within certain niche categories. I defined "niche" according to the substrate used, such as aquatic, semiaquatic, terrestrial, arboreal, etc., and the trophic strategy as folivore, grazer, frugivore, etc. It was then possible to see the manner in which niches were distributed over different geographic areas of the globe (Eisenberg 1981). Predictably, certain niche categories were absent from the temperate zone and present in the tropics, and this further led me to concentrate on two tropical areas that were comparable in geographic position and size: Panama and Malaya. When the frequency of size classes were plotted for the various trophic categories, a remarkable similarity could be demonstrated. Of course, there were some differences: The megafauna of Malaya is still intact, but Late Pleistocene extinctions in Panama have eliminated some of the larger forms (Webb 1985). The distribution of size classes over trophic categories might even be more similar if recently extinct forms were included in the tabulations. This result suggests that specialization for a certain niche sets certain constraints on body size, and that Darlington's principle of complementarity is a reality (Eisenberg 1981; see Figure 3.1).

Since I am addressing paleontologists, another way of looking at the problem is to study communities from different parts of the globe through time. Here paleontological data become important. For example, when one looks at mammalian predatory guilds through time and compares relative body proportions and size, one can discern some remarkable convergences in community structure. This was demonstrated by Van Valkenburgh (1984, 1985) when the carnivore guild of the Serengeti was compared with a carnivore guild in North America during the Pliocene. In a similar fashion, Larry Martin (1990) has been able to demonstrate that even when carnivore assemblages become extinct, the replacement forms are remarkably similar to their predecessors in size and body proportion. This again implies that there are some general rules governing form and function as well as community structure and dynamics, and that, given an appropriate range of mammalian stocks, selection will inevitably tend to produce similar forms of secondary consumers within those communities that are comparable in vegetative cover, primary productivity, and primary consumers.

A similar case could be made for the replacement of mammalian primary consumers such as the extinctions at the end of the Eocene, leading to the loss of the Uintatheriidae and subsequent replacement by the Brontotheriidae in North America (Osborn 1929).

Body size, trophic strategy, and population dynamics

In an attempt to unravel some rules or patterns of mammalian community structure, my colleagues and I have published a series of papers concerning body size, diet, population biomass, and the rate of reproduction (Eisenberg 1980; Eisenberg, O'Connell, & August 1979; Kinnaird & Eisenberg in press; Robinson & Redford 1986a,b). We chose Neotropical communities for analysis because we had a great deal of direct experience in the area; and, furthermore, there was a pressing need to develop management plans for some species that required fundamental information concerning tropical community structure and function. I will attempt to summarize briefly several general rules that we uncovered. Much of the following information is included in Eisenberg et al. (1979) and Eisenberg (1980, 1981). In all communities, herbivores – in particular, browsers and grazers – collectively comprise the largest proportion of the mammalian biomass within a community. In short, primary consumers, or those feeding directly upon primary productivity, are the standing-crop biomass dominants. Secondary and tertiary consumers not only exist at much lower densities but contribute a smaller proportion to the total mammalian biomass in a community. These results are broadly consistent with the global analysis presented by Damuth (1987).

On the other hand, the smaller forms, although not comprising a significant portion of the standing-crop biomass, are often the key forms in terms of annual productivity and may account for the major consumption of the primary producers. Eisenberg et al. (1979) demonstrated that in the llanos of Venezuela the cane rat *Zygodontomys brevicauda* (at 35 g) was not a biomass dominant, but in terms of annual production had a very significant impact on the community.

A large portion of the standing-crop biomass of a closed-canopy, multistratal, tropical evergreen forest tends to be comprised of arboreal species as opposed to terrestrial, because of the reduced primary productivity on the shaded forest floor of a mature rain forest. As one proceeds through a series of communities showing reduced vegetational cover, a more open canopy, or an earlier form of plant growth (second growth forest), the proportion of community biomass contributed by terrestrial forms increases until, in an almost treeless open savanna, the

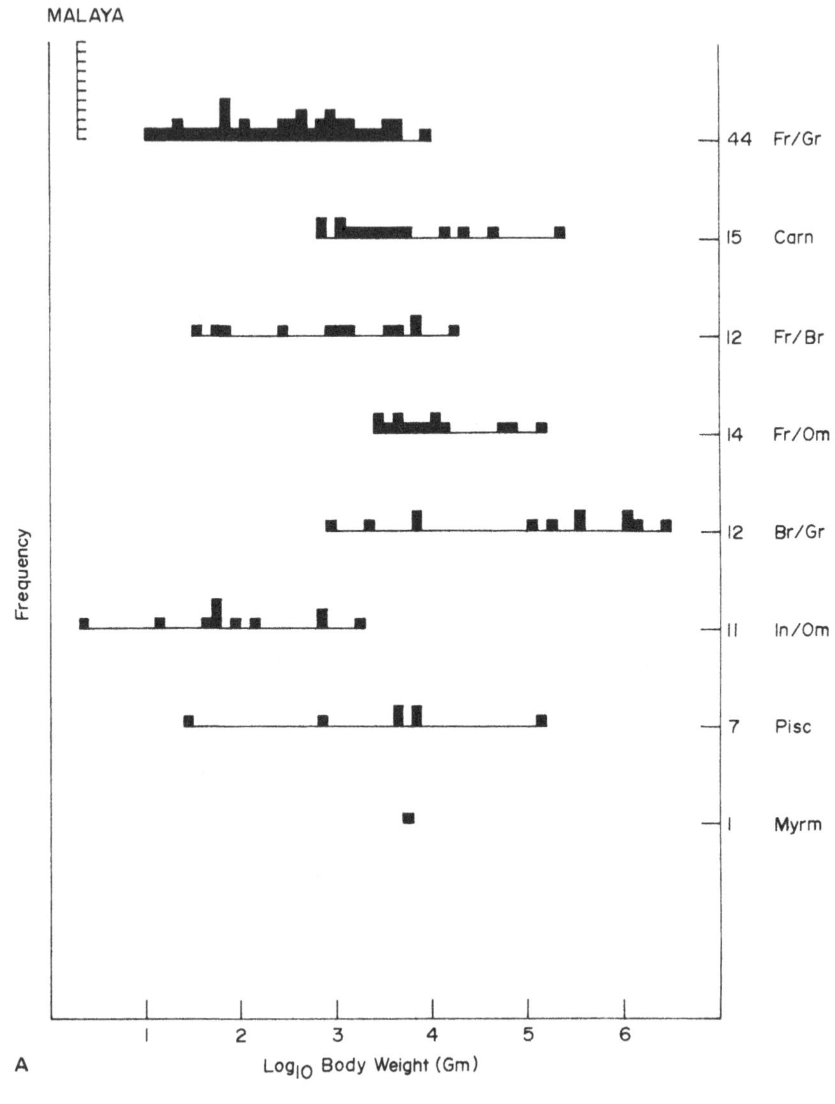

Figure 3.1. Frequency of size classes according to trophic strategy: Malaya (A) compared with Panama (B, opposite page). Log_{10} body weight (g) (abscissa) is plotted against frequency (ordinate). The number of genera within each trophic class is given to the right of each size series. Fr/Gr = frugivore/granivore; Fr/Om = frugivore/omnivore; In/Om = insectivore/omnivore; Fr/Br = frugivore/browser; Br/Gr = browser/grazer; Carn = carnivore; Pisc = piscivore; Myrm = myrmecophage. (Data in part based on Handley 1966 and Medway 1978.)

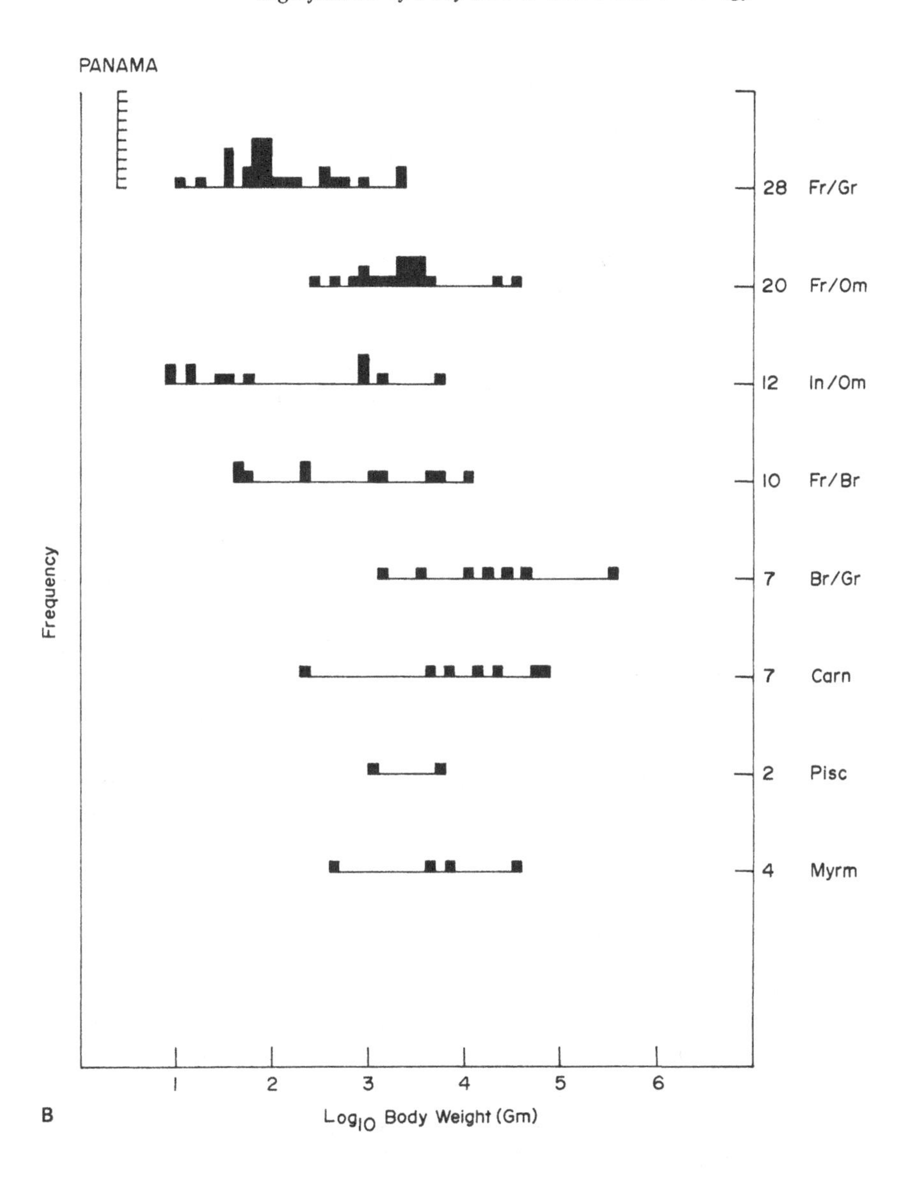

terrestrial components dominate completely over all other locomotor categories (scansorial or arboreal). This is intuitively obvious, but is nevertheless an important factor to bear in mind when trying to reconstruct extinct communities (Eisenberg and Seidensticker 1976).

The proportion of total standing-crop biomass contributed by different size classes bears the following relationship: If one plots average standing-crop biomass for each species against its average weight class, the relationship is positive. That is to say, larger forms contribute more to total standing-crop biomass than do smaller species (Kinnaird and Eisenberg in press). However, there is considerable scatter around the regression line, and in an attempt to account for the scatter, multiple regression analyses were performed. More than 70% of the variation can be accounted for by trophic strategy. For example, within any given size class, herbivores contribute more to standing-crop biomass than do secondary and tertiary consumers.

Population densities of a species are predictably related to their size and trophic specialization. Again, a plot of population densities against mean body size demonstrates a negative relationship. Smaller forms can exist at higher densities than larger ones. Again there is a great deal of scatter around the regression line. Almost all of this scatter can be accounted for when the trophic specializations are plotted separately. Herbivores within a given size class exist at higher densities than secondary or tertiary consumers (Robinson and Redford 1986a).

Although a species may be a biomass dominant, it does not necessarily follow that this species shows a high productivity value. Indeed, the intrinsic rate of population increase covaries negatively with body mass. This trend does not necessarily correlate with dietary specialization, and in fact, seems to be more closely tied to the phylogenetic history of the organism than any special adaptation for a particular diet (Robinson and Redford 1986b).

Body size, behavior, and life history

Body size has profound implications for how a mammal can locomote, whether it be on land, sea, or air. For example, there is a maximum limit on size if an animal becomes specialized for steady, powered flight rather than hovering flight. Such relationships have been studied not only for bats but of course for birds and insects (Pirlot 1977).

If we consider nonvolant mammals and terrestrial locomotion, then the following generalizations can be made: Maximum speed covaries positively with body size, even though step frequency covaries negatively with body size. That a large mammal can achieve a higher speed derives from the fact that the stride length is relatively greater as size increases (Calder 1984). Larger-bodied mammals have the capacity to use a larger area of space more efficiently because, although metabolic rate increases in all species as speed increases, the per-gram cost of rapid locomotion is higher for smaller than for larger forms (Taylor, Heglund, & Maloiy 1982). (This becomes apparent when metabolic rate during locomotion is plotted against speed, with large-bodied forms plotted separately from small-bodied ones [Calder 1984]: The slope is less for larger forms than for smaller forms.) It follows that larger forms can more efficiently exploit a larger home range; and, although the total energy consumed by a larger form is greater than that of a smaller one (and thus, ultimately, larger-bodied forms, whatever trophic specialization, need larger areas), it is a fact that home-range size increases with body size at a faster rate than do individual metabolic requirements (Harestad & Bunnell 1979; McNab 1963). Once again there is considerable scatter around the regression line which can in part be accounted for by trophic specializations. For any given size class, given comparable primary production values for the habitat, a herbivore has a smaller home range than a carnivore.

What a mammal can perceive often scales with body size. For example, in regard to sound production with vocal chords, larger-bodied forms produce a range of sounds having lower frequencies than those produced by smaller-bodied forms; and with respect to audition in terrestrial forms, the highest frequency perceived covaries negatively with body size (Calder 1984; Dooling 1980).

As indicated in the previous section, trophic strategy and body size are inextricably linked. For example, mammals feeding on small insects scattered widely in the forest floor litter are often of small body size. Those taxa specialized for feeding on social insects, such as ants or termites, are a special case. These "anteaters" may be of modest, to even large, body size. Granivores, mammals that harvest small seeds, are often of modest body size. Of course, man, who cultivates and processes cereal grains on a large scale, is an exception. Ruminants, or foregut fermenters, have a lower limit on the body size they can attain because of the necessity for ceasing feeding while fermentation takes place in the foregut, thus reducing the rate of passage for ingested herbaceous material (Janis 1976; Parra 1978). Constraints on niche occupancy as a function of body size have been extensively reviewed in Eisenberg (1981).

Body size profoundly affects life history strategies, a phenomenon expansively dealt with by Bonner in his 1965 book. For example, generation interval scales positively with body size, longevity scales positively with body size, and the reproductive strategy of a species is very much influenced by adult body size. Mammals that opt for a near semelparous mode of reproduction are almost always of modest body size and attain their adult growth within a year. On the other hand, those species showing extreme iteroparity, with a litter size of one and a long interbirth interval, are often of large body size (Eisenberg 1981). Adaptation for various foraging strategies has profound implications with respect to the form of social organization that a species exhibits (Eisenberg 1966; Jarman 1974).

In a previous publication (Eisenberg 1981), I attempted to arrive at some general correlations between trophic strategy and social structure. The argument is difficult to condense, but the rest of this section summarizes the major hypotheses. Throughout the evolution of the Mammalia there has been pressure from competitors (both intraspecific and interspecific), predators, pathogens, and parasites coupled with a fluctuating global climate, resulting in fluctuating levels of primary productivity. The role of pathogens and parasites in shaping mammalian communities has often been overlooked (Barbehenn 1978; Muul 1978). Yet I concur with Jerison (1973) that the history of the Mammalia on the "contiguous" continents (North America, Europe, Asia, Africa) has resulted in an overall increase in relative brain size. The corollary to this statement is that the neocortex has increased in size; thus the stage was set for a feedback loop between individually acquired experience and learning by descendents, enabling them to operate in a more and more unpredictable world as new niches were invaded. The evolution of complex, interdependent social groupings is not the only outcome of such processes; rather it is only one of four possible contemporary outcomes of the total, historical selective forces (as illustrated in Figure 156 of Eisenberg [1981]). Within the extant species of the class Mammalia, a relatively large brain tends to be associated with selective forces that have shaped a species that is *K-selected*; that is, the individual within the species has a relatively long reproductive life and a small litter size. Often, but not always, within a historical lineage such a species has a relatively large body size (Eisenberg 1981).

Using current data to project to the past

The above represents the argument for the important influence of trophic strategy upon the density a mammalian species can achieve within a community. Likewise the standing-crop biomass exhibited by a species (or a population of a species) within a community is also determined by trophic strategy and life history strategy. Life history strategies are constrained by trophic specializations, but demographic processes are ultimately the outcome of age at first reproduction, mean annual litter size, and age at last reproduction (Cole 1954). For a mammalian paleontologist, the genuine problem in reconstructing extinct communities is to decide on which trophic category to assign a species that one has never seen. Moreover, one confronts the problem that one will never see the species in nature. In the early years of community studies, the trophic categories allocated to secondary consumers were broad: herbivore, omnivore, carnivore.

Let us take the category of herbivore for closer scrutiny. In the original sense of the term, it implied feeding on plants or parts of plants. We now know that there are profound consequences for the life history of a species if it is selected to become a frugivore rather than a folivore, or a browser rather than a grazer. In short, our science is moving toward a finer definition of trophic specialization for primary and secondary consumers. Consider the work of Hoffmann and Stewart (1972) in establishing the correlations between the stomach structure and feeding habits of East African ungulates, correlating with the behavioral observations of Jarman (1974). The feeding habits of browsers and grazers in the East African ungulate community are a continuum; yet morphological and behavioral data also form a concordant continuum. Unfortunately, behavioral data and the anatomy of the "soft parts" are unavailable to the paleontologist. Yet, I feel we can make progress in paleoecology even with only skulls, teeth, and, hopefully, postcranial elements at our disposal.

The work of Kay and Cartmill (1977), Kay and Hylander (1978), and Cartmill (1972) justifies the previous assertion. In his article, Cartmill asked the question, What morphological specializations can one discern in the anatomy of extant arboreal species (both diurnal and nocturnal) that may have relevance for the interpretation of fragmented fossils from the Paleocene–Eocene boundary? Great insight was demonstrated in selecting the extant *Sciurus carolinensis* and *Didelphis virginiana* as two of several possible examples for comparison and contrast. Both species were readily observable and available in North America. Kay and Cartmill undertook a more ambitious project when they attempted to infer the habits of the Paleocene *Palaechthon nacimienti*. In this case, the task was to determine diet, and probable hunting strategy (i.e., major sensory inputs), from a set of pitiful fragments. The authors took many measurements from extant forms to establish such facts as relative size of orbit and activity cycle, relative size of the foramina, and vibrissal development to create a picture of the sensory world of *Palaechthon*. Kay and Hylander addressed, and answered, the question, Can the degree of feeding on leaves in extant primates be correlated with molar tooth measurements? The three articles cited are not indicative of a new trend in paleontological research. Undoubtedly, from the time the first fossil skull was discovered by a human possessing an inquiring mind, the habitus of the fossil was postulated based on the discoverer's own experience with extant forms. Indeed, all museum exhibitions of fossils represent "educated guesses" at natural history. I propose nothing new in one sense, but I do propose that as neontologists become more sophisticated, so should paleontologists. The dialogue established at the symposium should – no, *must* – continue.

Concluding remarks

As this brief review indicates, there are many rules of biological scaling that we can discern in extant forms that are probably excellent predictors for extinct species. Once the radiation of mammals had commenced in the Paleocene, and the major lines of descent had diverged, recognizable lineages developed which through time have become specialized for various niches. And even with extinction events, the remaining stocks of mammals have seemingly been able to adaptively radiate into similar niches. This is especially true for the period following the extinctions at the end of the Oligocene, when the Late Miocene and Pliocene communities emerged. These communities have more or less persisted through time, although shifting geographically depending on global climatic changes. Some cautionary words must be uttered, however, in that when any of these allometric relationships have been established, there usually is a great deal of residual scatter around the regression line. As I have indicated in this paper, and as McNab has indicated in his, much of this scatter can be accounted for if one knows something about the trophic strategy of the species. One should never discard or discount the residual variation; it often is of the greatest biological interest.

Let us consider the problem of scaling and regression analysis. Many biologists would hope to develop mathematical expressions of their measurements that would allow a predictive power emulating those equations derived by physicists (Lotka 1956). Yet, biology is a young science, as compared to astronomy, let alone physics. Furthermore, even though we revere physics for its precision and foundation in the experimental method, the discipline is as subject to the influence of political forces as any other science dependent on government subsidies (McCormmach 1982). Most biologists, in plotting a supposedly dependent variable against an independent variable, consider the scatter about a regression line as "noise." This is unfortunate, since the scatter may reflect our inability to appropriately categorize our data set into biologically meaningful subsets. Given the inertia of institutionally sanctioned jargon, I say, beware of "hardening of the categories."

For example, once again we may consider the regression analysis performed by Kinnaird and Eisenberg (in press) for Neotropical mammals. Standing-crop biomass for species was regressed against mean body weight. There was enormous scatter, but greater than 70% of the variance could be accounted for by trophic specialization. Similar results were established by Robinson and Redford (1986b) for species densities of Neotropical mammals. Scatter about a regression line should not be considered an annoyance, but rather a challenge.

References

Barbehenn, K. R. 1978. Discussion: concluding comments, from the worm's view "eco-pharmacodynamics" and 2000 A.D. In *Populations of Small Mammals under Natural Conditions,* ed. D. P. Snyder, pp. 231–236. Special Publication Series, Vol. 5. Linesville, Penna.: Pymatuning Laboratory of Ecology. University of Pittsburgh.

Bonner, J. T. 1965. *Size and Cycle.* Princeton, N.J.: Princeton University Press.

Calder, W. A., III. 1984. *Size, Function, and Life History.* Cambridge, Mass.: Harvard University Press.

Cartmill, M. 1972. Arboreal adaptations and the origin of the order Primates. In *The Function and Evolutionary Biology of Primates,* ed. R. Tuttle, pp. 97–122. Chicago/New York: Aldine-Atherton.

Cole, L. C. 1954. The population consequences of life history phenomena. *Q. Rev. Biol. 29*:103–137.

Damuth, J. 1987. Interspecific allometry of population density in mammals and other animals: the independence of body mass and population energy-use. *Biol. J. Linn. Soc. 31*:193–246.

Dooling, R. J. 1980. Behavior and psychophysics of hearing in birds. In *Comparative Studies of Hearing in Vertebrates,* ed. A. N. Popper & R. R. Fay, pp. 261–285. New York: Springer.

Eisenberg, J. F. 1966. The social organizations of mammals. *Handbuch der Zoologie, VIII* (10/7), Lieferung 39, 92pp.

Eisenberg, J. F. 1980. The density and biomass of tropical mammals. In *Conservation Biology: An Evolutionary–Ecological Perspective,* ed. M. E. Soule & B. A. Wilcox, pp. 35–55. Sunderland, Mass.: Sinauer Press.

Eisenberg, J. F. 1981. *The Mammalian Radiations.* Chicago: University of Chicago Press.

Eisenberg, J. F., & Seidensticker, J. 1976. Ungulates in southern Asia: A consideration of biomass estimates for selected habitats. *Biol. Conserv. 10*:293–308.

Eisenberg, J. F., O'Connell, M., & August, P. V. 1979. Density, productivity, and distribution of mammals in two Venezuelan habitats. In *Vertebrate Ecology in the Northern Neotropics,* ed. J. F. Eisenberg, pp. 187–207. Washington, D.C.: Smithsonian Institution Press.

Handley, C. O., Jr. 1966. Checklist of the mammals of Panama. In *Ectoparasites of Panama,* ed. R. L. Wenzel & V. J. Tipton, pp. 753–795. Chicago: Field Museum of Natural History.

Harestad, A. S., & Bunnell, F. L. 1979. Home range and body weight – a re-evaluation. *Ecology 60*:389–402.

Hoffmann, R. R., & Stewart, D. R. M. 1972. Grazer and browser: a classification based on the stomach structure and feeding habits of East African ruminants. *Mammalia 36*:226–240.

Janis, C. 1976. The evolutionary strategy of the Equidae and the origins of rumen and cecal digestion. *Evolution 30*:757–774.

Jarman, P. J. 1974. The social organization of antelopes in relation to their ecology. *Behavior 58*(3–4):215–267.

Jerison, H. J. 1973. *Evolution of the Brain and Intelligence.* New York: Academic Press.

Kay, R. F., & Cartmill, M. 1977. Cranial morphology and adaptations of *Palaechthon nacimienti* and other Paromomyidae (Plesiadapoidea, ?Primates), with a description of a new genus and species. *J. Human Evol. 6*:13–53.

Kay, R. F., & Hylander, W. L. 1978. The dental structure of mammalian folivores with special reference to primates and phalangeroids (Marsupialia). In *The Ecology of Arboreal Folivores,* ed. G. G. Montgomery, pp. 173–192. Washington, D.C.: Smithsonian Institution Press.

Kinnaird, M., & Eisenberg, J. F. in press. A consideration of body size, diet, and population biomass for Neotropical mammals. In *Advances in Neotropical Mammalogy,* ed. K. H. Redford & J. F. Eisenberg.

Lotka, A. J. 1956. *Elements of Mathematical Biology.* New York: Dover Press.

Martin, L. D. 1990. Fossil history of the terrestrial Carnivora. In *Carnivore Behavior, Ecology, and Evolution,* ed. J. L. Gittleman, pp. 536–538. Ithaca: Cornell University Press.

McCormmach, R. 1982. *Night Thoughts of a Classical Physicist.* Cambridge, Mass.: Harvard University Press.

McNab, B. K. 1963. Bioenergetics and the determination of home range size. *Am. Nat. 97*:133–140.

Medway, L. 1978. *The Wild Mammals of Malaya and Singapore,* 2nd ed. Oxford: Oxford University Press.

Muul, I. 1978. Review: Small mammal populations in zoonotic disease and toxicological studies. In *Populations of Small Mammals under Natural Conditions,* ed. D. P. Snyder, pp. 208–223. Special Publication Series, Vol. 5. Linesville, Penna.: Pymatuning Laboratory of Ecology, University of Pittsburgh.

Osborn, H. F. 1929. *The Titanotheres of Ancient Wyoming and Nebraska.* U.S. Geol. Survey Monogr. No. 55.

Parra, R. 1978. Comparisons of foregut and hindgut fermentation in herbivores. In *The Ecology of Arboreal Folivores,* ed. G. G. Montgomery, pp. 205–230. Washington, D.C.: Smithsonian Institution Press.

Pirlot, F. 1977. Wing design and the origin of bats. In *Major Patterns in Vertebrate Evolution,* ed. M. K. Hecht, P. C. Goody, & B. M. Hecht, pp. 375–422. New York: Plenum Press.

Robinson, J. G., & Redford, K. H. 1986a. Body size, diet and population density of Neotropical forest mammals. *Am. Nat. 128*:665–680.

Robinson, J. G., & Redford, K. H. 1986b. Intrinsic rate of natural increase in Neotropical forest mammals: relationship to phylogeny and diet. *Oecologia 68*:516–520.

Taylor, E. R., Heglund, N. C., & Maloiy, G. M. L. 1982. Energetics and mechanics of terrestrial locomotion. J. *Exp. Biol. 97*:1–21.

Webb, S. D. 1985. Late Cenozoic mammal dispersal between the Americas. In *The Great American Biotic Interchange,* ed. F. G. Stehli & S. D. Webb, pp. 357–386. New York: Plenum Press.

Van Valkenburgh, B. 1984. A morphological analysis of ecological separation within past and present predator guilds. Ph.D. dissertation. Baltimore: The Johns Hopkins University.

Van Valkenburgh, B. 1985. Locomotor diversity within past and present guilds of large predatory mammals. *Paleobiology 11*(4):406–428.

4

The functional anatomy of body mass

THEODORE I. GRAND

One central challenge for the paleontologist is to reconstruct body mass from fragments of bone and teeth. Estimates of volumes, whether of specific soft tissues or the entire body mass, are derived from linear measures. Moreover, this becomes a closed system: One makes successive approximations, but can never verify volumetric/mass statements about extinct animals. The paleontologist, unable to make direct analyses, looks to living species for guidance and the luxuries of choice and abundance. The anatomist, on the other hand, works relatively unconstrained: There are infinite opportunities to dissect; to make direct measurements of skin, adipose stores, muscle, bone, and body mass; to observe directly the biomechanics of motion, of physiological processes, of social behavior. How can the "loop" of linear measures of volume and body mass open up? What can the anatomy of living animals say to the paleontologist?

I have studied more than 150 genera of metatherian and eutherian mammals (from 10-g bats to 1000-kg whales). But a survey is not appropriate: The student of fossil carnivores has no direct interest in modern bats; the student of ungulate origins need not know about the anatomy of modern whales. One conclusion about mammalian body composition, however, is important: For the largest tissues of the body (skin, muscle, bone, adipose stores), the *dominating components* of weight vary with function and, to a degree, independently of one another. How does evolutionary process "tinker" with the component tissues, while weight itself remains constant?

In *Body Size in Mammalian Paleobiology: Estimation and Biological Implications,* John Damuth and Bruce J. MacFadden, eds.

To demonstrate that locomotor adaptation has a substantive effect on body composition, I will compare pairs of species of equal body mass: the potto, *Perodicticus potto,* with the greater galago, *Galago crassicaudatus,* at 1000 g; the red panda, *Ailurus fulgens,* with the raccoon, *Procyon lotor,* at 4000 g. Differences in the proportion of skin and muscle correlate directly with locomotor and antipredator behavior. Since this is true across my wide study sample, I suggest that an analysis of body mass must begin by interpreting functionally the largest tissues–for example, cutaneous adaptations such as quill or carapace, scaling effects on skin proportion, differences between arboreal and terrestrial locomotor behaviors, and patterns of adipose storage. Then, we have a basis for understanding body mass and the potential for better estimates among extinct species.

Methods

My published data (Grand 1977) on three *P. potto* and two *G. crassicaudatus* have been increased to five adult individuals of each species. Results of the studies of five adult *A. fulgens* and six *P. lotor* have not yet been published. The mass-related measurements have been described in a series of papers (Grand 1977, 1978, 1983).

Body composition: Skin, muscle, skeleton, and adipose stores were dissected and weighed separately, and each quantity was divided by total weight to determine its percentage of total body weight (%TBW).

Functional muscle groups: The weight of each muscle group (forelimb, hindlimb, back extensors, masticatory, tail) was divided by the total weight of all muscle.

The ratios of limb skeleton: The weight of the forelimb (scapula, humerus, radius and ulna, hand/paw) skeleton was divided by that of the hindlimb (femur, tibia and fibula, foot).

The mean values for each tissue and muscle group are given in the figures. Error bars show the 95% confidence intervals for these means.

Results

*Pair contrasts – 1000 g: the potto (*Perodicticus potto*) and the greater galago (*Galago crassicaudatus*)*

These animals represent the structural and behavioral extremes of the lorisoid radiation of African prosimians (Charles-Dominique 1977;

Grand 1977). Both are arboreal and nocturnal, and at 800–1200 g (Napier & Napier 1967) are equal in body mass. There the similarity ends. The potto is a slow-moving climber with a stumpy, almost flaplike tail. The extreme mobility of each limb joint provides the basis for the enormous reach of hands and feet within three-dimensional space; fingers and toes have a viselike grip; the vertebral column exhibits snakelike, sinusoidal flexibility. Pulling with the forelimb and pushing with the hindlimb enable the animal to glide across branch supports like a "slinky" toy. Because one or more limbs is always in contact with a branch, the potto is supported during the entire gait cycle. As a consequence, branches deform slowly, and rustling noises are reduced. This acoustic "crypsis" is a common antipredator strategy, and part of the potto's own stalking technique with regard to insects, nestling birds, and other live food.

By contrast, the galago is a long-legged, long-tailed vertical clinger and leaper (Napier & Walker 1967). As in the case of *Tarsius* (Grand & Lorenz 1968) hindlimb dominance is unequivocal: The great mass and length of each hindlimb segment, the elongate and elaborate fulcrum within the foot (which nevertheless retains prehensile digits) are apparent. The ability to leap and change direction at landing resembles the ricochetal hop and antipredator evasion of some small rodents, insectivores, and the macropods.

Body composition. The potto has 15% more skin, but only 75% of the muscle of the galago (Figure 4.1). Although the percentage of skeletal mass is similar, the ratios of forelimb to hindlimb differ significantly – 1:1 in the potto and 1:2 in the galago.

Functional muscle groups. In the potto almost 30% of all muscle is concentrated in the hindlimb; 40% is found in the forelimb (Figure 4.2). This muscular balance reflects the push–pull nature of potto climbing (Charles-Dominique 1977). Less than 10% of all muscle is concentrated in the back extensors; with the tail reduced, there is almost no sacral muscle.

In the galago a higher proportion of muscle is concentrated in the hindlimb (over 40% of all muscle) than in the forelimb (under 30%). Moreover, the back extensors constitute a higher proportion of muscle than in the potto. Tail muscle adds another 3%.

There are "local" muscle-group differences as well. The quadriceps femoris are larger (jumping is based upon powerful knee-joint extension) than those of the potto. However, the adductors, powerful climbing

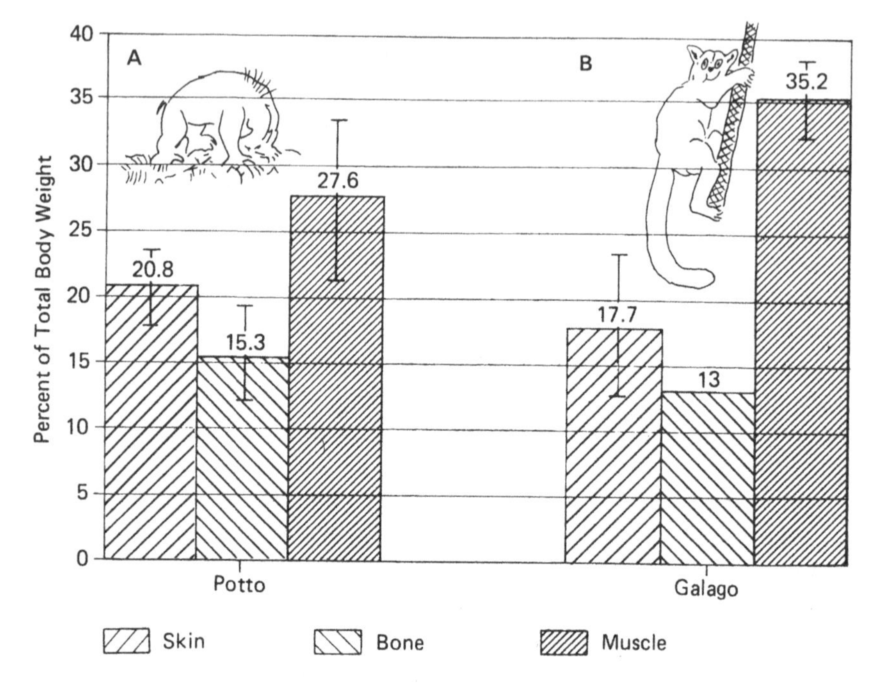

Figure 4.1. Body composition in (A) *P. potto* and (B) *G. crassicaudatus*: skin, bone, and muscle as percentages of total body weight.

and stabilizing muscles, are much smaller in the galago than they are in the potto.

The potto has reduced the bulk of the lumbar extensor muscles as well as the lengths of transverse and spinous processes. This complex of elements enhances flexibility of the lumbar region, a primary adaptation to slow climbing. The galago has much larger lumbar extensors, and the transverse and spinous processes are much more pronounced. The loss of side-to-side mobility, which makes hindlimb propulsion more effective, is like that of scansorial, hindlimb-dominant species such as the kangaroo rat (*Dipodomys*), the brown four-eyed opossum (*Metachirus*) (Grand 1983), and the elephant shrew (*Elephantulus*).

*Pair contrasts – 4 kg: the red panda (*Ailurus fulgens*) and the raccoon (*Procyon lotor*)*

The panda is an arboreal climber that can walk on the ground; the raccoon is a terrestrial quadruped that can climb (Ewer 1973; Lotze &

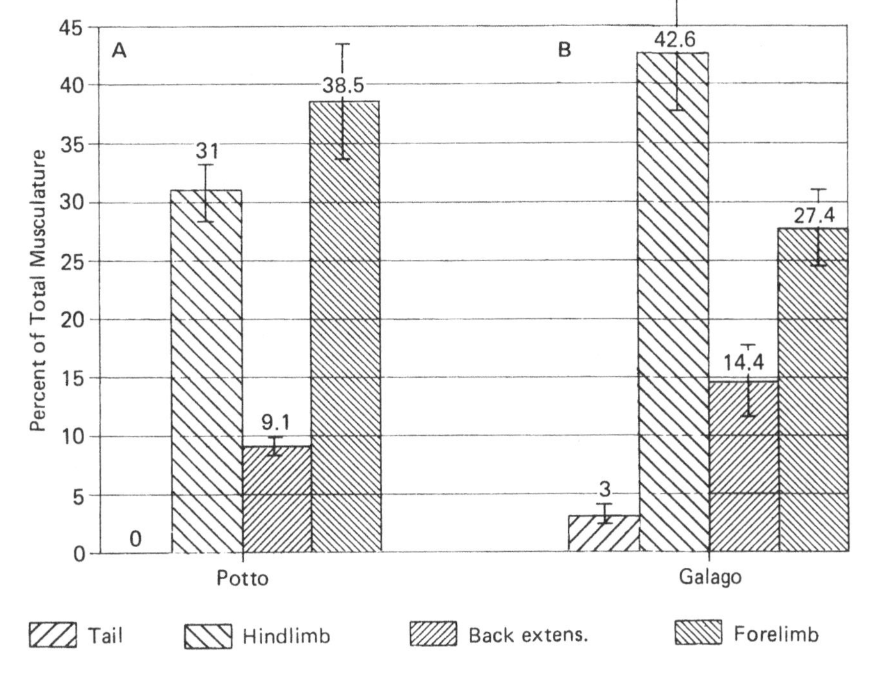

Figure 4.2. Functional muscle groups in (A) *P. potto* and (B) *G. crassicaudatus*: the mass of muscle groups in tail, hindlimb, back, and forelimb as percentages of total body musculature.

Anderson 1979; MacClintock 1988; Roberts & Gittleman 1984). Limb proportions are similar (Davis 1964), but the segments differ in relative mass (Grand, unpublished data): The thigh of the panda is two-thirds that of the raccoon; the tail is almost 50% heavier.

Body composition. The skin (16% TBW), adipose tissue (6% TBW), and bone (15% TBW) in both species are equal (Figure 4.3). Data on the raccoon (Grand, unpublished) show enormous seasonal fluctuations in adipose stores. In terms of muscularity, the two animals differ significantly. The red panda is 30% muscle; the raccoon, 35% muscle.

Proportions of the limb bones. The ratio of forelimb to hindlimb bone is 1:1 in the red panda; 0.75:1 in the raccoon.

Functional muscle groups. Less than 30% of all muscle in the red panda is concentrated in the hindlimb, slightly more than 30% in the forelimb

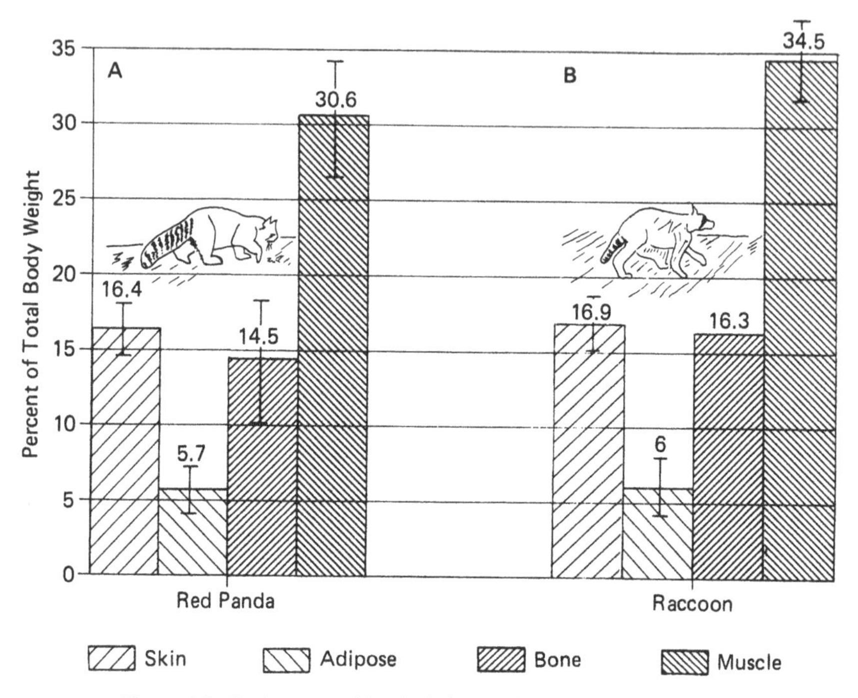

Figure 4.3. Body composition in (A) *A. fulgens* and (B) *P. lotor*: skin, adipose tissue, skeleton, and musculature as a percentage of total body weight.

(Figure 4.4). Two percent of all muscle is in the tail. The masticatory muscles represent more than 6% of all muscle. The sizes of the masseter and temporalis explain the relative distortions of the zygomatic arch, the nuchal and sagittal crests, and the coronoid and angular processes, described by Davis (1964).

The hindlimb of the raccoon is more heavily muscled (representing 36% of all muscle) than the forelimb (27%). The masticatory musculature represents 4%, the tail 1% of all muscle.

Discussion

In the natural history of a species, body mass is only one attribute among many. The potto and greater galago are the same size, but the proportions of the two largest tissues (skin and muscle) are significantly different. Locomotor adaptation is the interpretive framework that correlates directly with the differences *and* explains them. With red panda and raccoon, related carnivores of equal size and hotly debated

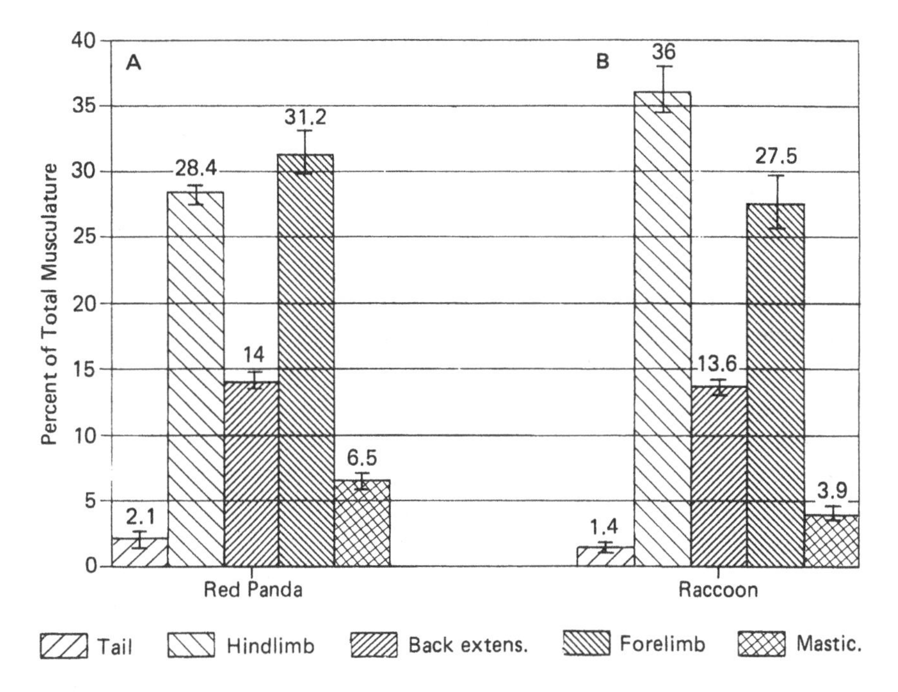

Figure 4.4. Functional muscle groups in (A) *A. fulgens* and (B) *P. lotor*: musculature in tail, hindlimb, back, forelimb, and masticatory apparatus as a percentage of total musculature.

kinship (Ewer 1973; Sarich 1976), muscularity has been sculptured primarily by locomotor pattern: in the case of the former, for arboreal climbing; in the case of the latter, for terrestrial quadrupedalism. If we think in terms of 1- or of 4-kg size classes, we might not seek out what is fundamentally different.

Thus, an obvious tension exists between the data and expectations of the paleontologist and those of the anatomist. The former moves from linear measures of bones and teeth to predictions about volumes. He or she tends to see weight as a numerical value in itself, unmindful of the essential variability of the major components of weight. The functional anatomist, however, travels comfortably (and comfortingly) among large data sets of many tissues and organs, and is compelled to see the functional bases of tissue variability.

These tissue masses vary tremendously, and to some degree, independently. Among moderately sized mammals, skin ranges from 5% to 20% TBW; muscle from less than 25% to more than 50% TBW; bone from 7% to 17% TBW. Skin may be elevated with antipredator adaptations such as quill or carapace; bone and muscle vary with locomotor skill. Arboreal climbers, regardless of taxonomic origin, are significantly less muscular than terrestrial species (Thompson & Grand, unpublished data). Climbers also tend to balance pushing and pulling motions, and thus the ratios of bone and muscle to the limbs. Terrestrial runners and jumpers (and the arboreal galago) become hindlimb dominant, and muscular ratios rise from 0.75:1 to 0.25:1 (the macropods) (Grand, unpublished data).

Muscle weight also responds to function with greater sensitivity than bone. If an animal breaks its leg, months later after the leg has healed, bony weights of the two sides will be equal, even though the muscle weights of the two sides show the asymmetry of disuse. In old age, an animal may lose muscle bulk even though the bony proportions remain the same. In these cases significant changes in the muscular components are masked by body weight as a number.

One might call particular attention to the thigh as a predictor of locomotor function and of body weight. In slow climbers, the thigh may represent only 3–4% TBW; in runners, jumpers, and walking man, it may rise to 8–12% TBW. The thigh seems to "track" muscularity and locomotor repertoire better than the other segments. The foot is too small and without muscle; the tail is too variable; the head is too complex (as it is the focus of brain evolution, masticatory adaptation, and intraspecific display such as horns or antlers). But even as one studies the thigh across species, one must know the local (taxonomic and functional) rules of body composition and regional muscle distribution.

Acknowledgments

I would like to thank John Damuth and Bruce MacFadden, co-organizers of the Workshop, for the opportunity to join them. John Eisenberg, Miles Roberts, Adrienne Zihlman, Mary Ellen Morbeck, and Sentiel Rommel commented on the manuscript at various stages in its evolution.

References

Charles-Dominique, P. 1977. *Ecology and Behavior of Nocturnal Prosimians.* New York: Columbia University Press.

Davis, D. D. 1964. The Giant Panda. A morphological study of evolutionary

mechanisms. *Fieldiana: Zool. Mem.*, Vol. 3. Chicago: Natural History Museum.
Ewer, R. F. 1973. *The Carnivores*. Ithaca, N.Y.: Cornell University Press.
Grand, T. I. 1977. Body weight: its relation to tissue composition, segment distribution, and motor function. I. Interspecific comparisons. *Am. J. Phys. Anthropol. 47*:211–240.
Grand, T. I. 1978. Adaptations of tissue and limb segments to facilitate moving and feeding in arboreal folivores. In *The Ecology of Arboreal Folivores*, ed. G. G. Montgomery, pp. 231–241. Washington, D.C.: Smithsonian Institution Press.
Grand, T. I. 1983. Body weight: its relationship to tissue composition, segmental distribution of mass, and motor function. III. The Didelphidae of French Guyana. *Austr. J. Zool. 31*:299–312.
Grand, T. I., & Lorenz, R. 1968. Functional analysis of *Tarsius bancanus* (Horsfield, 1821) and *Tarsius syrichta* (Linnaeus, 1758). *Folia Primatol. 9*: 161–181.
Lotze, J. H., & Anderson, S. 1979. *Procyon lotor*. Mammalian Species, No. 119, pp. 1–8.
MacClintock, D. 1988. The red pandas: In *A Natural History*. New York: Scribners.
Napier, J. R., & Napier, P. H. 1967. *A Handbook of Living Primates*. New York: Academic Press.
Napier, J. R., & Walker, A. C. 1967. Vertical clinging and leaping, a newly recognized category of locomotor behavior among primates. *Folia Primatol. 6*:180–203.
Roberts, M., & J. Gittleman 1984. *Ailurus fulgens*. Mammalian Species, No. 222, pp. 1–8.
Sarich, V. 1976. Transferrin. *Trans. Zool. Soc. Lond. 33*:165–171.

5

Estimating body mass and correlated variables in extinct mammals: travels in the fourth dimension

ROBERT A. MARTIN

Well, I do not mind telling you I have been at work upon this geometry of Four Dimensions for some time. Some of the results are curious. For instance, here is a portrait of a man at eight years old, another at fifteen, another at seventeen, another at twenty-three, and so on. All these are evidently sections, as it were, Three-Dimensional representations of his Four-Dimensional being, which is a fixed and unalterable thing.

We never learn the identity of Wells's Time Traveler, who above describes with reasonable accuracy the spatiotemporal continuity of a segment of an ontogeny or a phyletic sequence, but through his tale in *The Time Machine* we do vicariously experience the thrill of moving freely through the so-called fourth dimension, and fully appreciate why, after returning from a harrowing adventure, the Time Traveler is irrevocably drawn once again into the unknown.

During the Pleistocene, North America was populated by a fabulous array of creatures: giant capybaras and beaver the size of black bears, llamas and camels, dire wolves, cave bears and spectacled bears, giant jaguars, lions (or tigers), cheetahs, mammoths and mastodons, wild horses, ground sloths and glyptodonts, sabertooth and scimitar-tooth cats, huge vampire bats, and herds of peccaries. Paleo-Indians ranged among many of these beasts, and saw the last of them disappear from the continental United States. These ancient Indians could tell us the size of *Panthera atrox,* how many cubs were born to its females each spring, and how long the species lived. They saw dire wolves, probably

In *Body Size in Mammalian Paleobiology: Estimation and Biological Implications,* John Damuth and Bruce J. MacFadden, eds.

hunting in packs like their living counterparts, bring down llamas and horses. Paleo-Indians could tell us if the giant vampire *Desmodus stocki* preyed upon ground sloths and other megaherbivores. But the Paleo-Indians left no significant heritage, and these scenarios represent merely the fanciful reconstructions of an imaginative biologist. My purpose in this essay is to show that such reconstructions are not beyond our abilities if we can accept as scientific reasonable inferences about extinct species based upon extant ones. The key to this process is the observation that many linear dimensions, areas, and volumes of teeth and bones are highly correlated with body mass in extant mammals and that body mass is tightly correlated with many physiological and ecological variables. If length of the first lower molar can be used to estimate body mass accurately, then it can also be used to estimate metabolic rate or home range. In this chapter, I will consider the statistical basis of body mass estimation, discuss the assumptions and limitations of estimating biological variables for extinct mammals, review my previous studies with cotton rats, and make some predictions for future work in this relatively new field of inquiry.

Statistical considerations

Body mass treated as an unknown dependent variable is best related to an independent variable by the least-squares regression model (Gingerich, Smith, & Rosenberg 1982). Although this method may also be used in functional analyses of general scaling phenomena (e.g., tooth area as a function of mass), the major axis method is preferred for the latter purpose by some investigators (Gingerich et al. 1982; Jolicoeur 1973; Pilbeam & Gould 1974; Sokal & Rohlf, 1969).

The independent variable differs considerably among authors, and includes head–body length (Jerison 1973; Radinsky 1978), greatest width across the occipital condyles (Martin 1980; Wing & Brown 1979), femoral head width (Martin 1980; Wing & Brown 1979), tooth crown area (Gingerich et al. 1982), length of the first lower molar (Martin 1984), a variety of postcranial measurements (Scott 1983), and a composite of measurements from the postcranial skeleton (McHenry 1975, 1976). Gingerich et al. (1982) conclude that composite measurements, such as tooth area, yield more accurate estimates than single dimensions, but Damuth (this volume) suggests that tooth areas are likely to provide inaccurate estimates in archaic ungulates, and argues instead that the individual variable of tooth length is the more appropriate measurement for these forms. Although it is true that a product contains more information than either a length or width taken alone, it may also be another construct entirely, with its own evolutionary and scaling dynamic.

The particular independent variable chosen is of little significance as long as two criteria are satisfied: (1) It must be measurable in a fossil assemblage with enough regularity to be practical; and (2), in extant mammals, it must be highly correlated with body mass. McHenry (1974) also argues that the measure should be isometric with regard to mass, but this is of minimal concern as long as the correlation is high and the sample size is large. It may at first seem appropriate to constitute the data base with only closely related species of similar body mass. In dealing with fossil hominids, for example, McHenry (1974, 1975, 1976) limited the predictive equation to *Homo sapiens.* However there are a number of drawbacks to this approach. Small sample size and inherent variability can lead to spurious values for the slope. Also, if the data base is limited, ontogenetic scaling may have a pronounced statistical effect in spite of the best efforts made to limit the data set to adult specimens. A regression that will allow accurate estimation of mass must be defined on a group of closely related (and/or functionally related; see Janis, Fortelius, Damuth, this volume) species extended through a wide range of body mass.

Although "grand" predictive equations constructed for all mammals (e.g., Martin 1980, Wing & Brown 1979) may be useful in testing null hypotheses relating to the overall scaling of body mass to a particular variable, they are generally too crude to produce highly accurate results. Jerison's (1973) conclusions concerning brain/body size ratios in evolving ungulates and carnivores were shown by Radinsky (1978) to be in error, in part because Jerison combined data from both carnivores and ungulates to construct his predictive equation for body mass.

The level of intended resolution will also dictate the manner in which a regression is calculated. Limits of resolution range from estimating the mass of an individual represented by a fossil in hand to an average value of mass represented by a fossil considered to be of a particular average size. In either case, the estimated mass value is the same, but the 95% confidence interval is much wider in the former case, reflecting the lower probability of calculating an accurate mass value for a single individual (Zar 1984).

As will be seen below, body mass estimates of extinct species cannot be tested in the conventional sense. And there are good statistical arguments that a high correlation coefficient in itself cannot guarantee the highest accuracy (Smith 1980). Consequently, Van Valkenburgh (this volume) recommends calculating two additional parameters, the percent prediction error (%PE) and the percent standard error of the estimate (%SEE). In calculating the %PE, known body masses of specimens from an extant species are compared to those estimated from the predictive equation. Our confidence in these estimates would be further increased if the specimens used for comparison were independent of those used to construct the data base.

Assumptions and limitations

Assumption 1 is a blanket statement meant to cover body mass and all variables correlated with body mass. Assumption 2 is listed primarily to cover the rare circumstance in which a mammal demonstrates ectohomoiothermy, although other kinds of variation are discussed below.

Assumptions

Assumption 1: Scaling relations for all relevant variables correlated with body mass in extinct species were the same as in the extant ones used to generate body mass equations.

Assumption 2: There has been no significant change in temperature/energy relations within the study taxon since the time of its origin.

Estimates of body size in extinct mammals are taken from regressions of body mass (or, a less useful unit, head–body length) on some measurement of teeth or bones in an extant mammalian taxon. If it could be shown that the measurement scaled differently to body size in an extinct group than it does today, assumption 1 would be invalid. Damuth (this volume) documents just such a circumstance, in which M_1 area is not a very good predictor of body mass since on average, archaic ungulates had relatively wider molars than do extant ones. This sort of relationship also occurs in fossil muskrats, where an increase in length of M_1 is accompanied by a relatively great increase in width in archaic species, but width becomes inflexible after a certain length has been attained (Martin 1978). As Damuth (this volume) correctly notes, this bias can be removed by using dental length alone as the estimator. Investigators working with archaic groups that have no extant analogues always run the risk of introducing bias by violating assumption 1, but the bias can at least be reduced by paying careful attention to potential sources of error.

Energy relations differ among extant mammalian groups. We know, for example, that arvicoline rodents and weasels have higher metabolic rates than predicted by body size in the Kleiber equation (McManus 1974; McNab 1974), and for this reason I constructed a new equation to estimate metabolism from mass for these groups (Martin 1980). Sloths and marsupials, on the other hand, manifest lower metabolic rates than predicted by the Kleiber equation. In the former case, low metabolism seems to be related to folivory (McNab 1978), but in the latter case the causative factor remains uncertain. Whereas the metabolic rate of an extinct mammalian species generated from a mass estimate entered into the Kleiber equation runs a high risk of inaccuracy, one that is estimated from an equation limited to the genus or family (and feeding guild) of which the extinct species is a member may be as accurate as any that would be generated in extant species. I know of no mammalian genus or family in which there are pronounced differences in metabolic patterns once it has been partitioned according to food habits (Eisenberg, this volume). Yet the primary concern here is really not so much this taxonomic variation, which can always be refined, but whether or not the group demonstrates any radical differences in temperature/energy relations, such as the ectohomoiothermy seen in *Heterocephalus glaber,* the naked mole rat (McNab 1966). On the basis of currently available information, I would hazard the guess that assumption 2 is, for all practical purposes, satisfied for all mammalian taxa that are not fossorial.

Limitations

Unlike assumptions, which must be satisfied before the analysis can be performed, limitations reflect the potential inadequacy of an estimate, owing to technical difficulties.

Limitation 1: In constructing the predictive equation for body mass, numbers of measurements from extant species may be limited.

Limitation 2: Body mass data are not available on specimen tags of extant specimens from which measurements are taken.

Limitation 3: Body mass data from the literature are not available for a random sample of each species.

Limitation 4: Measurement error.

Limitation 5: Data for parameters other than body mass may be highly variable.

Limitation 6 (also known as Wells's rule): Without a time machine there is no way to validate results.

There are no studies in which large samples of extant specimens with associated tag mass data form the data base on which body mass regressions are made. In all of the studies reported in this symposium volume, only the data for cotton rats were derived from an equation based solely on specimens with associated tag mass data. Unfortunately, very few specimens in museums have associated tag mass data, and therefore it becomes necessary to use a mean of published mass data with either a mean of a measured sample for a species (Janis, this volume) or a single male and female chosen at random (Van Valkenburgh, this volume). In an earlier study (Martin 1980), where I constructed a single equation relating body mass to condylar width of the skull in mammals, I compared samples in which mass data were taken from the literature and from tag data. Regression slopes derived from the two sets were statistically identical ($p > .05$), suggesting that literature values may be substituted for tag data when the latter are unavailable.

Grand (this volume), Scott (this volume), and Van Valkenburgh (this volume) discuss the problems associated with mass data reported for "big game" animals. Sometimes the only mass data available are taken from the largest individuals of the species. Additionally, there are problems with "dressed weight" versus true wet weight.

In spite of the most consistent attempts to minimize measurement error, it is unavoidable when different individuals are involved in an analysis or calibrations and corrections are made. Nevertheless, I would argue that in most cases measurement error probably accounts for less than 1% (and certainly less than 5%) of total measurement variation (see Van Gelder 1968).

It is one thing to estimate body mass in an extinct species; it is another thing entirely to use that mass value in another equation to estimate a second parameter. All of the possible errors from the primary analysis are now compounded with inherent errors in the secondary analysis. But the situation is not hopeless. Consider population density. This variable in cotton rats is known to range over two orders of magnitude, depending upon a variety of (mostly unknown) environmental factors. However, the population density for *Sigmodon hispidus* estimated from Peters's (1983) temperate herbivore equation is statistically indistinguishable from an average value of population density from the literature on cotton rats (Martin 1986). Partly, this is a function of statistical testing: The null hypothesis cannot be rejected because the data are so variable. Yet, this also represents the biological reality; cotton rats (and other mammals) have highly variable population densities, but those densities still can be quantified for each species, and a reasonable estimate can result.

Also, it pays to remember that we are rarely dealing with the kind of predictive power that would allow any of these variables to be estimated for a single specimen (see Statistical considerations, above). I suppose that the predictive power of a mass-estimating equation for cricetine rodents might, when the sample size is large enough, be so robust that we could feel fairly comfortable in stating that a particular animal massed at 34.6 g ($\pm$ some value), but I certainly would not feel comfortable estimating the same animal's metabolic rate. However, there should be no compunction in estimating an average metabolic rate for the species. In the final analysis, the best we can hope for in estimating parameters from secondary equations are values that fall within species limits. But these can prove most instructive in a host of paleoecological investigations. Some scaling equations relating biological attributes to body mass in extant mammals are presented in Table 5.1.

There are, at least at face value, hypotheses in paleontology that are testable (e.g., stasis of morphology in a taxon), but body masses of extinct species, considered as hypotheses, cannot be tested. Strictly speaking then, body mass estimation of an extinct species falls under the umbrella of scientific inference, and thus conforms to Wells's rule. And yet the estimate of body mass in an extinct species is probably roughly analogous to extrapolation of data beyond the end of a regression (as in estimates of body size made by Gould [1974] of the giant Irish elk, *Megaloceros giganteus*). We know that both are unverifiable, but there is a very high (although essentially unknown) probability that they are correct within tolerable limits.

Estimating body mass and other variables in cotton rats

To this point I have reviewed the technical aspects of estimating body size and its correlated variables in extinct mammals. In this section, I provide an example of the analyses possible when accurate estimates of body mass can be made for an extinct taxon. Cotton rats (*Sigmodon*) are small, pastoral, runway-making rodents that are ubiquitous in Late Pliocene and Pleistocene deposits of the southern United States. They

Table 5.1. *Regression equations that scale selected physiological and ecological parameters to body mass in extant mammals*

Parameter	Equation	Correlation (r)	Source
Home range (H in hectares; W in g)			
Large herbivores	$H = 0.002\ W^{1.02}$	0.87	Harestad & Bunnell 1979
Large omnivores	$H = 0.05\ W^{0.92}$	0.95	As above
Large carnivores	$H = 0.11\ W^{1.36}$	0.91	As above
Basal metabolic rate (M_b in kcal/day; W in kg)			
Mammals	$M_b = 70\ W^{0.75}$	—	Kleiber 1961
Mass-specific basal metabolic rate (M_b/W in cc O_2/g-h; W in g)			
Mammals	$M_b/W = 3.4\ W^{-0.25}$	—	McNab 1970
Shrews, arvicoline rodents, weasels	$M_b/W = 0.59\ W^{-0.45}$	0.93	Martin (1980)
Cotton rats	$M_b/W = 6.61\ W^{-0.30}$	—	Recalculated from Bowers 1971[a]
Life-span (L in days, W in kg except as indicated)			
Artiodactyla	$L = 2.14 \times 10^3\ W^{0.22}$	0.93	Recalculated from Western 1979[a]
Carnivora	$L = 3.31 \times 10^3\ W^{0.17}$	0.71	As above
Primates	$L = 4.57 \times 10^3\ W^{0.24}$	0.88	As above
Mammals (W in g)	$L = 630\ W^{0.17}$	0.75	Blueweiss et al. 1978
Gestation time (G in days, W in kg)			
Artiodactyla	$G = 117.5\ W^{0.16}$	0.88	Recalculated from Western 1979
Artiodactyla	$G = 117.5\ W^{0.12}$	0.59	Recalculated from McDonald 1984[a]
Carnivora	$G = 51.29\ W^{0.12}$	0.60	Recalculated from Western 1979
Carnivora	$G = 47.86\ W^{0.17}$	0.72	Recalculated from McDonald 1964
Primates	$G = 4.57 \times 10^3\ W^{0.14}$	0.73	Recalculated from Western 1979
Perissodactyla	$G = 191\ W^{0.11}$	0.48	Recalculated from McDonald 1984
Age at first parturition (A in days; W in kg)			
Artiodactyla	$A = 257\ W^{0.24}$	0.69	Recalculated from McDonald 1984
Perissodactyla	$A = 195\ W^{0.32}$	0.93	As above
Carnivora	$A = 251\ W^{0.29}$	0.91	As above
Population density (D in No./km^2; W in kg except as indicated)			
Mammals	$D = 214\ W^{-0.61}$	0.80	Peters 1983
Mammals (W in g)	$D = 1.15 \times 10^{-4}\ W^{-0.78}$	0.80	Recalculated from Damuth 1987[a]

Table 5.1. (*cont.*)

Parameter	Equation	Correlation (r)	Source
Temperate herbivores	$D = 55\ W^{-0.90}$	0.81	Peters 1983
Tropical herbivores	$D = 16\ W^{-0.60}$	0.85	Peters 1983
Tropical herbivores (W in g)	$D = 1.20 \times 10^{-4}\ W^{-0.73}$	0.85	Recalculated from Damuth 1987
Carnivores	$D = 15\ W^{-1.16}$	0.82	Peters 1983
Vertebrate consumers (W in g)	$D = 2.95 \times 10^{-3}\ W^{-0.96}$	0.82	Recalculated from Damuth 1987

Note: W = body mass. Other abbreviations are defined in the table.
[a]If a regression equation was given in the primary literature as $\log y = a - b (\log x)$, it was here converted to its exponential form by the formula $y = (10^a)(x^{-b})$. In their conversion to an exponential equation, both Western (1979) and McDonald (1984) neglected the (10^a) step.

are ideal animals in which to test evolutionary and ecological hypotheses because fossil representatives are closely related structurally to extant species and because a large body of paleontological and neontological information is available for consultation (Martin 1979, 1984, 1986).

Methods

Measurements of first lower molar lengths were taken on 33 specimens of six cricetine species (Martin 1984). A least squares regression line was fitted to the $\log_{10}$ transformed data of body mass (W) plotted against M_1 length (Figure 5.1). Body mass data were then calculated for both extinct and extant species of *Sigmodon* from the predictive equation

$$W = 4.05\ L^{3.33} \quad (5.1)$$

where W is in grams and L is the length of the first lower molar. These mass data were subsequently used in a series of equations to estimate a variety of physiological and life history variables. These equations are as follows:

Metabolic rate. The equation, from Bowers (1971) is:

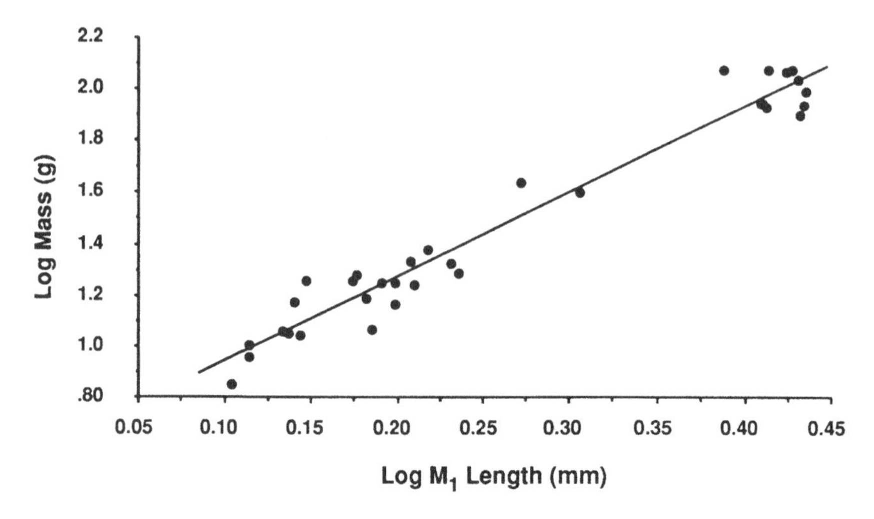

Figure 5.1. Least squares regression line running through scatter diagram of body mass plotted against length of the first lower molar in extant cricetines. Log Mass = 3.31 (Log M_1 length) + 0.611. (Data from Martin, 1984.)

$$M_b/W = 6.61\ W^{-0.304} \qquad (5.2)$$

where M_b/W is mass-specific metabolic rate in ccO_2/g-hr^{-1} and W is in grams. Total resting metabolism (M_b) in kcal/day was calculated as

$$M_b = (M_b/W)(W)(0.115) \qquad (5.3)$$

where 0.115 is a conversion factor including the value of 4.8 kcal/liter oxygen.

Caloric content (standing crop biomass). Caloric content (K) of *Sigmodon* tissue is about 5.0 kcal/g dry mass, and dry mass of cotton rat tissue averages approximately 30% of total adult wet mass (Fleharty, Krause, & Stinnett 1973). Thus, caloric content of an individual is calculated as

$$K = (5.0\ \text{kcal/g})(0.30)(W) \qquad (5.4)$$

and caloric content of the amount of *Sigmodon* biomass in a square kilometer is the product $(K)(D)$, where D = population density.

Population density. A number of methods have been published estimating population density in extant and extinct mammals (Damuth 1981,

1987; Martin 1981; Peters 1983); I found that Peters's equation for temperate herbivores yielded the most accurate results when compared to average values of *Sigmodon hispidus* populations reported in the literature. Consequently, the following equation from Peters (1983) was used to estimate population density in extinct cotton rats:

$$D = 214\ W^{-0.61} \tag{5.5}$$

where D is in numbers per square kilometer and W is in kilograms.

Home range. Home range (H) estimates were generated by McNab's (1963) equation for small mammal "croppers":

$$H = 6.76\ W^{0.63} \tag{5.6}$$

where H is in acres and W is in kilograms (1 hectare [ha] = 2.471 acres).

Results

When the natural logarithm of M_1 length is plotted against time, a general increase of tooth size is documented within the last 3.5 million years (myr; Figure 5.2). The slope of this line, 0.06, also represents the rate constant of change of M_1 length in darwins (Haldane 1949). This value is consistent with other reports of evolutionary rates in mammals (Gingerich 1983).

Comparison of body size changes from four depositional basins in North America (Figure 5.3) led to the conclusion that the increase in body size was the result not of gradual orthoselection, but rather of stochastic events introducing cotton rats of various body sizes (Martin 1986). Because at least one dwarfing speciation event was documented, Wright's rule may be satisfied (Gould & Eldredge 1977). However, large size seems to be adaptive in cotton rats, and linked with high levels of aggressive behavior. Consequently, the overall trend toward large size is quite likely the result of active selection against populations of small cotton rats as they randomly appear. This selection acts at the individual level, and therefore we have the interesting circumstance where Wright's rule may be operative but species selection is not indicated. That is, larger size does not appear to be manifested simply because species with large individuals speciate more rapidly than do those with small individuals.

Physiological and ecological variables for fossil samples of cotton rats were plotted by depositional basin through the geological study period

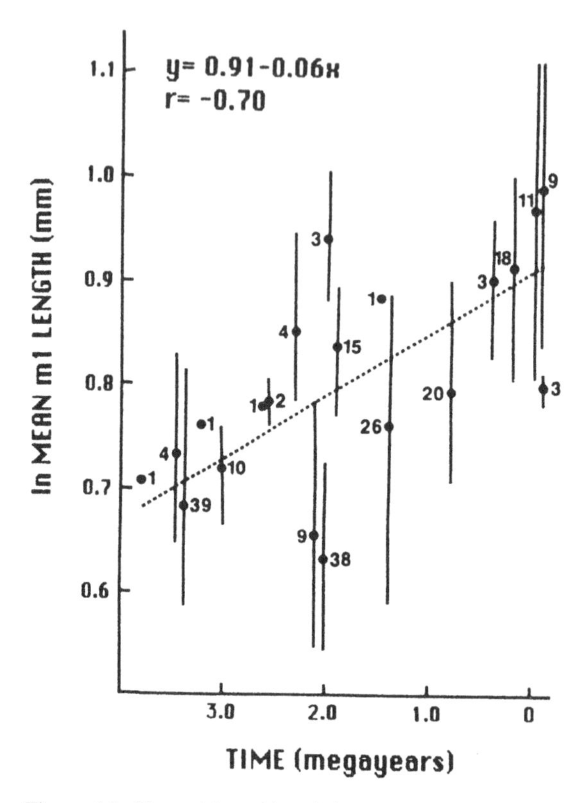

Figure 5.2. Natural logarithm (ln) of mean cotton rat lower first molar (M_1) length plotted against time in megayears (myr). Lines associated with each point represent the observed range of measurements. Numbers are sample sizes. The slope is negative because time on the abscissa decreases from the origin. See Martin (1986) for details. (From Martin 1986. With the permission of *Paleobiology*.)

(Figure 5.4). Although slight swings in body mass are accompanied by rather large changes in population density, cotton rat species of different body sizes were nevertheless considered to be energetically equivalent. This is represented in Figure 5.4 by only negligible change in population respiration (metabolism). Examination of relationships within and between extant *Sigmodon* species and their ecological analogues from different clades (such as *Microtus ochrogaster*) also led me to reject the hypothesis that large size conferred a thermoregulatory advantage among cotton rat species. Thus the physiological and ecological variables that were correlated with, and therefore predictable by, body mass appear to be neutral with regard to adaptation. At least within the cotton

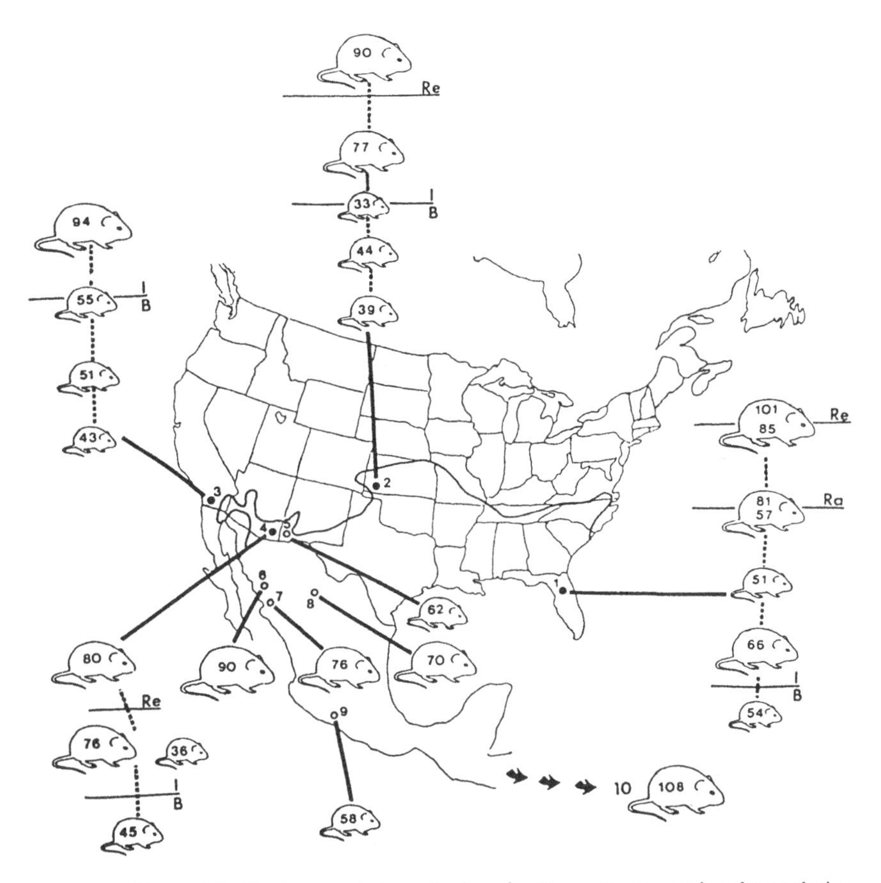

Figure 5.3. Replacement chronologies of cotton rats, comparing the evolution of body mass in *Sigmodon* with the distribution of body mass seen in modern species. The horizontal lines within each chronology separate land mammal ages. B = Blancan, I = Irvingtonian, Ra = Rancholabrean, Re = Recent. Solid circles represent fossil deposits; open circles are areas from which samples of extant species were captured. The solid black line on the U.S. map depicts the present limits of cotton rat distribution. Numbers within cotton rat icons are average mass values for the samples. Dashed lines are for continuity only. At each locality, replacement sequences are listed from oldest to youngest sample. 1 = northcentral Florida: *S. minor* (Haile XVA) – *S. curtisi* (Inglis IA) – *S. libitinus* (Haile XVIA) – *S. bakeri* (two samples; Coleman IIA and Williston IIIA) – *S. hispidus* (two samples; Reddick IA and Recent). 2 = Meade Basin: *S. minor medius* (Rexroad Loc. 3) – *S. minor medius* (Sanders) – *S. minor minor* (Borchers) – cf. *S. curtisi* (Kentuck) – *S. hispidus* (Recent). 3 = Vallecito Fish Creek beds: *S. m. medius* (Layer Cake) – *S. m. medius* (Arroyo Seco) – *S. m. medius* (transition zone between Arroyo Seco and Vallecito Creek faunas) – cf. *S. curtisi* (Vallecito Creek). 4 = San Pedro Valley: *S. m. medius* (Benson and Tusker combined) – *S. m. minor* (36 g; Curtis Ranch) and *S. curtisi* (76 g; Curtis Ranch) – *S. arizonae* (Recent). 5 = *S. fulviventer*. 6 = *S. leucotis*. 7 = *S. alleni*. 8 = *S. hispidus berlandieri*. 9 = *S. mascotensis*. 10 = *S. peruanus*, a large South American species. (From Martin 1986. With the permission of *Paleobiology*.)

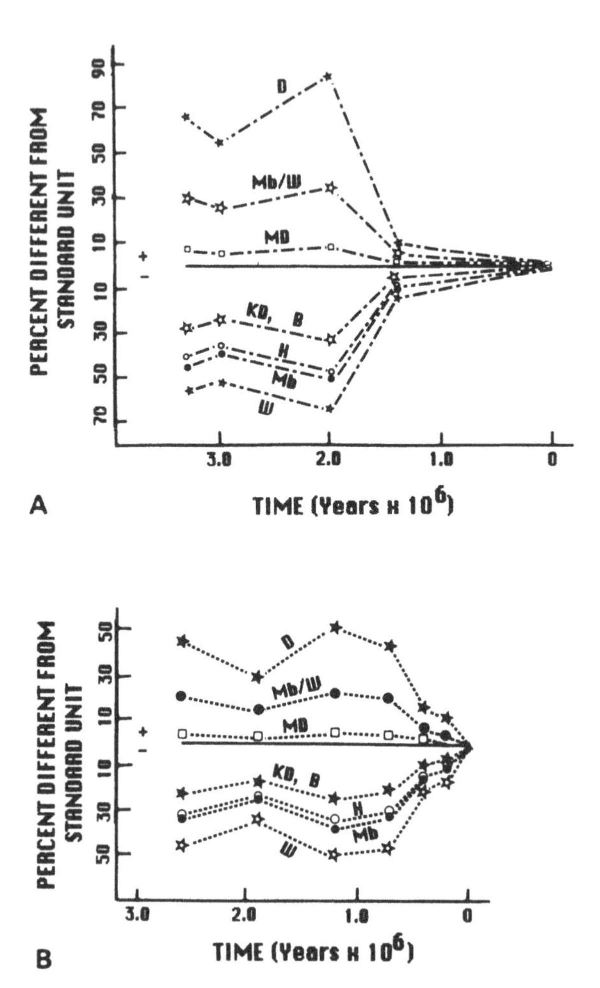

Figure 5.4. Variation in cotton rat body mass and some correlated energetic and ecological parameters during Late Pliocene and Pleistocene time in Meade Basin, Kansas (A) and northcentral Florida (B). D = population density, M_b/W = individual mass-specific metabolic rate, MD = population total metabolism, KD = population caloric content (production), B = biomass, H = home range, M^b = individual total metabolism, W = body mass. The solid horizontal line represents the set of standard values calculated for the latest-appearing, largest species, in this case *S. hispidus*. (From Martin 1986. With the permission of *Paleobiology*.)

rat clade, the major feature that has shaped their body size and diversity seems to be behavior.

Because little morphological evolution of consequence occurred during the 4-myr history of cotton rats in North America, I suggested

(Martin 1986) that most speciation events in a clade of organisms within the same adaptive zone were random and neutral with regard to adaptation. I referred to these as *second-order* speciation events. *First-order* speciation events, those that transfer an organism from one adaptive zone to another and sometimes from one higher taxonomic category to another, were considered to be rare. The first appearance of a cotton rat in New World grassland communities was a first-order event, and was followed, apparently relatively quickly, by a series of second-order events that produced the full species diversity of the cotton rat clade. This explosive proliferation has been known in the literature for many years as "adaptive radiation," but it is a different phenomenon from adaptive radiation at the higher taxonomic levels, the latter of which may be based almost exclusively on first-order speciation events. I consider adaptive radiation to be one of the more important macroevolutionary patterns, and hope that the above observations and suggestions will inspire further consideration of this topic.

Discussion

My recent work with cotton rats began as a study of the evolution of energetic strategies within a clade of organisms. It ended with the conclusion that behavior, not energetics, determined the fate of a given cotton rat species, although both behavior and metabolism may be correlated with body size. Without information on the behavior of modern species, this hypothesis would not have been even remotely substantiated. And, of course, as is the case with the body mass estimates that accompany it, the hypothesis cannot be tested. This does not make it unimportant, as it may, after all, be true. It merely points out another of the limitations of paleobiological reconstruction.

Fortelius (this volume) is disturbed by the "averaging" effect of body mass estimation. Indeed, we will never be able to generate a realistic momentary set of biological measurements for an extinct fauna; there is simply no way to inject the vagaries of environmental pressures on, for example, population density. But over long periods of time these variables have Gaussian distributions, and thus averages are biologically valid species parameters. However, I certainly agree that species originations and extinctions are often driven by stochastic environmental and hereditary events, and that only those events that radically alter the adaptive zone of a species or transfer a species from one adaptive zone to another have any significant effect on the direction of evolution. In a very important sense it is, as Fortelius (this volume) notes, the "outliers" that are the stuff of evolution. I suspect that the major contribution of body mass estimation in the future lies in its descriptive, rather than explanatory, power. In the following sections I examine the potential arena of this methodology.

Phylogenetic reconstruction

Although most biostatisticians might disagree, there is something inherently unsatisfying to me about pure measurement data, in spite of the elaborate statistical analyses that can be performed on them. It wasn't until I saw the body mass estimates for extinct *Sigmodon* species that I could comfortably agree with Cantwell (1969) and Eshelman (1975) that *Sigmodon minor* and *S. medius* were the same species (Martin 1986). Body size is a powerful character, and I am convinced that accurate estimates of body mass within clades can help solve a number of thorny taxonomic problems. I was most heartened to see Czaplewski's (1987) estimates of body size in the earliest *Sigmodon minor* (= *S medius*).

Community reconstruction

Maiorana (this volume) demonstrates that there is a polymodal distribution of body sizes in extant ground-living, foliage-feeding herbivorous mammals. It will be instructive to compare ancient mammalian communities to see if this was true in the past.

Modeling energetics of ancient predators

As noted in an earlier section of this chapter, it is relatively easy to determine the caloric content of an individual or population of an extinct species if body mass is known and a value for caloric content is available from a living analogue. This information can be used to reconstruct the energy content of past hominid diets directly from midden heaps (see Wing & Brown 1979), and also to approximate the energy available to predaceous birds, mammals, and reptiles in a given paleocommunity. These patterns can be followed through time, and may help to illuminate the manner in which predator guilds have evolved.

Testing evolutionary hypotheses

Stanley (1979, 1985) contends that most phyletic body size changes in lineages are the "fine tuning" of evolution, and are relatively unimportant from the standpoint of macroevolutionary processes. It will be instructive to estimate body mass for a clade within a stratigraphically controlled section of long duration (e.g., for artiodactyls and condylarths from the Eocene of Wyoming; Gingerich 1985). Because body mass is exponentially related to most measurement data, the biological significance of changes in linear dimensions may be greater than workers have suspected.

Modeling the evolution of faunal energetics

Damuth (1981) and I (Martin 1986) independently concluded that, at least theoretically, all mammalian species are potentially equivalent energetically, because small species have concomitantly larger population sizes than do large species. If the theory is valid, then a change in diversity of species in a mammalian community may indicate a change in available trophic energy. This could be due to modifications in climatic variables such as rainfall or temperature, or changes in incident solar radiation. Naturally, other taxa besides mammals compete for available energy, and an analysis of this sort is necessarily incomplete. However, if examination of extant mammalian communities demonstrates a tight correlation between solar radiation and mammalian species diversity, we will have a powerful tool for reconstructing past climatic regimes and, incidentally, the amount of energy striking the Earth's surface in times past (see Turner, Lennon, & Lawrenson 1988).

Analyzing functional scaling patterns in ancient taxa

Scaling has value and interest in its own right (e.g., Creighton 1980; Gould 1975) particularly for paleontologists interested in the functional morphology of extinct taxa. Estimates of body mass can provide important data regarding shape changes in evolving lineages.

It has been my good fortune to be involved in the early development of body mass estimation techniques for extinct mammals and, although the results of this symposium volume indicate that there is much work to be done, both in concept and method, there is every reason to believe that the technique will become a permanent and valuable tool for the reconstruction of mammalian community evolution.

Acknowledgments

I am most thankful to John Damuth and Bruce MacFadden for hosting the symposium on body size estimation at the University of Florida which inspired this contribution. My gratitude is also extended to the participants in the symposium, as this paper was greatly improved by their insights. John Damuth's editorial suggestions also proved most helpful. Financial support for my attendance at the symposium was provided by a Faculty Development Grant from Berry College.

References

Blueweiss, L. H., Fox, H., Nakashima, D., Peters, R., & Sams, S. 1978. Relationships between body size and some life history parameters. *Oecologia 37*:257–272.

Bowers, J. R. 1971. Resting metabolic rate in the cotton rat *Sigmodon*. *Physiol. Zool. 44*:137–148.

Cantwell, R. J. 1969. Fossil *Sigmodon* from the Tusker locality, 111 Ranch, Arizona. *J. Mammal. 50*:375–378.

Creighton, G. K. 1980. Static allometry of mammalian teeth and the correlation of tooth size and body size in contemporary mammals. *J. Zool., Lond. 191*:435–443.

Czaplewski, N. J. 1987. Sigmodont rodents (Mammalia; Muroidea; Sigmodontinae) from the Pliocene (Early Blancan) Verde Formation, Arizona. *J. Vertebr. Paleontol. 7*:183–199.

Damuth, J. 1981. Population density and body size in mammals. *Nature 290*: 699–700.

Damuth, J. 1987. Interspecific allometry of population density in mammals and other animals: the independence of body mass and population energy-use. *Biol. J. Linn. Soc. 31*:193–246.

Eshelman, R. E. 1975. Geology and paleontology of the Early Pleistocene (Late Blancan) White Rock fauna from north-central Kansas. C. W. Hibbard Memorial, Vol. 4, pp. 1–60. Ann Arbor: Mus. Paleontol., University of Michigan.

Fleharty, E. D., Krause, M. E., & Stinnett, D. P. 1973. Body composition, energy content, and lipid cycles of four species of rodents. *J. Mammal. 54*:426–438.

Gingerich, P. D. 1983. Rates of evolution: effects of time and temporal scaling. *Science 222*:159–161.

Gingerich, P. D. 1985. Species in the fossil record: concepts, trends and transitions. *Paleobiology 11*:27–41.

Gingerich, P. D., Smith, B. H., & Rosenberg, K. 1982. Allometric scaling in the dentition of Primates and prediction of body weight from tooth size in fossils. *Am. J. Phys. Anthropol. 58*:81–100.

Gould, S. J. 1974. The origin and function of "bizarre" structures: antler size and skull size in the "Irish elk," *Megaloceros giganteus. Evolution 28*: 191–220.

Gould, S. J. 1975. On the scaling of tooth size in mammals. *Am. Zool. 15*: 351–362.

Gould, S. J., & Eldredge, N. 1977. Punctuated equilibria: the tempo and mode of evolution reconsidered. *Paleobiology 3*:115–151.

Haldane, J. B. S. 1949. Suggestions as to quantitative measurement of rates of evolution. *Evolution 3*:51–56.

Harestad, A. S., & Bunnell, F. L. 1979. Home range and body weight – a reevaluation. *Ecology 60*:389–402.

Jerison, H. 1973. *Evolution of the Brain and Intelligence.* New York: Academic Press.

Jolicoeur, P. 1973. Imaginary confidence limits of the slope of the major axis of a bivariate distribution: a sampling experiment. *J. Am. Stat. Assoc. 68*:866–871.

Kleiber, M. 1961. *The Fire of Life: An Introduction to Animal Energetics.* New York: John Wiley.

Martin, R. A. 1979. Fossil history of the rodent genus *Sigmodon. Evol. Monogr. 2*:1–36.

Martin, R. A. 1980. Body mass and basal metabolism of extinct mammals. *Comp. Biochem. Physiol. 66A*:307–314.

Martin, R. A. 1981. On extinct hominid population densities. *J. Human Evol. 10*:427–428.

Martin, R. A. 1984. The evolution of cotton rat body mass. In *Contributions in Quaternary Paleontology: A Volume in Memorial to John E. Guilday,* ed. H. H. Genoways & M. R. Dawson, pp. 252–266. Pittsburgh: Carnegie Mus. Nat. Hist. Special Publ. No. 8.

Martin, R. A. 1986. Energy, ecology and cotton rat evolution. *Paleobiology 12*:370–382.

McDonald, J. N. 1984. The reordered North American selection regime and Late Quaternary megafaunal extinctions. In *Quaternary Extinctions: A Prehistoric Revolution,* ed. P. S. Martin & R. G. Klein, pp. 404–439. Tucson: University of Arizona Press.

McHenry, H. M. 1974. How large were the australopithecines? *Am. J. Phys. Anthropol. 40*:329–340.

McHenry, H. M. 1975. Fossil hominid body weight and brain size. *Nature 254*:686–688.

McHenry, H. M. 1976. Early hominid body weight and encephalization. *Am. J. Phys. Anthropol. 45*:77–84.

McManus, J. J. 1974. Bioenergetics and water requirements of the redback vole, *Clethrionomys gapperi. J. Mammal. 55*:30–44.

McNab, B. K. 1963. Bioenergetics and the determination of home range size. *Am. Nat. 97*:133–140.

McNab, B. K. 1966. The metabolism of fossorial rodents: a study of convergence. *Ecology 47*:712–733.

McNab, B. K. 1970. Body weight and the energetics of temperature regulation. *J. Exp. Biol. 53*:329–348.

McNab, B. K. 1974. The energetics of endotherms. *Ohio J. Sci. 74*:370–380.

McNab, B. K. 1978. Energetics of arboreal folivores: physiological problems and ecological consequences of feeding on an ubiquitous food supply. In *The Ecology of Arboreal Folivores,* ed. G. G. Montgomery, pp. 153–162. Washington, D.C.: Smithsonian Institution Press.

Peters, R. H. 1983. *The Ecological Implications of Body Size.* Cambridge: Cambridge University Press.

Pilbeam, D., & Gould, S. J. 1974. Size and scaling in human evolution. *Science 186*:892–901.

Radinsky, L. 1978. Evolution of brain size in carnivores and ungulates. *Am. Nat. 987*:815–831.

Scott, K. 1983. Body weight predictions in fossil Artiodactyla. *Zool. J. Linn. Soc. 77*:199–215.

Smith, R. J. 1980. Rethinking allometry. *J. Theor. Biol. 87*:97–111.

Sokal, R. R. & Rohlf, F. J. 1969. *Biometry.* San Francisco: W. H. Freeman.

Stanley, S. M. 1979. *Macroevolution: Pattern and Process.* San Francisco: W. H. Freeman.

Stanley, S. M. 1985. Rates of evolution. *Paleobiology 11*:13–26.

Turner, J. R. G., Lennon, J. J., & Lawrenson, J. A. 1988. British bird species distributions and the energy theory. *Nature 335*:539–541.

Van Gelder, R. G. 1968. The genus *Conepatus* (Mammalia, Mustelidae): variation within a population. *Am. Mus. Nat. Hist. Novitates 2322*:1–37.

Western, D. 1979. Size, life history and ecology in mammals. *Afr. J. Ecol. 17*:185–204.

Wing, E. S., & Brown, A. B. 1979. *Paleonutrition.* New York: Academic Press.

Zar, J. 1984. *Biostatistical Analysis,* 2nd ed. Englewood Cliffs, N.J.: Prentice Hall.

6

Evolutionary strategies and body size in a guild of mammals

VIRGINIA C. MAIORANA

Body size differences among related animals are frequently thought to indicate competitive displacement (Hutchinson 1959; MacArthur 1972). This is particularly true of animals that eat particulate and size-variable sorts of food, such as insects or seeds. Competing species can share such a resource by specializing on different parts of the size range available. Among feeders of particulate food, size differences in only the feeding organ itself sometimes exist. The resource for which such species compete is typically assumed to be food, but this may not always be the case (Maiorana 1978a).

Size displacement is a less obvious consequence of competition among species that feed on a food, such as terrestrial foliage, which is not handled in a size-specific way. A caterpillar and a deer both chew leaves of trees, but the former may consume only a part of one leaf a day whereas the latter may take several in one bite. Size differences are not obviously sorting the available food among species and thus are not often seen as a direct cause of size displacement. Whether herbivores are in fact food-regulated as a trophic level is still controversial and therefore so is the relative importance of competition for food in assembling species within a community. Studies are now beginning to demonstrate that body size influences the species of plants on which a herbivore can feed efficiently; they suggest that body size may be an important component in any competitive interactions over a limiting food supply (e.g., Belovsky 1984). Body size can also affect competitive ability in other ways, as I discuss below. Alternatively or in addition,

In *Body Size in Mammalian Paleobiology: Estimation and Biological Implications,* John Damuth and Bruce J. MacFadden, eds.

predators specializing on prey of different sizes can influence the size of their prey and thereby the shape of foliage-feeding assemblages (see below).

Studies of species-size displacement within a guild have typically been restricted to the relatively few species that are sympatric in a local region (typically fewer than ten). The meaning of the observed size differences is unclear because of lack of knowledge about the size distribution of the guild globally (see Connor & Simberloff [1986] and Wiens [1982] for discussion of some of the problems involved). Moreover, the evolutionary significance of the global size distribution of a guild is important to resolving the controversy over competitive versus neutral assembling of species in local communities. If the global distribution is itself a product of competitive size displacement, then random selection of species will often, but not always, produce assemblages that are competitively compatible.

Unfortunately, the sporadic analyses of global size distributions that have been made over the past several decades have focused on taxonomic groups that contain ecologically diverse species (e.g., Caughley & Krebs 1983; Griffiths 1986; Hutchinson & MacArthur 1959; Kirchner 1980; May 1978). Such approaches obscure the ecology and thus the evolutionary significance of the shape of the body size distributions.

Focusing on the global as well as local body-size distributions of species in a single guild may uncover some new patterns that need explaining as well as provide a different perspective from which to attack the old problems and controversies about size differences of species in communities. I propose to do this for one part of the guild of herbivores that consume green foliage. I restrict my analysis here, for several reasons, to ground-living mammals that feed on exposed parts of plants. This group of mammalian herbivores has the widest size range of any mammalian guild one can define, because there are few constraints on body size imposed by ground living. Consequently, the body size distribution of the guild is more likely to reveal the effects of competition or predation as selective forces if they are important. In contrast to the viruses, nematodes, insects, and mites which also exploit exposed green foliage, the geographic and temporal ranges of mammals are better known. It is possible to compile nearly complete distributions for virtually any spot in the world. A good fossil record will enable one to document the body size distributions of this group of herbivores throughout its history once reliable techniques for estimating body sizes of species from fragmentary fossils become available. No mammalian species has a naturally global distribution, so body size distributions of mammals from different parts of the world are assemblages of different species. Thus, any similarities among the distributions likely reflect convergent or parallel evolution at the community level (which does not imply selection of communities per se), and thus potentially provide evidence for selective shaping of the distributions. In contrast to other guilds of mammals, foliage feeders lack direct competition from other taxa of equal size, in particular birds; thus their size distributions reflect essentially all species in this guild and size range. Finally, many aspects of the biology of all the species in this guild are relatively well known for many areas; therefore, one can explore in more depth the significance of the size distribution, its origin, and its adaptive consequences, on the biology of the included species.

Following a description of the current body-size distributions of ground-dwelling mammalian foliage feeders in various regions of the world and a discussion of several hypotheses on their origin and differences, I investigate one consequence suggested by the observed distributions: Predation pressure is a nonmonotonic function of body size and is predicted to cause a nonmonotonic response in the size-specific life-history adaptations of the herbivores.

Polymodality in the distribution of foliage-feeding animals

When one considers the animals that one can find feeding on the grasses and forbs in a meadow, they seem to come in a variety of nonoverlapping groups with respect to body size. Deer, rabbits, and voles are clearly distinct, and even the largest insects are considerably smaller than voles. Insects themselves seem to come in a variety of sizes, ranging from relatively large caterpillars and grasshoppers to tiny aphids. Mites, nematodes, and viruses are essentially invisible to the unaided eye, and each of these taxa may also come in a variety of sizes. To quantify these impressions of the body sizes of foliage feeders, I have plotted in Figure 6.1 the body sizes of the North American mammals in this guild and the insects that were collected from collard plants (Root 1973). The relative number of insects in relation to mammals is not of importance here, merely the size distribution of this small but perhaps representative sample.

The distributions suggest a polymodality of sizes for both insects and

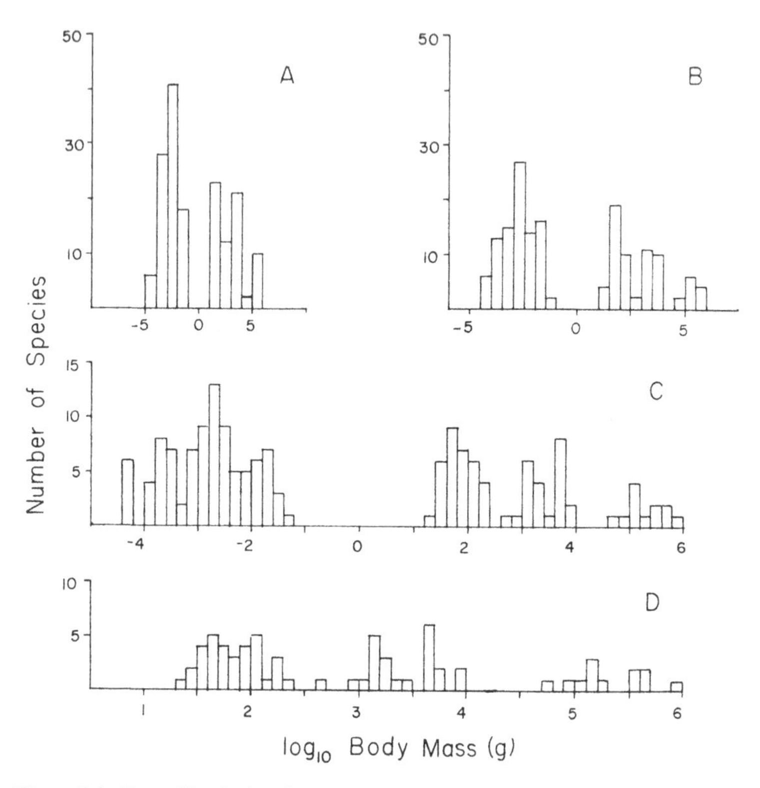

Figure 6.1. Logarithmic size distributions for foliage-feeding mammals of North America (Burt and Grossenheider 1976) and herbivorous insects on collard plants (Root 1973) at different scales of resolution. Insects are less than 1 g in body mass; mammals, greater. A through D represent successive doubling of the scale of the size axis. In D, only mammals are shown. The dry weights given for insects by Root (1973) were converted to wet weight by assuming that dry weights were 20% of wet weight.

mammals. However, note that the distinctness and even the number of clusters vary with the scale on which they are plotted (Figure 6.1). If one is to theorize about the significance of a polymodal distribution of body sizes, it is important to specify what scale one is using to find clusters. Because the biological meaning of such periodicity depends on its magnitude, the scale at which distributions are examined is central to the conclusions reached. Indeed, reducing or expanding the scale can eliminate the periodicity and thus the problem (see Caughley [1987] for an example). Ideally, the scale chosen for analysis should not be finer than the accuracy of the size data used.

Geographic variation in body size distributions

Results from a preliminary survey of mammalian body-size distributions are illustrated in Figure 6.2. The data were collected from regional handbooks that contained information on size and diet. If masses were given, I used those in the handbook from which I extracted the list of species; if not, I used the masses from another source or estimated them from related species of a similar length. I used only those rodents that eat mostly green vegetation and very few seeds or fruits. Species that feed mainly on the underground parts of the plants were not included nor were beavers and porcupines, both of which eat a lot of bark.

Figure 6.2 suggests the following conclusions: For North America (A) the three distinct clusters of body size found in more local areas (states, B–E) characterize the continent as well. In contrast, the three clusters that are typical of a European country (F–H) are indistinctly seen in the size distribution of the entire fauna. In Europe there are relatively fewer medium-sized species (large rodents and lagomorphs) globally and the average size of large species is smaller than that in North America. Both factors produce a distribution that approaches bimodality. The size distribution for Pakistan (I) is also more bi- than trimodal. In Africa (J–P), where the greatest number of larger mammals have survived, the size distribution appears somewhat polymodal with approximately four to five clusters (two to three among ungulates). One can tentatively conclude that in all areas except North America, lagomorphs and large rodents have widespread distributions and so are disproportionately represented in local faunas. In North America many of the medium-sized herbivores have restricted ranges, and they are as proportionately represented in local faunas as the other two size clusters. For example, California contains 33%, 39%, and 33% of the small, medium, and large species of North American foliage-feeders, respectively.

The cluster of medium-sized foliage feeders in North America is bimodal in shape and that of the small species is somewhat bimodal when the size scale is expanded (Figure 6.1D). Small ungulates are missing in North America and Europe, producing a larger gap between the medium and large cluster of body sizes. In contrast, small–medium rodents (100–300 g in weight) are scarce in the Old World, which produces a larger gap between small and medium herbivores than between medium and large. An examination of the body sizes of the foliage-

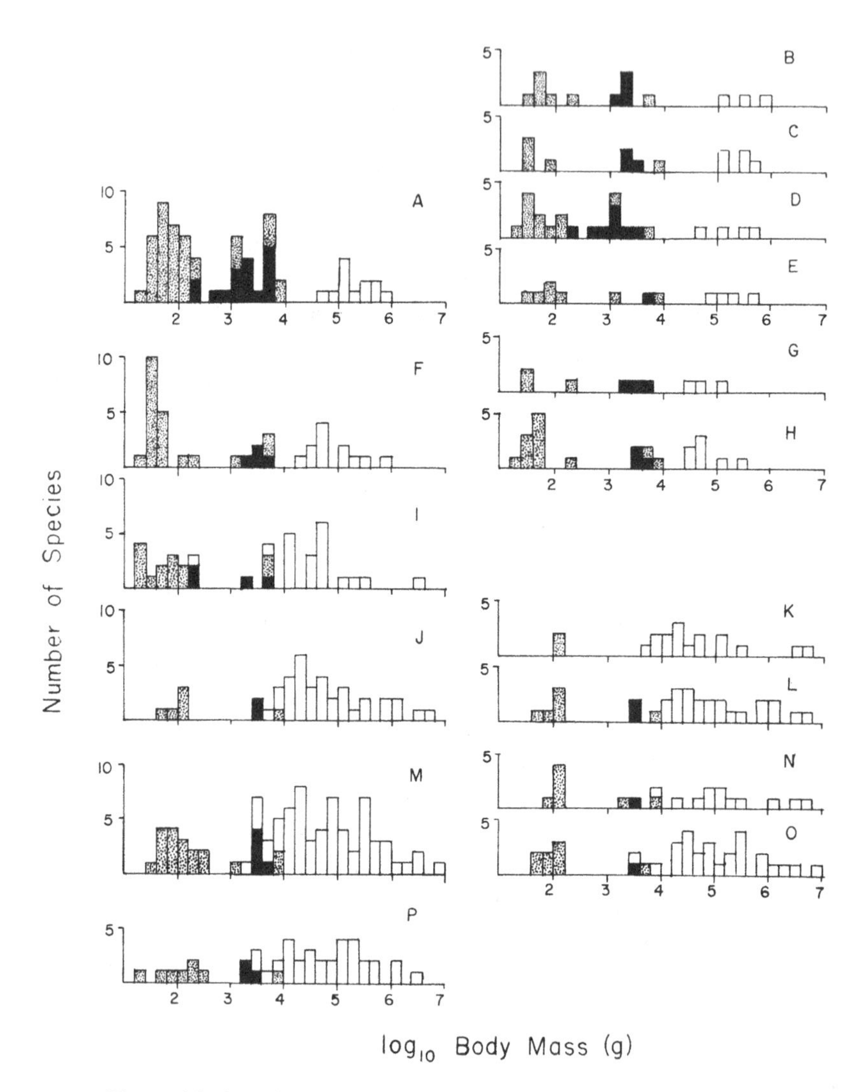

Figure 6.2. Logarithmic size distributions of foliage-feeding mammals from various areas of the world. Stippled bars represent rodents; solid bars, lagomorphs; and open bars, other foliage-feeders, which are mostly ungulates. Sources of information follow. (A) North America: Burt and Grossenheider (1976); (B) Virginia: Bailey (1946): (C) Michigan: Burt (1948); (D) California: Ingles (1947); (E) Northern Alaska: Bee and Hall (1956); (F) Europe: Corbet (1980): (G) England: Corbet and Southern (1977): (H) Italy: Toschii (1965); (I) Pakistan: Roberts (1977); (J–L) Nigeria: Happold (1987); (K) Nigeria, rainforest; (L) Nigeria, savannah; (M) East Africa: Kingdon (1971–1982): (N) Uganda, Rwenzori Park: Delany and Happold (1979): (O) Serengeti: Hendrichs (1970); (P) Botswana: Smithers (1971).

feeding guild through the Pleistocene might help clarify the reasons for the differences observed now in the body size distributions of these mammals.

The various size-distributions from regions of diverse geographic extent suggest that the number of large species seems not to influence directly the number of smaller species. In Africa, where there is by far the greatest number of large species, there are just more possible clusters, which are approximately equally spaced apart (see below). Also, the more local faunas retain nearly the entire size range of the more global regions; clusters are more frequently reduced than eliminated. Even more locally, the foliage-feeders characteristic of a particular type of habitat rarely overlap in size. Such a pattern of decreasing overlap in body size of guild members in more local areas, in which the species are more likely to compete, is typically taken as evidence of competitive size displacement. Such a pattern is also expected most of the time from a random sampling of the more global distribution. It does not really matter whether the local sets of foliage feeders are randomly or selectively assembled, if the global distribution reflects competitive size displacement.

An explanation for polymodality involving competition

The distributions illustrated here are clearly not unimodal on a logarithmic scale, as most distributions of large taxonomic groups have been observed to be (e.g., Hutchinson & MacArthur 1959; Kirchner 1980; May 1978). A polymodal distribution of sizes can be interpreted as a consequence of competitive displacement if one can explain the magnitude of the observed differences (see below) but noncompetitive factors can also cause disruptive selection (cf., Griffiths 1986).

The suggestion of a regular periodicity in the size distribution of African ungulates (also discussed by Griffiths [1986] and Jarman [1974]) is reinforced by plotting the data on an expanded scale (Figure 6.3). For this analysis I plotted the maximum body masses of larger African terrestrial foliage feeders for the entire continent and three smaller regions. In these comparisons I used the masses obtained from Haltenorth and Diller (1980) for all African species. This procedure means that the same species will be in the same position in all of the distributions. Thus, the distributions do not reflect intraspecific geographic variation in size which is often significantly variable in mammalian spe-

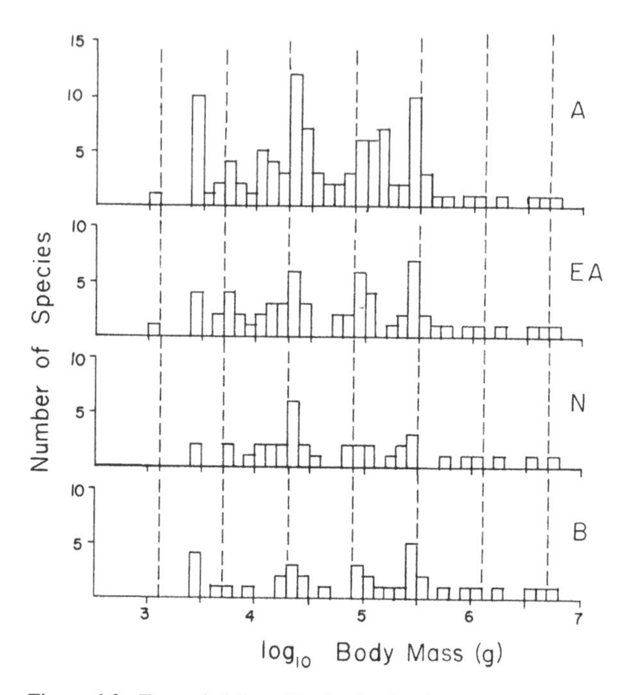

Figure 6.3. Expanded logarithmic distributions of the maximum body masses of larger (greater than 1 kg) foliage-feeding mammals of Africa (A) and three subregions (EA, East Africa; N, Nigeria; B, Botswana). The masses given by Haltenorth and Diller (1980) were used for all regions. The dashed lines indicate the position of successive peaks on the assumption of a regular fourfold difference in body mass.

cies over the distances involved here. However, at the level of clumping used here, geographic size variation is relatively unimportant.

The results suggest a striking periodicity with successive peaks visually occurring at approximately four times the mass of the previous peak. The lines indicate my estimate of best fit of a fourfold periodicity to the actual distributions. If the suggested periodicity is biologically meaningful, it should also fit the peaks of smaller mammals in this guild as well as the size distributions from other regions of the world. An expanded plot of the size distributions of all North American and East African terrestrial foliage feeders does suggest that the actual size clusters of smaller species are concordant with the periodicity derived from larger species (Figure 6.4). There are gaps, as was discussed above, but the position of the small rodents in this guild centers near a predicted size for a peak. The suggested bimodality in these rodents has a peri-

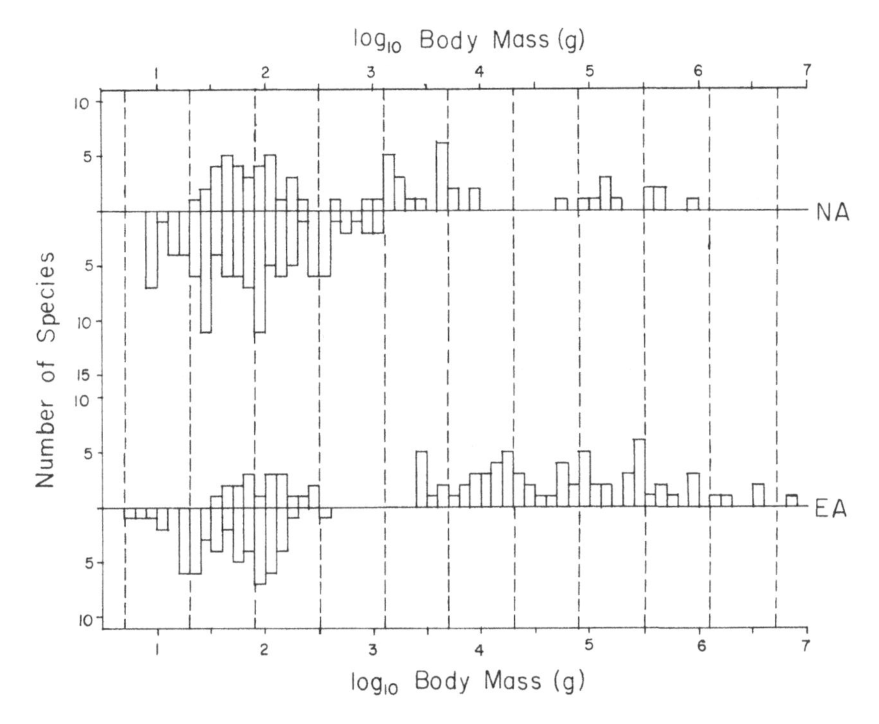

Figure 6.4. Expanded logarithmic size distributions of the entire foliage-feeding guild of North America and East Africa above the axis and the distribution of other terrestrial rodents from the same region below the axis. The sizes from the original sources were used. The dashed lines indicate the position of peaks at fourfold intervals as predicted from the distributions in Figure 6.3.

odicity somewhat smaller than expected. However, the size distribution of other terrestrial rodents (omnivores, seed-eaters, and insectivores), which as a group are considerably more numerous than foliage feeders, fits the predicted fourfold periodicity remarkably well (Figure 6.4).

The magnitude of the periodicity should have a biological interpretation if it is real and not an accidental pattern. Unfortunately, polymodality in an arbitrary distribution is difficult if not impossible to test statistically (see Roth 1979, 1981). Periodicity itself, though, can be tested, but existing tests distinguish merely between periodicity and randomness, ignoring the poorly defined but large and realistic set of cases of nonperiodic and nonrandom variation. This problem and others have become notorious in the controversy on periodicity of mass extinctions (see Hoffman [1989] and references therein). I have therefore not used such tests here.

Nevertheless, one can use statistical techniques to determine which periodicity fits the data most closely. The closest fit is statistically the "real" periodicity, but it is not statistical proof of the existence of any periodicity. For the type of data I have, I use a procedure for finding the best periodicity from a range of possible ones that blends techniques from Roth (1979) and L. M. Van Valen (personal communication).

Seven periodicities from two to eight times the body mass were chosen. These range from three to nine units on the mass distributions under consideration and represent the closest alternative periodicities consistent with the accuracy of the data. I first collapsed the distributions into one period, modulo the periodicity according to the method of Roth (1979), except that for each of the three figures I used the lowest value in each as the starting point for all distributions in that figure. Since a fourfold periodicity was visually expected, I tested three periodicities on either side. What I compared was the standard deviation of the collapsed distributions. Because the standard deviation is sensitive to the range of values, I expanded the distributions below log 4 and contracted those above to have the same log range for each periodicity. The midpoint of each bar was used in the calculation. Additionally, I shifted the collapsed distributions to have the form with least variance, which implies the best fit for a periodicity. Because the distributions are collapsed, the origin is arbitrary and the right end of each distribution abuts the left end. Despite the collapsing and expansion, uniform distributions with fewer bars have lower standard deviations than those with more because of the positions of the midpoints of the bars. To eliminate this small bias, I divided the observed standard deviations by the standard deviation of the uniform distribution for each distribution and periodicity. I call this ratio the "adjusted standard deviation." The results of nine distributions for seven periodicities are summarized in Table 6.1.

On two criteria – mean "adjusted standard deviation" and number of cases in the lowest and next-to-lowest category – a periodicity of four is the best. Thus, there may be a periodicity of body mass among terrestrial, herbivorous mammals that lies near four times the body mass. Because this result is based on faunas of two continents with virtually no overlap in species and even little in families, the results suggest an ecological rather than a phylogenetic explanation.

A regular polymodal distribution of the type observed for this guild of mammals can be explained (although perhaps not uniquely so) by

Table 6.1. *"Adjusted standard deviations" for the mass distributions of mammals illustrated in Figures 6.3–6.5 for superimposed periodicities of different magnitudes*

		Superimposed periodicities						
Distribution	No.	log 2 (0.3)	log 2.5 (0.4)	log 3.2 (0.5)	log 4 (0.6)	log 5 (0.7)	log 6.3 (0.8)	log 8 (0.9)
Large foliage feeders[a]								
Africa	106	0.934	0.921	0.881	0.940	1.016	0.885	0.869
E. Africa	65	0.952	0.912	0.879	0.764	0.876	0.949	0.980
Nigeria	38	0.940	0.980	0.844	0.864	0.981	0.874	0.853
Botswana	36	0.923	0.827	0.873	0.834	0.981	0.885	0.851
All foliage feeders[b]								
E. Africa	88	0.971	0.957	0.963	0.922	0.914	0.971	1.040
N. America	68	0.977	0.865	0.861	0.927	0.927	0.972	1.014
All herbivores[b]								
E. Africa	142	0.991	0.978	0.956	0.946	0.888	0.960	0.978
N. America	172	0.987	0.986	0.935	0.938	0.980	1.012	0.983
Number of ranges of large foliage feeders[c]								
E. Africa	169	0.998	0.980	0.962	0.915	0.918	0.944	0.984
Mean adjusted SD		0.964	0.934	0.906	0.895	0.942	0.939	0.949
No. distributions								
Lowest SD		0	1	3	2	2	0	1
Next lowest SD		0	1	1	4	2	0	1

Note: The lowest value for each distribution is underscored.
[a]Figure 6.3; [b]Figure 6.4; [c]Figure 6.5.

competitive displacement. Species in adjacent peaks can (by assumption) competitively coexist on the basis of size alone, whereas those that fall into the same peak presumably differ in some other dimension of niche space. The distance between peaks is the hypothetical average difference in size necessary for the coexistence of species that are otherwise ecologically indistinguishable. A precise mechanism for fourfold mass differences affecting competition is unclear (but see discussion below); the existence of a pattern may stimulate development of theory. The fourfold mass difference in mammals is compatible with the approximate 1.6 linear difference (which translates to a fourfold mass difference) in bird communities that was noted by Griffiths (1986). The periodicity in mammals may be even more general (Maiorana, unpublished data). However, if such a mechanism exists, why are the peaks separated by a magnitude of four times the body mass?

Hutchinson's (1959) famous ratio for minimum separation of potentially competing congeners suggests a twofold difference in mass or a 1.3 ratio in length. I have shown that the Hutchinsonian ratio may represent the difference in size necessary to produce zero overlap in the size distributions of adjacent congeneric species (Maiorana 1978b). The magnitude of the size separation or ratio then depends on how variable the species are. More variable species have larger ratios than less variable ones. Are mammals unusually variable, so that a fourfold difference in mass (or a 1.6 ratio of length) is necessary for competitive coexistence? No, they tend to be average in this respect. For the average mammalian species the largest individual is about double the mass of the smallest (see Figure 6.5). The fourfold difference, which is twice what is expected from the average species' size range, may be a result of diffuse competition. Such competition can cause a spreading out of peaks, but a priori the magnitude is unknown.

In relatively local areas there may be more than one species in each size cluster. Although these species of nearly the same size differ ecologically in other ways, they still exert some competitive influence that may lead to minor shifts in size. As a consequence, the total size range of all the species that fall into one cluster is larger than the size range of any single species. The fourfold difference reflects competitive displacements of sets of species rather than single species. Some support is given to this idea by the body mass distribution of insects collected from collards (Figure 6.1). The distribution suggests a sixfold difference in mass between peaks, which translates to double the typical species' size range (Maiorana, unpublished data).

To examine whether the idea explains the observed polymodality of mammals, I plotted the size ranges for the East African large, terrestrial foliage feeders and summed the number of ranges that fell into each size interval (Figure 6.5). The distribution of size ranges shows the same periodicity as the distribution of maximum sizes. The position of the peaks is shifted to a smaller size, as expected. If one divides the distance between peaks in half (units of twofold magnitude), there should be and are distinct valleys and peaks with respect to the number of species whose size ranges are centered in the size unit (Figure 6.5). The position chosen was arbitrary; moving one unit to the right does not substantially influence the results.

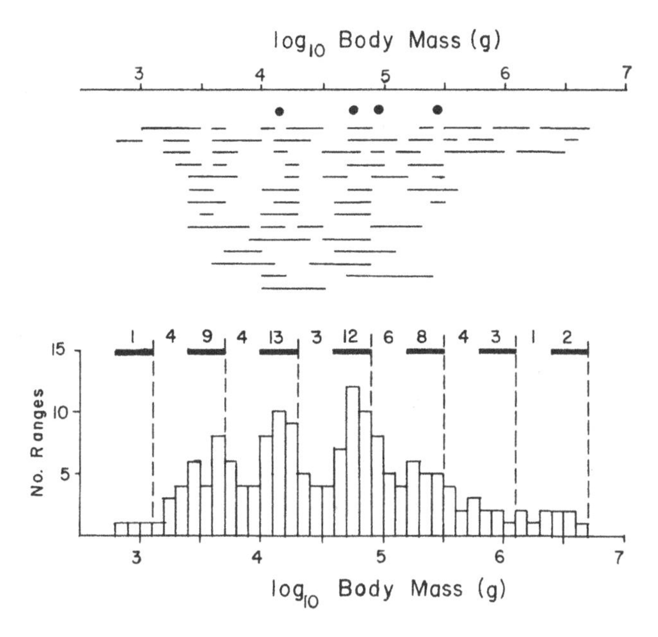

Figure 6.5. Logarithmic size ranges of species of East African larger foliage-feeders (above) with a distribution of the number of species' ranges (data from Kingdon 1971–1982) that fall in each unit of mass (below). The dashed lines indicate the predicted periodicity derived from the modes of maximum body masses in Figure 6.3. The numbers above the bars and spaces between the arrows are the number of species for which 50% or more of their body size ranges fall into that size interval. The bars designate predicted peaks; the spaces, predicted valleys.

In conclusion, foliage-feeding, terrestrial mammals cluster together in distinct size classes which are separated by distances that are compatible with an interpretation of competitive size displacement. In all regions surveyed, there are gaps in the distributions that cannot be explained adequately by competition. Other factors that may have shaped the size distributions of this guild of mammals are discussed below.

Origin and ecological significance of the body size distributions

Clearly, the body sizes of foliage-feeding mammals fall into distinct clusters in approximately the same position on the body mass distributions in all parts of the world, despite differences in the species that occur in these different regions. Different regions vary with respect to which of the expected clusters are present. This result suggests that these observed modal sizes are adaptive for some reason. The cause seems not to be simply competitive displacement because this predicts only a constant distance between successive clusters and not necessarily the same position of the peaks on the size axis as one goes from one community or region to another. Nor is it a taxonomic constraint, because each taxon can be (and is in various parts of the world) composed of organisms of very different sizes. There have even been rhinoceros-sized rodents, pony-sized elephants, insects with a wingspread of nearly a meter, and nematodes as long as a rabbit. A historical perspective would be of help here in confirming the relative stability of the position of these clusters, but it is not yet available. Knowledge of size change through time is essential for understanding the origin and adaptive significance (if any) of the current distributions. Such information needs to be assembled before serious theorizing of causes. Here, I merely describe some possibilities.

Any competitive model for the coexistence of different-sized species requires a source of heterogeneity that gives each size a place in "niche space" where it is competitively superior. Some possibilities include the following:

Digestibility of plant food: Larger foliage-feeding mammals can digest older and tougher plant material more efficiently than smaller species can because the food stays in the intestinal tract for longer periods of time (Demment & van Soest 1985; McNaughton & Georgiadis 1986). Although larger species may prefer younger or more succulent vegetation, smaller species may be able to harvest this at a faster rate because they can get to it faster. Small mammals appear to cover a greater proportion of their annual home range daily than do larger ones (Maiorana, unpublished data; Swihart, Slade, & Bergstrom 1988) and thus can be more selective about what they eat since they encounter all that is available more or less each day. Thus the many voles that inhabit the home range of a single deer can keep the more succulent vegetation grazed and thus largely unavailable to the deer, but they cannot survive on the tougher vegetation.

Climatic fluctuations: The frequency at which adverse periods of climate occur may be relative to body size. That is, periods of severe weather that adversely affect population growth or survival of small species are more frequent than those that affect larger species, because small species have less physiological buffering capacity. However, small species have a greater reproductive potential to spring back after such an interval than do large species. When very infrequent periods of adverse climate occur, large species may go extinct at least locally and probably globally more frequently than small species, a disadvantage of large size that helps explain why small species continue to exist. During these very severe but infrequent periods of adverse climate, smaller species have the advantage of not needing absolutely as much food as larger species to survive although as individuals they starve more rapidly. It is conceivable that the actual periodicities of climatic fluctuations are such that mammals of the different size clusters observed are at different optima for different severities.

Rather than trying to explain the adaptive significance of the observed modal sizes, one might ask why there are so few foliage-feeding species in the areas of the gaps. It could be that there are adaptive thresholds that keep the mammals clustered around certain sizes. The following possibility applies to the broad trimodality that characterizes the distribution of North American and European foliage feeders, and the gap between 500 and 1000 g that occurs in all regions.

The major alternative to competition is predation. Small mammals are nocturnal and burrow to escape predators. They must be active during part of the day (and thus accessible to predators) because they cannot store enough to last from one night to the next (see Schmidt-Nielsen 1984). Somewhat larger mammals also burrow but can remain entirely nocturnal. Large mammals can be diurnal and run to escape their predators. In temperate climates, however, small ungulates may be excluded for physiological reasons. They cannot burrow to escape cold weather and are too small to be buffered adequately by their size alone. The large cluster of small ungulates that is typical of Africa and Pakistan drops out in North America and Europe. It may be that the three major modes of adaptation are separated by thresholds (perhaps spectral, *sensu* Van Valen [1969:195]) that inhibit the occurrence of masses close to their size. If so, such adaptive valleys should inhibit mammals from evolutionarily changing their body size across them, so clades should be concentrated in single size classes when some adaptive factor such as predation or physiology is important in creating the valleys.

Mammalian (and avian) carnivores, which are the major predators of these mammals, are of intermediate size relative to the herbivores. For example, in North America 70% of the carnivoran species cluster between 1 and 20 kg, the size range corresponding to the middle cluster of foliage feeders. Over evolutionary time carnivores may have selected for divergence in prey species size at both extremes and a loss of species in the middle. Smaller species have greater reproductive potential to compensate for loss to predators and individually are less rewarding to a predator than are larger species. Larger species might be able to get too large to be caught when adult and in good condition. Medium-sized species have little alternative but to perfect their defenses such as hiding or armor. In fact, the more extreme types of morphological and behavioral defenses are usually found in medium-sized species (e.g., porcupines, beavers, and armadillos, although New World Neogene glyptodonts were a counterexample).

Predation does not explain very well the fact that medium-sized species are about as abundant as large ones locally and in North America even regionally. However, it can explain the pattern of small mammals increasing and large ones decreasing in body size on islands, which often have reduced predation, a topic I discuss more fully at the end of the next section. It is at least equally likely that the size distribution of predators has been selected for by that of their prey. Nevertheless, the predators existing at the present time are expected to exert greater pressure on medium-sized species than on either larger or smaller prey. This pressure should be evident in the life-history adaptations of the species, a prediction I investigate in the next section.

Predation and a dip in the *r–K* continuum for mammalian herbivores

Defense against predation can take the form either of rendering the individual difficult to capture or of having it reproduce rapidly enough to compensate for loss to predators (see Maiorana 1979). Intermediate-sized mammalian herbivores show more elaborate defenses of both varieties than do smaller or larger relatives. Here I present evidence for greater *r*-adaptation among medium-sized herbivores. *r*-Adaptation as opposed to *K*-adaptation as a defense involves rapid pre- and postnatal growth, large and frequent litters, little maternal care, and early maturity as a consequence of low expected survivorship. Because body size has been considered a correlate of the *r–K* dichotomy, smaller species have been seen as more *r*-selected than larger species (Pianka 1970). However, the problem I want to analyze is whether medium-sized mammals are more *r*-selected than would be expected for their size. To do so, one needs a technique for removing body size from the definition of *r–K* selection.

Removing body size from the r–K *continuum*

Body size influences life history traits because of the disproportionate scaling of metabolic rate on body size. Metabolic rate reflects how fast organisms are processing energy and thus sets limits on how fast they can grow and reach maturity or how much they can put into reproduction (e.g., Blueweiss et al. 1978; Calder 1983). Given metabolic rate and either the number or relative size of offspring, one can derive scaling factors for all of the other major life-history parameters. In this derivation of expected scaling factors, I assume that metabolic rate scales as the 0.75 power of body mass (Kleiber 1961) for all mammalian taxa. Available data (Elgar & Harvey 1987; McNab 1986) support this value for the herbivorous taxa examined here (Table 6.2). The model is summarized in Figure 6.6; I discuss briefly how the scaling of each life-history parameter is derived.

Productivity and litter mass. Kleiber (1961) noted that the secondary productivity of individual mammals scaled as the 0.75 power of body mass. If all species on the average put the same proportion of their available energy into reproduction, then litter mass would be expected to scale as the 0.75 power of adult mass, holding number of litters per lifetime constant.

Birth mass and litter size. If birth mass is proportional to adult mass and if litter mass scales as the 0.75 power of body mass, then litter size scales as the −0.25 power of birth or adult mass. The proportionality of birth and adult mass is an assumed value in the model that seems reasonable and in fact is close to what is observed among mammals (Table 6.2).

Pre- and postnatal growth rate. Growth rate is expected to be proportional to metabolic rate because it is constrained by the same factors that influence metabolic rate. For mammals postnatal growth rate is directly proportional to adult metabolic rate (see Appendix in this chap-

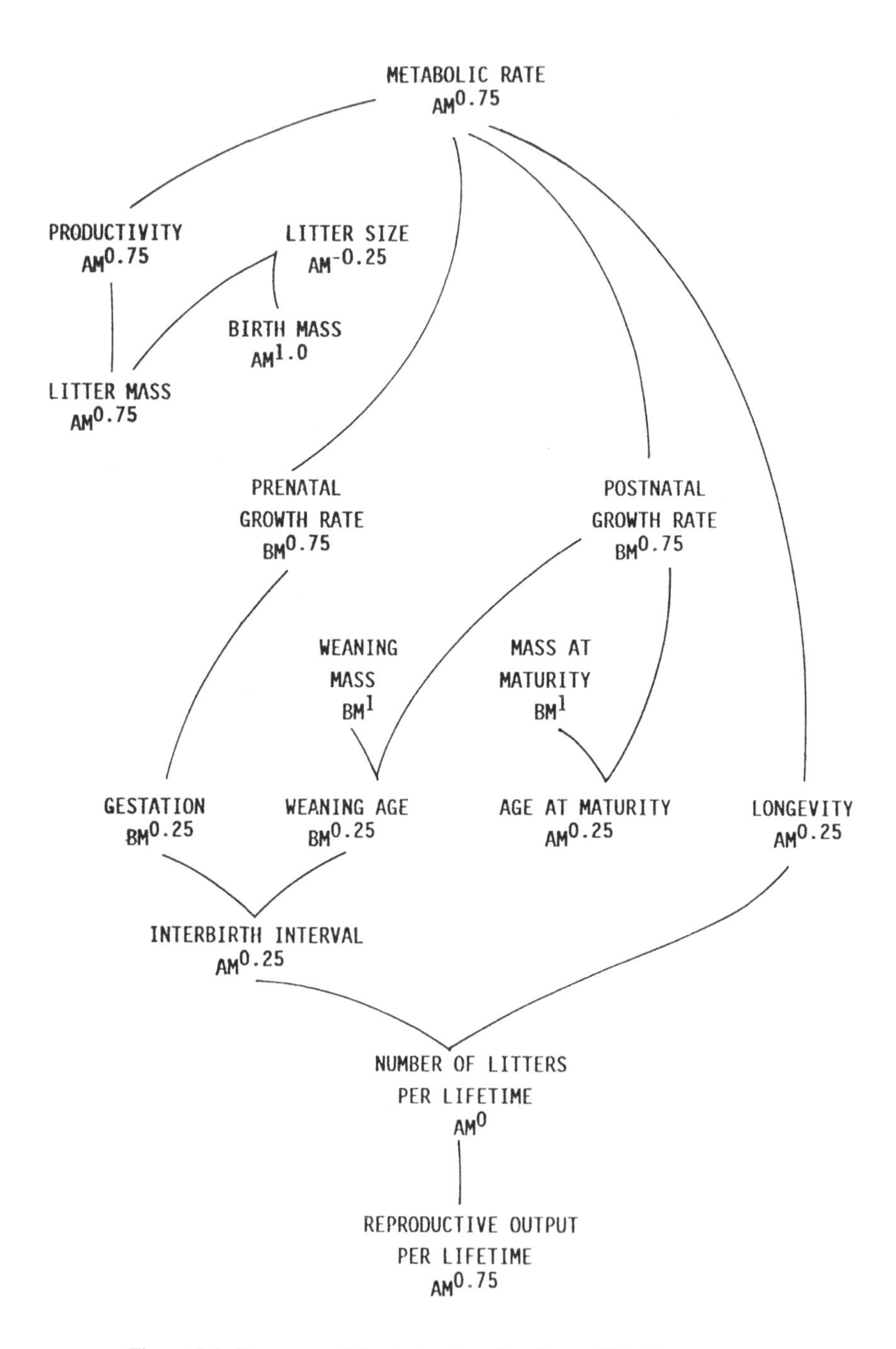

Figure 6.6. Summary of the derivation of scalings of life-history traits on body mass. AM, adult mass; BM, birth mass.

ter) and is thus expected to scale similarly. Because postnatal growth slows as maturity approaches, I use the period from just after birth to about 40% of adult size; this represents a period of approximately linear increase in mass and for many mammals it is also the period of nursing.

Gestation time. If prenatal growth rate is assumed to scale as the 0.75 power of body mass, then the period of gestation can be estimated from a knowledge of mass at birth because all mammals begin life at the same size (that of the fertilized egg). Gestation time is expected to scale as birth mass^1/birth mass$^{0.75}$, or as the 0.25 power of birth mass.

Mass and age at weaning. If mammals nurse their young until they reach a certain proportion of adult mass, which is some multiple of birth mass in those cases where birth mass is proportional to adult mass, then weaning age scales as other time factors – that is, as birth mass^1/weaning mass$^{0.75}$ or as birth or adult mass$^{0.25}$.

Age at maturity. If mass at maturity scales proportional to birth mass, then age at maturity is expected to scale as gestation and weaning age.

Longevity. Metabolic rate determines how fast an organism lives and thus how long it takes to reach certain points in its life such as birth, weaning, age at maturity, or death from senescence (Schmidt-Nielsen 1984). Thus, longevity is expected to scale as the 0.25 power of body mass like the other time factors discussed above.

Number of litters per life-span. Minimum interbirth intervals are the same as either gestation or the age to weaning or the sum of the two periods. The number of litters per life-span is expected to be independent of size if litters are produced as fast as they can be: life-span ($M^{0.25}$)/ interbirth interval ($M^{0.25}$) = litters/life-span (M^0), where M is adult mass. Seasonality presumably exerts some constraints on when mammals can breed, so the number of litters per year or per life-span may not be a simple linear function of body size.

Reproductive effort per life-span. If the number of litters per life-span is constant, then reproductive effort per lifetime is expected to scale as the 0.75 power of adult mass since reproductive effort per litter is expected to scale as the 0.75 power of adult mass. Reproductive effort after birth is assumed to be proportional to the energy demands of the young during the weaning period. This is then proportional to: growth rate of each young ($M^{0.75}$) × litter size ($M^{-0.25}$) × weaning period ($M^{0.25}$) = reproductive effort ($M^{0.75}$), where M again is adult mass. If litter mass also scales as 0.75, total reproductive effort will as well.

Brain size, basal metabolic rate, and life history. A growing realization that many life-history traits of mammals are better explained by relative brain size than by body size alone (e.g., Eisenberg 1981; Harvey & Bennett 1983; Harvey, Martin, & Clutton-Brock 1987; Sacher 1959; Sacher & Staffeldt 1974) has helped clarify some of the observed scalings that have been obtained (see below). Because the development and maintenance of brain tissue require more energy than do those of somatic tissue (Armstrong 1983), a larger brain draws energy away from reproduction and growth. Development of a proportionately larger brain, which presumably increases the ability of an individual to survive at the cost of reducing its reproductive output, can be considered a K-adapted trait. Relatively large-brained species should have higher basal metabolic rates, all else being equal (Lewin 1982; Martin 1981). Therefore, if larger brains are considered a K-adaptation, they carry with them a higher basal rate of metabolism, which is also a K-adaptation. For mammals both brain size and metabolic rate scale approximately as the 0.75 power of body mass (Martin 1981). Scalings lower than this value indicate that larger species are more r-adapted with respect to these traits than are their smaller relatives.

Body size and life histories of herbivorous mammals: observation of the expected with an interesting deviation

The model described above derives the expected scalings of life-history traits from known scalings of metabolic rate and birth mass. If species do scale as expected, their life-history traits reflect the average physiological potential for their size and they cannot really be considered either more K- or r-adapted. This is not to say that these life-history differences have no ecological or adaptive consequences, only that they have not evolved independently from their physiological correlates in response to selection for particular ecological strategies (see below). For the six groups of herbivorous mammals I examined, the observed scalings of various life-history traits are close to what the above model predicted them to be (Table 6.2). Larger mammals are really not more K-adapted than expected for their size. In fact, two of the three signif-

Table 6.2. *Comparison of the scaling factors (slopes of the linear regression) of various life-history traits on body mass, to those expected*

Trait	Expected scaling	Direction if *K*-adapted	Observed scaling (95% confidence interval)	Direction of deviation
BM/AM	1.0	Higher	1.04 (0.87–1.21)	0
LM/AM	0.75	Lower	0.86 (0.82–0.90)	+
LS/AM	−0.25	Lower	−0.19 (−0.34 – −0.04)	0
PNG/BM	0.75	Lower	0.71 (0.52–0.90)	0
G/BM	0.25	Higher	0.30 (0.18–0.41)	0
WA/BM	0.25	Higher	0.24 (0.07–0.41)	0
AFM/AM	0.25	Higher	0.31 (0.15–0.48)	0
L/AM	0.25	Higher	0.19 (0.14–0.24)	–
BR/AM	0.75	Higher	0.65 (0.57–0.73)	–
BMR/AM	0.75	Higher	0.75 (0.70–0.80)	0

Note: The regressions use the mean values of six herbivorous mammalian taxa (see Appendix, this chapter). Significant deviations from the model (beyond the 95% confidence interval of the observed slope) are indicated by +, if the direction is that predicted for *K*-adaptation; −, if for *r*-adaptation. See Appendix (this chapter) for the regression statistics.

Abbreviations: BM, birth mass; AM, adult mass; LM, litter mass; LS, litter size; PNG, postnatal growth rate; G, gestation period; WA, weaning age; AFM, age at female maturity; L, maximum longevity; BR, brain mass; BMR, basal metabolic rate.

icant deviations are in the *r*-direction. The acquisition of large size in mammals is typically accompanied by a life-history pattern expected for their size. It is the deviations from the expected that are of interest and the focus of this section.

Indeed, allometries themselves evolve. One of the kinds of evidence for the pervasiveness of adaptation is the common difference (Gould 1966) between within-species and among-species allometries. To some extent the scalings discussed here are themselves adaptive and selected for. For example, one can argue that the expected life-history parameters of large mammals are basic to their survival or competitive success; mammals tend to evolve toward larger size because of the superior life-history traits they acquire as a consequence. The analyses discussed here cannot refute such a contention since they demonstrate merely a set of correlations between body mass and life history and not a causal mechanism. A more detailed analysis of the taxa involved may reveal such a mechanism. Although the results reported here cannot clearly distinguish among predation, competition, and physiological advantage because the predictions of all three coincide to a large extent, a differential role of the three potential mechanisms is suggested.

Given a higher predation pressure on medium-sized herbivores for the reasons outlined in the previous section, one can predict the following rules about the scaling of life-history traits within and among taxa.

1 Larger species in taxa with small to medium average body sizes are expected to be more r*-adapted than smaller relatives.* The slopes of the various life-history traits are expected to deviate from the predicted slopes in the *r*-direction. In contrast, species in taxa with large average body size are expected to have slopes that deviate in the *K*-direction because the smaller members are expected to be more *r*-selected. The results from six herbivorous taxa are summarized in Table 6.3. The data tend to support the prediction with respect to the directions of the deviations from the expected scalings.

The data used in this analysis have numerous problems, which caution against accepting the hypothesis at this stage. By using secondary compilations, I repeat any errors made by the compiler. The sample size for any particular trait may be relatively small. There is a lack of geographic control both among and within taxa. Data on body mass and a particular parameter may not be from the same study and may thus be misleading because many species vary geographically in body size. The same sets of species were not used for each trait, and these species did not necessarily cover the entire size range of the taxon (the smaller the size range, the less reliable the slope).

However, errors introduced by the factors above are presumably random with respect to the predictions, and some parameters are measured with reasonable reliability. Nevertheless, many of the slopes in Table 6.3 do not differ significantly from the expected value, and even some of those that do may not reflect real biological differences. The variation in the scaling of life-history traits revealed in the present analysis suggests that it would be worthwhile to pursue the collection of more reliable information on these parameters.

Some notable deviations of the observed from the expected include the following: The deviation of litter mass in a direction opposite of that expected reflects the fact that litter size is approximately independent of body mass for the taxa used in this analysis. Consequently, litter mass shows about the same scaling as birth mass. Litter mass, however, may be a poor indicator of relative *r*- or *K*-adaptation. More relevant is the mass in all the litters per unit time or per lifetime, since this

Table 6.3. *The scaling factors (slopes of the linear regression) of various life-history traits on body mass for six mammalian taxa*

Trait	Observed scalings among taxa	Observed scalings within taxa					
		Bovids	Cavio.	Lago.	T. Sci.	Cricet.	Micro.
BM/AM	1.04	0.95 (+3)	0.88 (−5)	0.59 (−15)*	0.63 (−14)*	0.65 (−13)*	0.68 (−12)*
LM/AM	0.86	0.93 (−2)	0.80 (+2)*	0.64 (+8)*	0.56 (+12)*	0.51 (+14)*	0.77 (+3)*
LS/AM	−0.19	0.00 (−11)*	0.07 (−15)*	0.05 (−14)*	0.07 (−15)*	0.06 (−14)*	0.08 (−16)*
PNG/BM	0.71	0.48 (+9)*	0.76 (+2)	0.94 (−8)*	1.18 (−15)*	0.90 (−7)	1.26 (−17)*
G/BM	0.30	0.18 (−6)	0.18 (−6)	0.18 (−6)	0.02 (−15)*	0.17 (−6)*	0.11 (−10)*
WA/BM	0.24	0.43 (+10)*	0.26 (+2)	−0.08 (−18)*	−0.05 (−16)*	0.09 (−8)	0.00 (−13)*
AFM/AM	0.31	0.21 (−5)	0.16 (−8)	0.24 (−4)	0.25 (−3)	0.26 (−2)	0.21 (−5)
L/AM	0.19	0.17 (−1)	0.04 (−9)*	0.01 (−10)*	0.15 (−2)	0.07 (−7)*	0.03 (−9)*
BR/AM	0.65	0.56 (−4)*	0.61 (−2)	0.62 (−1)	0.36 (−13)*	0.34 (−14)*	0.36 (−13)*
BMR/AM	0.75	0.84 (+3)*	0.84 (+3)*	0.64 (−4)*	0.89 (+5)*	0.69 (−2)*	0.63 (−5)*
Net number of traits for which larger species are significantly *K*-adapted		+1	0	−5	−4	−5	−8
Sum of the deviations from observed scalings		−4	−36	−72	−76	−59	−97

Note: In parentheses are the direction and amount that a scaling factor for a single taxon deviates from the observed scaling among taxa. An asterisk denotes that the slope falls beyond the 95% confidence interval of the observed slope (see Table 6.2). The deviations are arctan differences between the two slopes.

Abbreviations: Life history traits as in Table 6.2; Cavio., caviomorph rodents; Lago., lagomorphs; T. Sci., terrestrial sciurid rodents; Cricet., cricetine rodents; Micro., microtine rodents. See Appendix (this chapter) for regression statistics.

measures the relative energetic investment in offspring more accurately than does the mass of a single litter. Unfortunately, too few data are available to say much about what is one of the more indicative traits of an *r–K* adaptive strategy.

Length of gestation also deviates notably from the model. Its slope is lower than expected for all groups and is not consistently lower for those taxa for which the larger species are predicted to be more *r*-selected. A possible explanation involves the fact that brain size tends to scale lower than predicted for most of the taxa examined here. Because the length of gestation is better estimated from neonatal brain mass than from body mass (Sacher & Staffeldt 1974), reduced brain size of larger species effectively shortens their gestation time and depresses the scaling below expected.

Age at maturity is another important trait that strongly influences the ability of a population to expand rapidly but for which good comparable data are unavailable to make any strong conclusions. A basic problem with the data is comparability of definitions of this age. The set of data I analyzed here (Wooton 1987) uses the age at first reproduction for females. In natural populations the age at which a female can physiologically reproduce may not be the age at which she actually reproduces. Seasonality frequently delays actual reproduction long beyond the onset of capability. The original sources from which the data are compiled, which themselves are frequently secondary rather than original sources of data, do not always clearly specify which age is relevant. Conclusions from this analysis are particularly tenuous as a consequence. The unexpected result that the two taxa for which I expected particularly low scalings, lagomorphs and terrestrial sciurids, have the highest, might possibly be explained by the relatively small size of their offspring. Smaller, more altricial young require more time to reach maturity than more precocial ones, all else being equal. However, relatively faster postnatal growth rates of altricial young should reduce the time it takes to reach adult size. The interaction of birth mass and postnatal growth may thus produce a scaling that does not deviate from expected. Further analysis may resolve this issue, although more consistently comparable data are needed.

Maximum longevity tends to increase considerably less than expected in all taxa. As for age at maturity, values for maximum longevity tend to be extremely variable in what they measure. To avoid this problem as best I could, I used the maximum captive longevity reported in Nowak and Paradiso (1983). Because maximum potential longevity like gestation is more influenced by brain size than by body size (Sacher 1959), the low scalings obtained for this trait may reflect in part the low scalings of brain to body mass (Table 6.3). However, the observed scalings for these two traits suggest a negative correlation rather than the expected positive one. A multivariate analysis is again required to resolve the issue, but an adequate one is extremely difficult at present. Obtaining values of all traits for the same set of species is difficult if not impossible, but for some analyses it is required to obtain meaningful results.

Despite the problems and qualifications noted above, a general trend emerges: For traits that can be measured with the greatest accuracy or reliability, medium-sized herbivores tend to be more *r*-adapted on the average than either their smaller or larger relatives.

2 *Among taxa, those with medium-sized species are expected to be more* r-*adapted than are taxa with either larger or smaller species.* To examine this I did pairwise regressions among the six taxa examined, using the mean values of body mass and life-history trait for each taxon (see Figure 6.7). The results are summarized in Table 6.4 and Figure 6.8. The comparisons use the three foliage-feeding taxa as the referent taxon against which to compare relative *r*- and *K*-adaptation of either larger taxa or those that are primarily seed-eaters.

The results indicate that only lagomorphs are more *r*-adapted than are bovids, and that seed-eaters are more *K*-adapted than are foliage feeders. Both cricetine and microtine rodents, for their size, appear to be no more *r*-selected than their largest relatives, the bovids. Lagomorphs, as the medium-sized foliage feeders, are also more *r*-adapted than microtines, as was predicted. Although the individual values that have gone into this analysis are not very reliable, as discussed above, the size range over which the comparisons are made is sufficiently large that it would require enormous change in the value of a trait to change the slope by much. Furthermore, the use of the mean value for each taxon in this analysis gives more confidence in the scaling values than those for each individual taxon that are summarized in Table 6.3.

In conclusion, this analysis of size-specific life-history adaptations suggests that predation may have been an important selective factor on herbivorous mammals. It may have selected for increasing or decreasing size at the extremes but has not eliminated the intermediate-sized species. An intermediate size may be optimal for obtaining and processing energy because whatever defense used by medium-sized species is en-

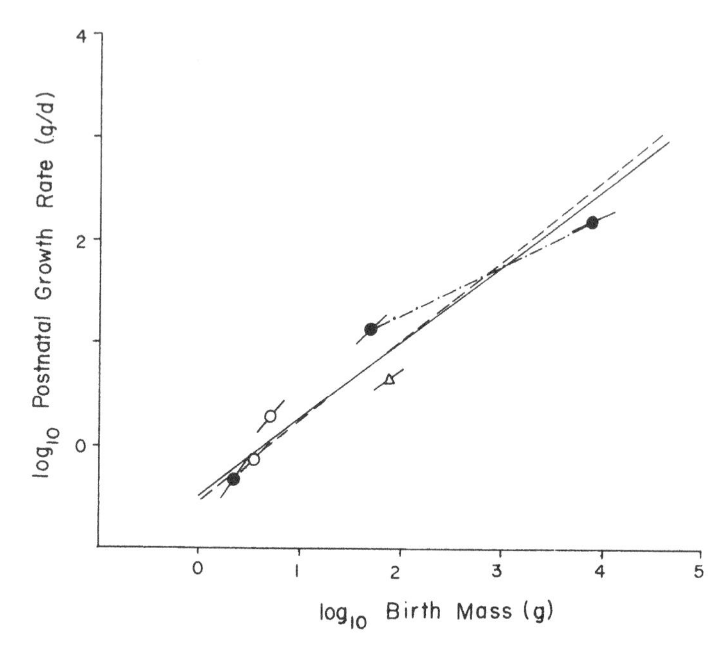

Figure 6.7. Postnatal growth rate as a function of birth weight illustrates the various scaling factors (slopes of the linear regression) used in this analysis. The long solid line is the slope for the mean values of the taxa (solid circles: foliage feeders [microtines, lagomorphs, bovids]; open circles: seed eaters [cricetines, terrestrial sciurids]; triangle: caviomorphs). The dashed line has the slope predicted from the model (Table 6.2). The short solid lines have slopes based on species within a taxon (Table 6.3). The dash-dotted line has the slope between mean values of paired taxa (Table 6.4).

ergetically expensive. One does not expect persistence of the middle size if becoming larger or smaller is a superior defense.

An alternative explanation for the persistence of intermediate-sized species despite what appears to be potential for great predation pressure involves a potential physiological advantage of their size: These species can shunt proportionally more energy into reproduction than can either smaller or larger ones. Smaller species expend more energy maintaining a metabolic rate high enough to avoid torpor (Schmidt-Nielsen 1984), whereas larger species cannot hide from either inclement weather or predators and so expend proportionally more energy on running away or generating heat, etc. Physiological superiority of the middle size suggests that species will evolve toward this size if external selective pressures to become very large or small are absent. Such reduction in predatory and competitive pressures are thought to occur on islands

Table 6.4. *The scaling factors (slopes of the linear regression) of means of various life-history traits, on body mass, for pairs of mammalian taxa*

			Observed scalings between paired taxa: Referent taxon						
			Bovids			Lagomorphs		Microtines	
Trait	Observed scaling among taxa	Lago.	T. Sci.	Cricet.	Micro.	Cricet.	T. Sci.	Lago.	T. Sci.
BM/AM	1.04	1.14 (−3)	1.19 (−4)	1.01 (+1)	1.04 (0)	0.83 (+6)*	1.29 (−6)*	0.91 (−4)	0.48 (−20)*
LM/AM	0.86	0.83 (−1)	0.88 (+0)	0.86 (0)	0.84 (−1)	0.91 (+1)*	1.01 (+4)*	0.84 (−1)	0.66 (+8)*
LS/AM	−0.19	−0.34 (−8)	−0.30 (−6)	−0.15 (+2)	−0.20 (−0)	0.06 (+14)*	−0.23 (−2)	−0.06 (−8)	0.17 (−21)*
PNG/BM	0.71	0.50 (−8)*	0.63 (−3)	0.73 (+1)	0.73 (+1)	1.15 (+14)*	0.92 (+8)*	1.09 (−12)*	1.51 (−22)*
G/BM	0.30	0.38 (+5)	0.30 (0)	0.28 (−0)	0.29 (−0)	0.09 (−11)*	0.10 (−10)*	0.16 (−7)*	0.26 (−1)
WA/BM	0.24	0.43 (−10)*	0.18 (+3)	0.26 (−2)	0.28 (−3)	−0.10 (+19)*	−0.42 (+36)*	0.00 (−13)*	1.42 (+42)*
AFM/AM	0.31	0.25 (+3)	0.12 (+10)*	0.30 (+0)	0.35 (−2)	0.35 (−2)	−0.43 (+40)*	0.47 (+8)	0.89 (+25)*
L/AM	0.19	0.21 (−1)	0.17 (+1)	0.17 (+1)	0.21 (−1)	0.10 (+5)*	−0.03 (+13)*	0.22 (−1)	0.34 (+8)*
BR/AM	0.65	0.76 (−4)*	0.68 (−1)	0.65 (0)	0.67 (−1)	0.50 (+6)*	0.45 (+9)*	0.55 (−4)*	0.63 (−1)
BMR/AM	0.75	0.77 (−1)	0.77 (−1)	0.78 (−1)	0.72 (+1)	0.79 (−1)	0.78 (−1)	0.66 (−4)*	0.34 (−34)*
Net number of traits for which taxon is significantly *K*-adapted		−3	+1	0	0	+6	+4	−4	0
Sum of the deviations from observed scaling		−28	−1	+2	−6	+51	+91	−46	−16

Note: See Appendix (this chapter) for the mean values used. Not all possible paired comparisons are shown. Conventions and abbreviations as in Tables 6.2 and 6.3.

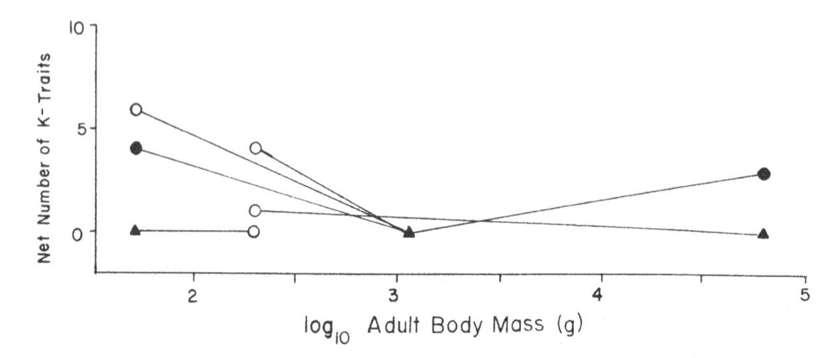

Figure 6.8. Summary of the relative degree of *K*-adaptation for pairs of mammalian taxa as measured by the net number of traits (Table 6.4) for which a taxon (solid circles: foliage feeders; open circles: seed eaters) was significantly more *K*-selected than the referent taxon (triangles). For clarity, the cricetid and microtine comparisons with bovids (both a net of zero) are not shown.

where, in fact, small mammals often increase and large species decrease in size (Foster 1964). One does not have to evoke specific selective factors favoring size change such as increased breadth of diet, less food required per individual, earlier age at reproduction, all of which may vary with the islands and the taxonomic groups involved (see, e.g., Case 1978b; Heaney 1978; Lawlor 1982; Lomolino 1985; Melton 1982; and Van Valen 1973, for examples of the variety of causes advanced to explain the "island rule"). The intrinsic physiological and reproductive advantages of being of middle size may be sufficient in itself to explain this pattern of size change in mammals when disruptive selective pressures are reduced as they often seem to be on islands.

Summary

Body size distributions for species of ground-dwelling, foliage-feeding mammals are distinctly polymodal in various regions of the world. Although similarities and differences in the size distributions from more local to more global areas suggest that they are the result of selection, the actual mechanism among several potential ones is not clear. An examination of the body size distributions of fossil assemblages may help clarify the factors shaping the size distributions observed among extant mammals. The observed size distributions may influence life-history adaptations and in turn be explained by them. For example, the size of predators relative to herbivores suggests that intermediate-sized herbivores may experience greater predation pressure than either smaller or larger species and thus be more *r*-selected. Although medium-sized mammalian herbivores are more *r*-adapted than smaller or larger relatives, it is not clear that predation or even competition is the reason for this pattern. Alternatively, the physiological advantages of this size range may be sufficient to account for the observed pattern; these advantages may also explain such phenomena as the "island rule" of mammalian size evolution and the persistence of the intermediate size class in herbivore faunas.

Appendix

The table below gives regression statistics for life-history traits analyzed in this paper. All values are $\log_{10}$ transformed. Masses are in grams, and times are in days. The data were obtained from the following sources: Case (1978a), Eisenberg (1981), Elgar and Harvey (1987), Hemmer (1976), Mace, Harvey, and Clutton-Brock (1981), McNab (1986), Millar (1977, 1981), Nowak and Paradiso (1983), Oboussier and Schliemann (1966), Western (1979), and Wooton (1987).

Taxon	*N*	Slope	Intercept	*r*	Mean *x* (SD)	Mean *y* (SD)
Birth mass on adult mass						
Bovids	24	0.945	−0.834	.892	4.785 (0.580)	3.688 (0.615)
Cavio.	28	0.882	−0.733	.955	2.974 (0.632)	1.889 (0.584)
Lago.	11	0.589	−0.090	.848	3.048 (0.531)	1.704 (0.369)
T. sci.	16	0.631	−0.699	.954	2.330 (0.516)	0.771 (0.341)
Cricet.	35	0.651	−0.532	.953	1.668 (0.520)	0.555 (0.355)
Micro.	19	0.678	−0.677	.955	1.663 (0.438)	0.450 (0.311)
Litter mass on adult mass						
Bovids	24	0.931	−0.758	.884	4.785 (0.580)	3.697 (0.432)
Cavio.	28	0.802	−0.195	.970	2.973 (0.632)	2.190 (0.523)
Lago.	11	0.644	0.294	.832	3.048 (0.531)	2.256 (0.411)
T. sci.	16	0.555	−0.235	.922	2.330 (0.516)	1.528 (0.311)
Cricet.	34	0.508	−0.156	.845	1.673 (0.527)	1.006 (0.317)
Micro.	19	0.578	−0.187	.891	1.665 (0.436)	1.087 (0.375)
Litter size on adult mass						
Bovids	24	NA	NA	NA	4.790 (0.587)	NA
Cavio.	17	−0.074	0.511	−.285	2.986 (0.810)	0.289 (0.211)
Lago.	11	0.050	0.396	.142	3.139 (0.481)	0.554 (0.170)
T. sci.	17	−0.066	0.900	−.267	2.310 (0.506)	0.747 (0.125)
Cricet.	33	−0.058	0.570	−.225	1.688 (0.528)	0.472 (0.137)
Micro.	19	0.085	0.498	.256	1.666 (0.436)	0.639 (0.144)

Taxon	N	Slope	Intercept	r	Mean x (SD)	Mean y (SD)
Postnatal growth rate on birth mass						
Bovids	14	0.478	0.370	.777	3.841 (0.455)	2.200 (0.278)
Cavio.	19	0.757	−0.772	.926	1.868 (0.582)	0.641 (0.476)
Lago.	7	0.938	−0.486	.952	1.724 (0.325)	1.131 (0.321)
T. sci.	11	1.175	−0.642	.810	0.755 (0.217)	0.245 (0.314)
Cricet.	32	0.903	−0.704	.807	0.583 (0.305)	−0.178 (0.342)
Micro.	9	1.257	−0.804	.850	0.365 (0.179)	−0.344 (0.264)
Gestation period on birth mass						
Bovids	22	0.177	1.647	.806	3.688 (0.641)	2.301 (0.141)
Cavio.	25	0.176	1.652	.623	1.869 (0.597)	1.980 (0.169)
Lago.	9	0.181	1.231	.769	1.643 (0.384)	1.533 (0.092)
T. sci.	11	0.021	1.439	.168	0.866 (0.374)	1.457 (0.046)
Cricet.	30	0.172	1.343	.738	0.570 (0.378)	1.441 (0.088)
Micro.	17	0.113	1.295	.574	0.422 (0.298)	1.343 (0.059)
Weaning age on birth mass						
Bovids	5	0.428	0.576	.918	3.706 (0.500)	2.162 (0.233)
Cavio.	9	0.262	1.154	.804	1.742 (0.719)	1.610 (0.234)
Lago.	7	−0.079	1.375	−.270	1.580 (0.394)	1.250 (0.115)
T. sci.	12	−0.048	1.650	−.010	0.714 (0.229)	1.616 (0.108)
Cricet.	19	0.090	1.295	.396	0.598 (0.311)	1.349 (0.070)
Micro.	11	0.004	1.251	.012	0.458 (0.371)	1.253 (0.117)
Weaning mass on birth mass						
Lago.	7	0.721	1.006	.871	1.561 (0.384)	2.131 (0.317)
T. sci.	12	0.995	1.025	.800	0.714 (0.229)	1.736 (0.285)
Cricet.	19	0.987	0.730	.927	0.557 (0.310)	1.261 (0.318)
Micro.	10	0.917	0.761	.984	0.494 (0.370)	1.214 (0.345)
Age at female maturity on adult mass						
Bovids	43	0.206	1.883	.639	4.858 (0.516)	2.884 (0.166)
Cavio.	19	0.161	1.903	.490	3.050 (0.756)	2.405 (0.249)
Lago.	9	0.236	1.707	.454	3.056 (0.494)	2.427 (0.256)
T. sci.	26	0.250	1.963	.665	2.617 (0.595)	2.616 (0.223)
Cricet.	29	0.258	1.506	.588	1.588 (0.464)	1.916 (0.204)
Micro.	21	0.208	1.437	.436	1.689 (0.422)	1.788 (0.201)
Longevity on adult mass						
Bovids	32	0.169	2.998	.717	4.981 (0.621)	3.838 (0.146)
Cavio.	17	0.044	3.370	.172	3.225 (0.953)	3.512 (0.245)
Lago.	4	0.013	3.392	.135	3.003 (0.581)	3.430 (0.055)
T. sci.	7	0.155	3.039	.592	2.604 (0.660)	3.441 (0.172)
Cricet.	10	0.075	3.174	.298	1.794 (0.544)	3.308 (0.135)
Micro.	4	0.031	3.100	.083	1.768 (0.311)	3.155 (0.116)
Brain mass on adult mass						
Bovids	35	0.560	−0.453	.980	4.725 (0.576)	2.192 (0.329)
Cavio.	15	0.608	−0.911	.958	2.962 (0.629)	0.890 (0.399)
Lago.	13	0.622	−0.988	.990	2.995 (0.486)	0.873 (0.305)
T. Sci.	34	0.361	−0.266	.936	2.370 (0.529)	0.590 (0.204)

Taxon	*N*	Slope	Intercept	*r*	Mean *x* (SD)	Mean *y* (SD)
Cricet.	26	0.335	−0.343	.958	1.700 (0.456)	0.227 (0.160)
Micro.	25	0.363	−0.459	.958	1.670 (0.424)	0.148 (0.161)
Basal metabolic rate (O_2/hr) on adult mass						
Bovids	8	0.835	0.196	.955	4.889 (0.357)	4.278 (0.312)
Cavio.	8	0.840	0.239	.932	3.223 (0.454)	2.948 (0.409)
Lago.	6	0.644	0.906	.960	3.064 (0.559)	2.880 (0.375)
T. Sci.	9	0.890	0.267	.969	2.008 (0.415)	2.054 (0.381)
Cricet.	20	0.694	0.626	.975	1.735 (0.505)	1.832 (0.359)
Micro.	24	0.633	0.893	.920	1.636 (0.378)	1.929 (0.260)
Regressions based on the mean values of all herbivorous taxa						
BM/AM	6	1.038	−1.340	.989	2.745 (1.167)	1.510 (1.225)
LM/AM	6	0.858	−0.396	.999	2.746 (1.166)	1.961 (1.002)
LS/AM	6	−0.186	0.963	−.807	2.763 (1.172)	0.450 (0.270)
PNG/BM	6	0.710	−0.465	.971	1.523 (1.292)	0.616 (0.945)
G/BM	6	0.295	1.230	.946	1.510 (1.214)	1.678 (0.379)
WA/BM	6	0.239	1.190	.841	1.466 (1.220)	1.540 (0.347)
AFM/AM	6	0.314	1.455	.900	2.810 (1.192)	2.339 (0.417)
L/AM	6	0.187	2.904	.971	2.896 (1.186)	3.447 (0.229)
BR/AM	6	0.650	0.960	.994	2.737 (1.133)	0.820 (0.741)
BMR/AM	6	0.748	0.591	.998	2.759 (1.242)	2.654 (0.931)
PNG/BMR	6	0.984	−0.996	.970	2.654 (0.931)	0.616 (0.945)

Abbreviations: Cavio., caviomorph rodents; Lago., lagomorphs; T. Sci., terrestrial sciurid rodents; Cricet., cricetine rodents; Micro., microtine rodents. BM, birth mass; AM, adult mass; LM, litter mass; LS, litter size, PNG, postnatal growth rate; G, gestation period; WA, weaning age; AFM, age at female maturity; L, maximum longevity; BR, brain mass; BMR, basal metabolic rate.

Acknowledgments

I thank Leigh Van Valen for various help with the manuscript. My ideas were further clarified by discussions with Jeff Ihara, Monte Lloyd, Joann White, and many of the participants of the workshop for which this paper was prepared. In particular, discussions with John Damuth have stimulated my interest in how body size influences the ecology and behavior of animals. Comments from John Eisenberg, Bruce MacFadden, Brian McNab, Bob Martin, and Louise Roth were also helpful in developing this paper.

References

Armstrong, E. 1983. Relative brain size and metabolism in mammals. *Science* *220*:1302–1304.

Bailey, J. W. 1946. *The Mammals of Virginia.* Richmond: Williams Printing.
Bee, J. W., & Hall, E. R. 1956. Mammals of Northern Alaska. *University of Kansas Mus. Nat. Hist. Misc. Publ. 8*:1–309.
Belovsky, G. E. 1984. Moose and snowshoe hare competition and a mechanistic explanation from foraging theory. *Oecologia 61*:150–159.
Blueweiss, L., Fox, H., Kudzma, V., Nakashima, D., Peters, R., & Sams, S. 1978. Relationships between body size and some life history parameters. *Oecologia 37*:257–272.
Burt, W. H. 1948. *The Mammals of Michigan.* Ann Arbor: University of Michigan Press.
Burt, W. H., & Grossenheider, R. P. 1976. *A Field Guide to the Mammals,* 3rd ed. Boston: Houghton Mifflin.
Calder, W. A., III. 1983. Ecological scaling: mammals and birds. *Annu. Rev. Ecol. Syst. 14*:213–230.
Case, T. J. 1978a. On the evolution and adaptive significance of postnatal growth rates in the terrestrial vertebrates. *Q. Rev. Biol. 53*:243–282.
Case, T. J. 1978b. A general explanation for insular body size trends in terrestrial vertebrates. *Ecology 59*:1–18.
Caughley, G. 1987. The distribution of eutherian body weights. *Oecologia 74*:319–320.
Caughley, G., & Krebs, C. J. 1983. Are big mammals simply little mammals writ large? *Oecologia 59*:7–17.
Connor, E. F., & Simberloff, D. 1986. Competition, scientific method, and null models in ecology. *Am. Sci. 74*:155–162.
Corbet, G. B. 1980. *The Mammals of Britain and Europe.* London: Collins.
Corbet, G. B., & Southern, H. N. (eds.). 1977. *The Handbook of British Mammals,* 2nd ed. Oxford: Blackwell Scientific Publ.
Delany, M. J., & Happold, D. C. D. 1979. *Ecology of African Mammals.* London: Longman.
Demment, M. W., & van Soest, P. J. 1985. A nutritional explanation for body-size patterns of ruminant and non-ruminant herbivores. *Am. Nat. 125*:641–647.
Eisenberg, J. F. 1981. *The Mammalian Radiations.* Chicago: University of Chicago Press.
Elgar, M. A., & Harvey, P. H. 1987. Basal metabolic rates in mammals: allometry, phylogeny and ecology. *Funct. Ecol. 1*:25–36.
Foster, J. B. 1964. The evolution of mammals on islands. *Nature 202*:234–235.
Gould, S. J. 1966. Allometry and size in ontogeny and phylogeny. *Biol. Rev. 41*:587–640.
Griffiths, D. 1986. Size-abundance relations in communities. *Am. Nat. 127*: 140–166.
Haltenorth, T., & Diller, H. 1980. *A Field Guide to the Mammals of Africa Including Madagascar* (English trans. by R. W. Hayman). London: Collins.
Happold, D. C. D. 1987. *The Mammals of Nigeria.* Oxford: Clarendon Press.
Harvey, P. H., & Bennett, P. M. 1983. Brain size, energetics, ecology and life history patterns. *Nature 306*:314–315.
Harvey, P. H., Martin, R. D., & Clutton-Brock, T. H. 1987. Life histories in comparative perspectives. In *Primate Societies,* ed. B. B. Smuts, D. L.

Cheney, R. M. Seyfarth, R. W. Wrangham, & T. T. Struhsaker, pp. 181–196. Chicago: University of Chicago Press.
Heaney, L. R. 1978. Island area and body size of insular mammals: evidence from the tri-colored squirrel (*Callosciurus prevosti*) of Southeast Asia. *Evolution 32*:29–44.
Hemmer, H. 1976. Gestation period and postnatal development in felids. *The World's Cats 3*(2):143–164.
Hendrichs, H. 1970. Schaetzungen der Huftierbiomass in der Dornbuschsavanne noerdlich und der anderen pflanzenfressenden Tierarten. *Säugetierk. Mitt. 23*:161–199.
Hoffman, A. 1989. *Arguments on Evolution.* Oxford: Oxford University Press.
Hutchinson, G. E. 1959. Homage to Santa Rosalia, or why are there so many kinds of animals? *Am. Nat. 93*:145–159.
Hutchinson, G. E., & MacArthur, R. H. 1959. A theoretical ecological model of size distributions among species of animals. *Am. Nat. 93*:117–125.
Ingles, L. G. 1947. *Mammals of California.* Stanford: Stanford University Press.
Jarman, P. J. 1974. The social organisation of antelope in relation to their ecology. *Behaviour 48*:215–267.
Kingdon, J. 1971–1982. *East African Mammals,* 3 vols. in 7. London: Academic Press.
Kirchner, T. B. 1980. *Community Structure in Relation to Body Size of Species.* Ph.D. dissertation, Colorado State University, Colorado.
Kleiber, M. 1961. *The Fire of Life.* New York: Wiley.
Lawlor, T. E. 1982. The evolution of body size in mammals: evidence from insular populations in Mexico. *Am. Nat. 119*:54–72.
Lewin, R. 1982. How did humans evolve big brains? *Science 216*:840–841.
Lomolino, M. V. 1985. Body size of mammals on islands: the island rule reexamined. *Am. Nat. 125*:310–316.
MacArthur, R. H. 1972. *Geographical Ecology.* New York: Harper & Row.
Mace, G. M., Harvey, P. H., & Clutton-Brock, T. H. 1981. Brain and ecology in small mammals. *J. Zool., Lond. 193*:333–354.
Maiorana, V. C. 1978a. Difference in diet as an epiphenomenon: space regulates salamanders. *Can. J. Zool. 56*:533–535.
Maiorana, V. C. 1978b. An explanation of ecological and developmental constants. *Nature 273*:375–376.
Maiorana, V. C. 1979. Nontoxic toxins: the energetics of coevolution. *Biol. J. Linn. Soc. 11*:387–396.
Martin, R. D. 1981. Relative brain size and basal metabolic rate in terrestrial vertebrates. *Nature 293*:56–60.
May, R. M. 1978. The dynamics and diversity of insect faunas. In *Diversity of Insect Faunas,* ed. L. A. Mound & N. Waloff, pp. 188–204. Oxford: Blackwell Scientific Publications.
McNab, B. K. 1986. The influence of food habits on the energetics of eutherian mammals. *Ecol. Monogr. 56*:1–19.
McNaughton, S. J., & Georgiadis, N. J. 1986. Ecology of African grazing and browsing mammals. *Annu. Rev. Ecol. Syst. 17*:39–65.

Melton, R. H. 1982. Body size and island *Peromyscus:* a pattern and a hypothesis. *Evol. Theory 6*:113–126.
Millar, J. S. 1977. Adaptive features of mammalian reproduction. *Evolution 31*:370–386.
Millar, J. S. 1981. Pre-partum reproductive characteristics of eutherian mammals. *Evolution 35*:1149–1163.
Nowak, R. M., & Paradiso, J. L. 1983. *Walker's Mammals of the World,* 2 vols., 4th ed. Baltimore: Johns Hopkins University Press.
Oboussier, H., & Schliemann, H. 1966. Hirn-Körpergewichtsbeziehungen bei Boviden. *Z. Säugetierk. 31*:464–471.
Pianka, E. R. 1970. On *r* and *K* selection. *Am. Nat. 106*:581–588.
Roberts, T. J. 1977. *The Mammals of Pakistan.* London: Ernest Been.
Root, R. B. 1973. Organization of a plant–arthropod association in simple and diverse habitats: the fauna of collards *(Brassica oleracea). Ecol. Monogr. 43*:95–124.
Roth, V. L. 1979. Can quantum leaps in body size be recognized among mammalian species? *Paleobiology 5*:318–336.
Roth, V. L. 1981. Constancy in the size ratios of sympatric species. *Am. Nat. 118*:394–404.
Sacher, G. A. 1959. Relation of lifespan to brain weight and body weight in mammals. In *CIBA Foundation Colloquia on Aging,* Vol. 5: *The Lifespan of Animals,* ed. G. E. W. Wolstenholme & M. O'Connor, pp. 115–133. London: Churchill.
Sacher, G. A. & Staffeldt, E. E. 1974. Relation of gestation time to brain weight for placental mammals: implications for the theory of vertebrate growth. *Am. Nat. 108*:593–615.
Schmidt-Nielsen, K. 1984. *Scaling.* Cambridge: Cambridge University Press.
Smithers, R. H. N. 1971. *The Mammals of Botswana.* Museum Memoir, No. 4. Salisbury: National Museum of Rhodesia.
Swihart, R. K., Slade, N. A., & Bergstrom, B. J. 1988. Relating body size to the rate of home range use in mammals. *Ecology 69*:393–399.
Toschii, A. 1965. *Fauna d'Italia,* Vol. 7: *Mammalia.* Bologna: Acad. Nazionale Italianadi di Entomol. & Italiana Unione Zoologica.
Van Valen, L. 1969. Variation genetics of extinct animals. *Am. Nat. 103*: 193–224.
Van Valen, L. 1973. Pattern and the balance of nature. *Evol. Theory 1*:31–49.
Western, D. 1979. Size, life history and ecology in mammals. *Afr. J. Ecol. 17*:185–204.
Wiens, J. A. 1982. On size ratios and sequences in ecological communities: are there no rules? *Ann. Zool. Fennica 19*:297–308.
Wooton, J. T. 1987. The effects of body mass, phylogeny, habitat, and trophic level on mammalian age at first reproduction. *Evolution 41*:732–749.

7

Problems and methods in reconstructing body size in fossil primates

WILLIAM L. JUNGERS

Body mass, more than any other single descriptive feature, is the primary determinant of ecological opportunities, as well as of the physiological and morphological requirements of an animal.

Lindstedt and Calder (1981:2)

Introduction

The significance of body size as a major feature of an animal's overall adaptive strategy has long been appreciated by primate paleontologists and paleoanthropologists (Fleagle 1978). "Size" is one of those rare attributes of a fossil that can be evaluated even from fragmentary remains. In addition to such obvious pragmatic considerations, paleontology's concern with size in the fossil record can be traced to the frequent demonstration that body size must be taken into explicit account in any comparative attempt to understand an animal's adaptation and evolution. In many ways, size itself can be characterized as an "ecological variable" (Calder 1984; Peters 1983; Smith 1985), and accurate estimates of body size in fossils can have very important implications for a wide variety of other related biological variables. For example, assessment of relative brain size or degree of encephalization depends critically upon reliable measures of both cranial capacity and body size (Conroy 1987; Gingerich 1977; Hofman 1983; Jerison 1973; Radinsky 1974). Similarly, reconstruction of dietary adaptations (e.g., Kay 1984, 1985) and positional repertoires (Aiello 1981; Jungers 1982, 1984) also require relatively accurate predictions of body size if fossils

In *Body Size in Mammalian Paleobiology: Estimation and Biological Implications,* John Damuth and Bruce J. MacFadden, eds.

are to be placed into a comparative, extant context. A complete list of size-related variables of potential interest to paleontologists would be a long one indeed.

Despite its obvious importance, size can be a very slippery and sometimes contentious concept. Some morphometricians believe that size is best represented as some linear combination of dimensions or a "general factor" (Humphries et al. 1981; Mosimann 1970). In some situations, this statistical construct has great advantages (e.g., Darroch & Mosimann 1985), but it is rarely the variable of choice in the biological contexts outlined above. More commonly, the most relevant size variable is body mass (weight). Although far from perfect and subject to various sources of variability (Ford & Corruccini 1985), it permits meaningful comparisons among animals of different shape (Jungers 1984; Schmidt-Nielsen 1977) and has been linked both theoretically and empirically to many other biological relationships of interest (e.g., Calder 1984; Schmidt-Nielsen 1984). Finally, and especially pertinent to paleontology, body mass can be predicted for fossils from many different, often fragmentary sources.

The perennially favorite source for such estimates among primate paleontologists has been tooth size (Blumenberg 1984; Conroy 1987; Fleagle & Kay 1985; Gingerich 1977; Gingerich, Smith, & Rosenberg 1982; Kay & Simons 1980). Because of their highly durable material properties, teeth are much better represented in most terrestrial vertebrate fossil assemblages than are other body parts, and their popularity, therefore, as a basis for estimating body mass is quite understandable. The higher frequency of teeth, however, is not an adequate rationale for their common use in body size estimation (also cf. Smith 1985). Along with other craniofacial measurements (e.g., occipital condyle size [Martin 1981]; palate breadth, etc. [Steudel 1980]), dental measurements are risky at best and probably inappropriate in general for a relatively simple reason. Following the logical argument put forth by Hylander (1985), because primates and most other vertebrates do not routinely, if ever, transmit body weight through their skulls or teeth, there exists no biomechanical reason to expect a direct or especially predictable relationship between such variables and body weight.

It follows, therefore, that weight-bearing portions of the locomotor skeleton should make better candidates for reliable estimators of body mass. Diaphyseal diameters or circumferences represent suitable alternatives to teeth for some predictions (Aiello 1981; McHenry 1976, 1988; Rightmire 1986). Cross-sectional geometrical variables such as cortical area or area moment of inertia also possess great logical appeal for sound biomechanical reasons (Ruff 1987). Because the loads borne by articular elements of the postcranial skeleton are closely linked mechanically to body weight (Alexander 1980), measurements of postcranial joint size should also be reliable predictors of body weight (Jungers 1988, in press; McHenry 1976; Rightmire 1986).

Joint size is used in the analyses to follow in order to provide new estimates of body size in several extinct primates. Contrasts are made to tooth-derived predictions of size in each case. The first case centers on a representative of the earliest recognized fossil hominid lineage, *Australopithecus afarensis*. The second case focuses on an enigmatic Miocene hominoid primate from Italy, *Oreopithecus bambolii*. The third and final example involves several species of recently extinct ("subfossil") giant prosimian primates from Madagascar.

Case 1: *Australopithecus afarensis* (A.L.288–1)

The partial skeleton of a small, presumably female *A. afarensis* (Figure 7.1) from the Pliocene of the Hadar Formation, Ethiopia (Johanson et al. 1982) provides a rare opportunity in human paleontology to compare directly different estimates of body mass drawn from teeth and postcranial joint size in the same australopithecine specimen (McHenry 1984). This individual is known popularly as "Lucy" (Johanson & Edey 1981) and preserves a variety of hindlimb articular surfaces. Six joint dimensions are included in this analysis: femoral head diameter (FHD), acetabulum height (ACH), anteroposterior diameter of medial tibial condyle (MTB), anteroposterior diameter of the lateral tibial condyle (LTB), anteroposterior diameter of the distal tibia (DTB), and width of the lumbosacral joint (LSW).

A specimen of this nature permits one to combine information from more than one joint surface, and this situation is relatively rare in the fossil record of early hominid evolution. More commonly, a single articular surface is preserved (e.g., femoral head). Least squares regression estimates from each individual measurement are compared to that based on a multiple regression analysis that includes all six variables. Logarithmic transformation of variables serves to reduce residual autocorrelation and improves correlation coefficients. Two groups of extant hominoid primates serve as the samples with known body mass and with measured joint size on which the predictor equations are based: The sex-specific means for (1) seven species of living hominoids

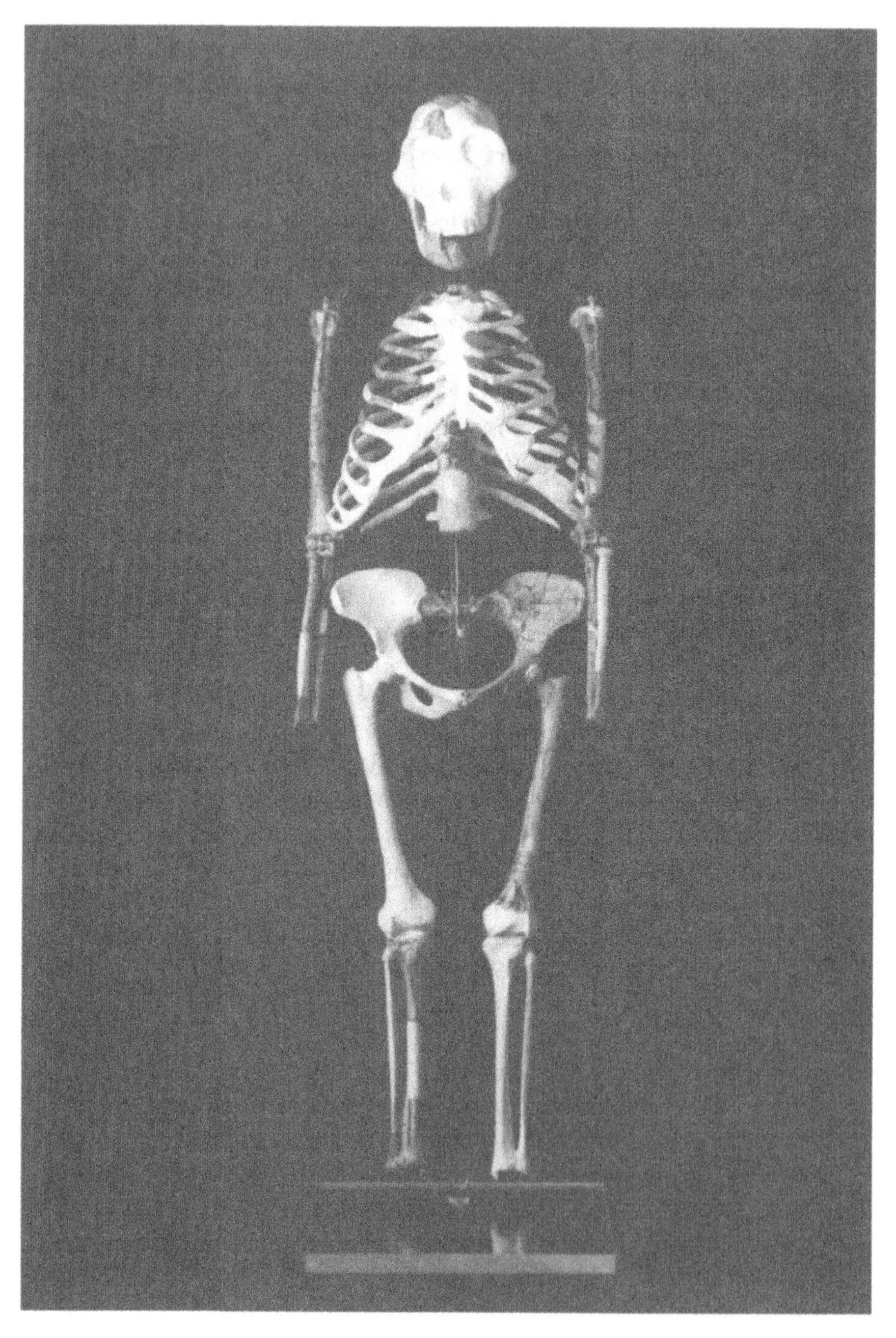

Figure 7.1. A reconstruction of a female *Australopithecus afarensis* (A.L.288-1, "Lucy") by P. Schmid (1983). (Photograph courtesy of Dr. P. Schmid.)

Table 7.1. *Estimates of body mass in A.L.288–1 (*Australopithecus afarensis*) from articular dimensions*

Articular dimension(s)	ALLHOM[a] (kg)	NONHOM[a] (kg)
Simple linear regression		
Femoral head diameter (FHD)	27.4[b]	29.6
Acetabulum height (ACH)	32.3	36.0
A–P diameter of medial tibial condyle (MTB)	32.2	35.5[b]
A–P diameter of lateral tibial condyle (LTB)	24.3	26.2
A–P diameter of distal tibia (DTB)	26.3	28.5
Width of lumbosacral joint (LSW)	35.1	39.8
Multiple regression		
FHD, ACT, MTB, LTB, DTB, and LSW	30.4	30.4

[a]ALLHOM, all extant hominoids included in sample; NONHUM, all nonhuman hominoids included in sample.
[b]Lowest standard error of estimate and lowest mean absolute percentage predicted error.

(ALLHOM: gorillas, common chimpanzees, bonobos or pygmy chimpanzees, orangutans, siamang, lar gibbons, and modern humans) and (2) the same assemblage minus the humans (NONHUM) were used. This precaution has been taken because modern humans possess hindlimb and lumbosacral joints that are exceptionally large for their body mass, and early human precursors do not appear to share this proportional distinction – despite a clear-cut commitment to bipedalism (Corruccini & McHenry 1978; Jungers 1988; Napier 1964). For this reason, and because extrapolation far outside the limits of the human sample would be necessary, intraspecific regression estimates based solely on modern humans were not attempted.

Estimates of body mass for A.L.288–1 are summarized in Table 7.1. The multiple regression estimates are virtually identical at 30.4 kg regardless of which extant sample is used. This value accords well with new estimates derived from the subtrochanteric circumference of the femoral diaphysis (McHenry 1988). Values based on individual articular variables range from 26.3 to 35.1 kg for ALLHOM regressions versus 26.2 to 39.8 kg for the NONHUM equations. Using the lowest standard error of estimate and lowest mean absolute percentage prediction errors as the criteria for selecting the "best" single variable predictors, femoral head diameter (27.4 kg) is best among the ALLHOM alternatives, whereas anteroposterior diameter of the medial tibial condyle (35.5 kg) is selected from the NONHUM possibilities. The former value is close to that favored by those workers who originally described the specimen (Latimer, Ohman, & Lovejoy 1987). On balance, a value of approximately 30 kg is regarded here as most reasonable. The considerable range of values possible for a single individual when different isolated variables are used is somewhat sobering and argues for combining as much information as possible into estimates (e.g., multiple regression) whenever feasible.

It is widely accepted that australopithecines (*A. afarensis* included) are characterized by relatively very large postcanine teeth for their body size (Kay 1985; McHenry 1984; Wolpoff 1973). A comparison of the results reported here for Lucy's mass with that based on tooth size demonstrates the accuracy of this conclusion in unequivocal fashion. Using first mandibular molar area (more or less the standard for dental estimates) and the regression equation for all primates reported by Gingerich et al. (1982), one predicts a body mass for A.L.288–1 of nearly 53 kg. This value is roughly 75% greater than the more reasonable value of 30 kg. Complex evolutionary scenarios for early hominid evolution predicated on body mass estimates derived from tooth size probably require substantial rethinking and reformulation (e.g., Blumenberg 1984).

When compared to other hominoids of similar body mass, Lucy's unique interlimb proportions are better appreciated. Relative hindlimb length is quite similar in A.L.288–1 to small-bodied African apes such as the bonobo (*Pan paniscus*) and is much shorter than that observed in any modern human sample, African pygmies included (Jungers 1982; Jungers & Stern 1983; Susman, Stern, & Jungers 1984). Relative and absolute elongation of the hindlimb, therefore, represents a later modification in human evolution, occurring well after the acquisition of bipedalism. Relative forelimb length in Lucy, however, is more similar to modern humans than to any pongid, and probably is related to a reorientation of the body's center of gravity closer to the hip joint (Jungers & Stern 1983; Preuschoft 1978). This interpretation is consistent with other aspects of australopithecine locomotor morphology, such as the closer approximation of the acetabulum to the sacroiliac articulation (Stern & Susman 1983). Considerable reorganization of the postcranial skeleton in *A. afarensis* as an adaptation to bipedal locomotion and posture is apparent, but pointed differences from modern humans (as well as from *Homo erectus*) refute the conclusion that this early hominid adaptation was already complete and functionally equivalent to that seen later in human evolution.

Case 2: *Oreopithecus bambolii* (IGF 11778)

The partial skeleton of *Oreopithecus bambolii* (IGF 11778) from the Late Miocene lignitic locality of Baccinello in central Italy is rare among hominoid fossils in that it also preserves dental and postcranial remains. The locomotor skeleton (Figure 7.2) exhibits clear-cut affinities with extant hominoids (see Harrison [1986] for a review), and this latter group (NONHUM of above) again serves as a useful baseline for making estimates of body mass. Considerable postmortem distortion via flattening is evident in the specimen, but four joint measurements could be taken reliably from several regions: femoral head diameter (FHD), width of the lateral femoral condyle (LCW), articular breadth of the patella (PAT), and maximum width of the distal radial articular surface (RAD). All four of these variables are highly correlated with body mass in nonhuman hominoids (Jungers in press). Again because more than one source of articular size information is available on this specimen, multiple regression was used in the estimation. A value of 31.8 kg results from this analysis. Femoral head diameter is again the best single predictor variable, and a value of 32.0 kg is obtained from FHD alone. Accordingly, 32 kg appears to be a reasonable value for this presumably male individual of *Oreopithecus*.

These results contrast with prior attempts to estimate body size in this individual. Schultz (1960) had confidently asserted that IGF 11778 had a mass of no less than 40 kg, and Aiello (1981) offered a regression-based value centered on 48.6 kg. If the formula for maxillary molars from Gingerich et al. (1982) is enlisted in an attempt to calculate the mass of the same specimen, one obtains values from 17 kg (M^2) to 19 kg (M^1). Therefore, one can conclude with some degree of confidence that *Oreopithecus* possessed relatively very small postcanine teeth – precisely the opposite condition seen in *A. afarensis*. Although the dietary implications of this finding are far from obvious, they serve again as a caveat against exclusive and uncritical estimation of body size from tooth size.

IGF 11778 can now be placed into a "narrow allometric" (*sensu* Smith 1980) context of similarly sized primates in order to examine its body proportions and probable locomotor adaptations (Table 7.2). Long bone lengths for IGF 11778 are taken from Straus (1963; confirmed on the original fossils by W.L.J.), and ilium length was measured on the original by the author. Five apes plus one baboon, all with known body mass at death, are approximately the same size as this individual of *Oreo-*

Figure 7.2. The partial skeleton of *Oreopithecus bambolii* (IGF 11778) following preparation. (Photograph courtesy of Dr. R. L. Susman.)

Table 7.2. *A comparison of skeletal dimensions in* Oreopithecus bambolii *(IGF 11778) and living anthropoids of the same size*

Variable	IGF 11778 (32 kg)	Orangs (2) (32.7 kg)	Bonobos (1) (31 kg)	Chimps (2) (31.5 kg)	Baboons (1) (30.9 kg)
Humerus length (mm)	298	327–329	282	251–271	244
Radius length (mm)	283	323	261	235–242	262
Femur length (mm)	255	244–251	285	252–273	281
Tibia length (mm)	235	215–219	238	214–227	259
Ilium length (mm)	145	150–152	174	167–172	137

pithecus. Only the highly suspensory orangutans have forelimbs longer than the fossil. Hindlimb length in IGF 11778 is not quite as short as in the orangutan, but it is shorter than in all but one of the remaining extant specimens – that is, one of the two common chimpanzees. Ilium length is intermediate between that of the baboon and the two orangutans.

Overall similarity among these seven specimens can be summarized using standard numerical taxonomic methods (Sneath and Sokal 1973). In this case, a 7 × 7 symmetrical matrix of average taxonomic distances was computed from the five measurements, and a UPGMA clustering algorithm was employed to summarize its phenetic structure (Rohlf, Kishpaug, & Kirk 1986). The resulting phenogram is seen in Figure 7.3. The cophenetic correlation (0.811 in this case) measures the strength of association between the original distance matrix and that implied by the clustering. *Oreopithecus* clusters with the two orangutans; the African apes form a separate group that is closer to the baboon than to the *Oreopithecus*/orangutan cluster. The members of the chimpanzee/baboon group share quadrupedalism as a significant component of their locomotor repertoires. The union of IGF 11778 with orangutans suggests a similarity in overall body proportions (Biegert & Maurer 1972), and may indicate considerable overlap in locomotor and postural repertoires (e.g., a predominance of climbing and brachiation over pronograde quadrupedalism).

Case 3: Giant subfossil lemurs from Madagascar

Recently extinct (subfossil) species constitute a significant component of the adaptive radiation of prosimian primates in Madagascar. Radio-

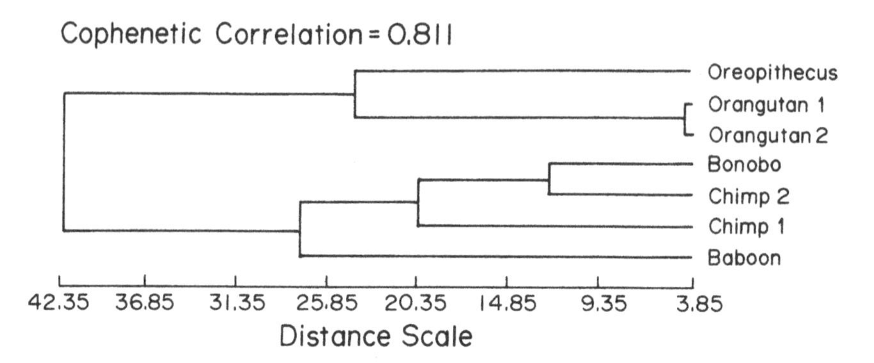

Figure 7.3. A phenogram of UPGMA clustering based on the average taxonomic distance matrix of body proportions in *Oreopithecus* and similarly sized extant primates.

carbon dates of most of the fossil-bearing localities span a relatively narrow period of 1,000 to 3,000 years B.P. (Tattersall 1982). In comparison with the living members of this diversified group, several of the subfossil species were truly "giants" (Jungers 1978, 1980; Tattersall 1973; Walker 1974). Included in this category are three species of *Megaladapis* (*M. edwardsi, M. grandidieri,* and *M. madagascariensis*), two species of *Palaeopropithecus* (*P. ingens* and *P. maximus*), and the monospecific *Archaeoindris fontoynonti* (= "*Lemuridotherium*").

Impressionistic, qualitative estimates of body size have been offered on several occasions for some of the subfossils. For example, *Megaladapis* (species usually unspecified) has been likened to a calf-sized animal (Kurten 1971; Lavauden 1931) or perhaps a Saint Bernard dog (Simons 1972). *Archaeoindris* was described as larger than a human (Lavauden 1931) or gorilla-sized (Lamberton 1934; Vuillaume-Randriamanantena 1988). Based on the enormous femur originally attributed to "*Lemuridotherium*" (Standing 1910) but now recognized as pertaining to *Archaeoindris* (Vuillaume-Randrianamanantena 1988; Walker 1974) Lavauden is reported to have proposed that this species represented the largest of known primates, even surpassing the gorilla in body size (Villette 1911). Although it shares closer functional affinities with sloths and orangutans (Carleton 1936; Jungers 1980; Walker 1974), *Palaeopropithecus* has recently been compared to the common chimpanzee in size (Vuillaume-Randriamanantena 1988).

A more specific estimate of perhaps 60 kg has recently been suggested for *Palaeopropithecus* (Burney et al. 1987). Somewhat earlier, Jungers

Table 7.3. *Body mass estimates in subfossil Malagasy primates based on tooth size and joint size*

	Predictor variable		
Species	M^1 area (kg)	M_1 area (kg)	Femoral head diameter (kg)
Megaladapis edwardsi	156.5	111.2	67.5
Megaladapis grandidieri	81.7	51.4	55.7
Megaladapis madagascariensis	66.2	47.3	36.0
Archaeoindris fontoynonti	230.5	176.8	244.1
Palaeopropithecus maximus	110.4	98.4	47.4

(1978) had predicted body mass values for two of the *Megaladapis* species from approximate trunk length; *M. edwardsi* was said to range from 50 to 100 kg, and *M. grandidieri* was bracketed between 40 and 75 kg. Because these extinct species are so much bigger than living prosimians, it is difficult to decide on the most appropriate living analogies for body size estimation. For five of these giant lemurs (Table 7.3), body mass has been estimated here from two sources: (1) maxillary and mandibular molar size, using the all-primate equations of Gingerich et al. (1982); and (2) femoral head diameter, using the NONHUM regression employed above. Nonhuman hominoids are the only extant primates with members in the probable size range of the subfossils, and hominoid locomotor analogies are not uncommon for some of the species.

Table 7.3 results point once again to large discrepancies between dental and articular estimates. For *M. edwardsi,* tooth-derived estimates are 1.6 (M_1) to 2.3 (M^1) times the value predicted by FHD. As Figure 7.4 illustrates, this species possessed a very large craniodental skeleton for its body size, an observation that helps explain the marked discrepancy in body size reconstructions reported here. The differences are also substantial for *M. madagascariensis,* but mandibular molar size and FHD predict similar masses for *M. grandidieri.* The 67.5-kg value for *M. edwardsi* and the 55.7-kg value for *M. grandidieri* based on FHD are in the middle of the ranges suggested earlier (Jungers 1978). Articular size predicts a body size for *Palaeopropithecus* much less than that implied by tooth size; the 47.4-kg value falls within the range of extant chimpanzees, but only male gorillas exceed the values predicted by molar size (100+ kg). There exists fairly good agreement between predictions from M^1 and FHD for *Archaeoindris,* and a value well in excess of 200 kg appears plau-

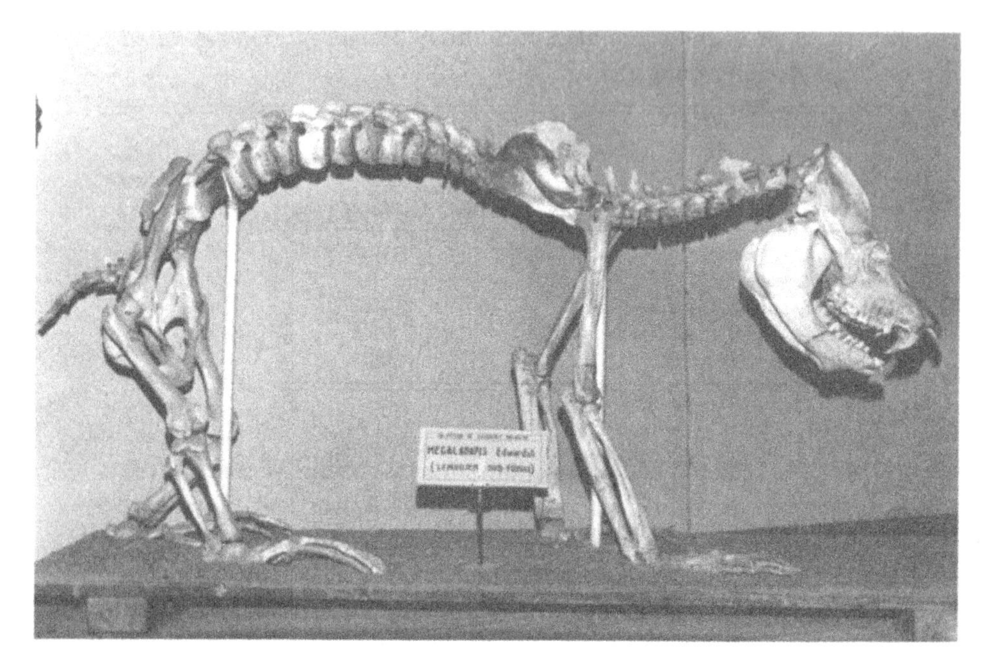

Figure 7.4. A reconstruction of *Megaladapis edwardsi* on display at the Académie Malgache in Antananarivo (Madagascar).

sible. If accurate, this estimate implies that this poorly known subfossil lemur was significantly larger than most male gorillas, and it was almost certainly terrestrial (Jungers 1980). Perhaps only *Gigantopithecus blacki* among fossil primates was larger than *Archaeoindris* (Simons 1972). It also implies that the entire radiation of Malagasy prosimians (subfossil plus extant species) spans the known range of body sizes for living primates outside of Madagascar (Fleagle 1978; Jungers 1984).

Acknowledgments

I wish to express my sincere appreciation to John Damuth and Bruce MacFadden for the invitation to contribute to this volume and to participate in the symposium at the Florida State Museum upon which it is based. Special thanks go to Peter Schmid for the photograph of his reconstruction of "Lucy" (A.L.288-1) and to Randall Susman for the photograph of the *Oreopithecus* skeleton. Leslie Jungers drew the phenogram (Figure 7.2), and Joan Kelly typed the bibliography. This research was supported by NSF grants BNS 8519747 and BNS 8606781.

References

Aiello, L. C. 1981. Locomotion in the Miocene Hominoidea. *In Aspects of Human Evolution,* ed. C. B. Stringer, pp. 63–97. London: Taylor & Francis.

Alexander, R. McN. 1980. Forces in animal joints. *Eng. Med. 9*:93–97.

Biegert, J., & Maurer, R. 1972. Rumpfskelettlange, Allometrien und Körperproportionen bei catarrhinen Primaten. *Folia Primatol. 17*:142–156.

Blumenberg, B. 1984. Allometry and evolution of Tertiary hominoids. *J. Human Evol. 13*:613–676.

Burney, D. A., MacPhee, R. D. E., Dewar, R. E., Wells, N. A. Andriamanantena, E. H., & Vuillaume-Randriamanantena, M. 1987. L'environment au cours de l'Holocène et la disparition de la mégafaune à Madagascar: quel rapport avec la conservation de la nature? In *Proceedings of the IUCN/SCC Symposium on Species Conservation Priorities in Madagascar,* ed. R. Mittermeier. Antananarivo. *Occasional Pap., IUCN Species Survival Commission 2*:137–143.

Calder, W. A., III. 1984. *Size, Function, and Life History.* Cambridge, Mass.: Harvard University Press.

Carleton, A. 1936. The limb bones and vertebrae of the extinct lemurs of Madagascar. *Proc. Zool. Soc. Lond. 110*:281–307.

Conroy, G. C. 1987. Problems of body-weight estimation in fossil primates. *Int. J. Primatol. 8*:115–137.

Corruccini, R. S., and McHenry, H. M. 1978. Relative femoral head size in early hominids. *Am. J. Phys. Anthropol. 49*:145–148.

Darroch, J. N., & Mosimann, J. E. 1985. Canonical and principal components of shape. *Biometrika 72*:241–252.

Fleagle, J. G. 1978. Size distributions of living and fossil primate faunas. *Paleobiology 4*:67–76.

Fleagle, J. G., & Kay, R. F. 1985. The paleobiology of catarrhines. In *Ancestors: The Hard Evidence,* ed. E. Delson, pp. 23–36. New York: Alan R. Liss.

Ford, S. M., & Corruccini, R. S. 1985. Intraspecific, interspecific, metabolic, and phylogenetic scaling in platyrrhine primates. In *Size and Scaling in Primate Biology,* ed. W. L. Jungers, pp. 401–435. New York: Plenum Press.

Gingerich, P. D. 1977. Correlation of tooth size and body size in living hominoid primates, with a note on relative brain size in *Aegyptopithecus* and *Proconsul. Am. J. Phys. Anthropol. 47*:395–398.

Gingerich, P. D., Smith B. H., & Rosenberg, K. 1982. Allometric scaling in the dentition of primates and prediction of body weight from tooth size in fossils. *Am. J. Phys. Anthropol. 58*:81–100.

Harrison, T. 1986. A reassessment of the phylogenetic relationships of *Oreopithecus bambolii* Gervais. *J. Human Evol. 15*:541–583.

Hofman, M. A. 1983. Encephalization in hominids: evidence for the model of punctuationalism. *Brain Behav. Evol. 22*:102–117.

Humphries, J. M., Bookstein, F. L., Chernoff, B., Smith, G. R., Elder, R. L., & Poss, S. G. 1981. Multivariate discrimination by shape in relation to size. *Syst. Zool. 30*:291–308.

Hylander, W. L. 1985. Mandibular function and biomechanical stress and scaling. *Am. Zool. 25*:315–330.

Jerison, H. J. 1973. *Evolution of the Brain and Intelligence*. New York: Academic Press.

Johanson, D. C., & Edey, M. 1981. *Lucy: The Beginnings of Humankind*. New York: Simon & Schuster.

Johanson, D. C., Lovejoy, C. O., Kimbel, W. H., White, T. D., Ward, S. C., Bush, M. E., Latimer, B. M., & Coppens, Y. 1982. Morphology of the Pliocene partial hominid skeleton (A.L.288–1) from the Hadar Formation, Ethiopia. *Am. J. Phys. Anthropol. 57*:403–451.

Jungers, W. L. 1978. The functional significance of skeletal allometry in *Megaladapis* in comparison to living prosimians. *Am. J. Phys. Anthropol. 19*:303–314.

Jungers, W. L. 1980. Adaptive diversity in subfossil Malagasy prosimians. *Z. Morphol. Anthropol. 71*:177–186.

Jungers, W. L. 1982. Lucy's limbs: skeletal allometry and locomotion in *Australopithecus afarensis*. *Nature 297*:676–678.

Jungers, W. L. 1984. Aspects of size and scaling in primate biology with special reference to the locomotor skeleton. *Yb. Phys. Anthropol. 27*: 73–97.

Jungers, W. L. in press. Scaling of postcranial joint size in hominoid primates. In *Gravity, Posture and Locomotion in Primates*, ed. F. K. Jouffroy, M. H. Stack, & C. Niemitz. Firenze: Il Sedicesimo.

Jungers, W. L. 1988. Relative joint size and hominoid locomotor adaptations with implications for the evolution of hominid bipedalism. *J. Human Evol. 17*:247–265.

Jungers, W. L., & Stern, J. T., Jr. 1983. Body proportions, skeletal allometry, and locomotion in the Hadar hominids: a reply to Wolpoff. *J. Human Evol. 12*:673–684.

Kay, R. F. 1984. On the use of anatomical features to infer foraging behavior in extinct primates. In *Adaptations for Foraging in Nonhuman Primates,* ed. P. S. Rodman & J. G. H. Cant, pp. 21–53. New York: Columbia University Press.

Kay, R. F. 1985. Dental evidence for the diet of *Australopithecus*. *Annu. Rev. Anthropol. 14*:315–341.

Kay, R. F., & Simons, E. L. 1980. The ecology of Oligocene African Anthropoidea. *Int. J. Primatol. 1*:21–37.

Kurten, B. 1971. *The Age of Mammals*. London: Weidenfeld & Nicolson.

Lamberton, C. 1934. Contribution à la connaissance de la faune subfossile de Madagascar. *L'Archaeoindris fontognonti* Stand. *Mém. Acad. Malgache 17*:9–39.

Latimer, B., Ohman, J. C., & Lovejoy, C. O. 1987. Talocrural joint in African

hominoids: implications for *Australopithecus afarensis*. *Am. J. Phys. Anthropol.* *74*:155–175.
Lavauden, L. 1931. Animaux disparus et légendaires de Madagascar. *Rev. Scientifique* *10*:297–308.
Lindstedt, S. L., & Calder, W. A. III. 1981. Body size, physiological time, and longevity of homeothermic animals. *Q. Rev. Biol.* *56*:1–16.
Martin, R. A. 1981. On extinct hominid population densities. *J. Human Evol.* *10*:427–428.
McHenry, H. M. 1976. Early hominid body weight and encephalization. *Am. J. Phys. Anthropol.* *45*:77–84.
McHenry, H. M. 1984. Relative cheek-tooth size in *Australopithecus*. *Am. J. Phys. Anthropol.* *64*:297–306.
McHenry, H. M. 1988. New estimates of body weight in early hominids and their significance to encephalization and megadontia in robust australopithecines. In *The Evolutionary History of the Robust Australopithecines,* ed. F. E. Grine, pp. 133–148. New York: Aldine DeGruyter.
Mosimann, J. E. 1970. Size allometry: size and shape variables with characteristics of the lognormal and generalized gamma distributions. *J. Am. Stat. Assoc.* *65*:930–945.
Napier, J. R. 1964. The evolution of bipedal walking in the hominids. *Arch. Biol. (Liege)* *75*:673–708.
Peters, R. H. 1983. *The Ecological Implications of Body Size.* Cambridge: Cambridge University Press.
Preuschoft, H. 1978. Recent results concerning the biomechanics of man's acquisition to bipedality. In *Recent Advances in Primatology,* ed. D. J. Chivers & K. A. Joysey, Vol. 3, pp. 435–458. London: Academic Press.
Radinsky, L. 1974. The fossil evidence of anthropoid brain evolution. *Am. J. Phys. Anthropol.* *41*:15–27.
Rightmire, G. P. 1986. Body size and encephalization in *Homo erectus*. *Anthropos* *23*:139–150.
Rohlf, F. J., Kishpaug, J., & Kirk, D. 1986. *NT-SYS: Numerical Taxonomic System of Multivariate Statistical Programs.* Stony Brook: State University of New York.
Ruff, C. 1987. Structural allometry of the femur and tibia in Hominoidea. *Folia Primatol.* *48*:9–49.
Schmid, P. 1983. Eine Rekonstruktion des Skelettes von A.L.288–1 (Hadar) und deren Konsequenzen. *Folia Primatol.* *40*:283–306.
Schmidt-Nielsen, K. 1977. Problems of scaling: locomotion and physiological correlates. In *Scale Effects in Animal Locomotion,* ed. T. J. Pedley, pp. 1–21. New York: Academic Press.
Schmidt-Nielsen, K. 1984. *Scaling. Why Is Animal Size So Important?* Cambridge: Cambridge University Press.
Schultz, A. H. 1960. Einige Beobachtungen und Masse am Skelett von *Oreopithecus* im Vergleich mit anderen catarrhinen Primaten. *Z. Morphol. Anthropol.* *50*:136–149.
Simons, E. L. 1972. *Primate Evolution: An Introduction to Man's Place in Nature.* New York: Macmillan.

Smith, R. J. 1980. Rethinking allometry. *J. Theor. Biol. 87*:97–111.
Smith, R. J. 1985. The present as a key to the past: body weight of Miocene hominoids as a test of allometric methods for paleontological inference. In *Size and Scaling in Primate Biology,* ed. W. L. Jungers, pp. 437–447. New York: Plenum Press.
Sneath, P. H. A., & Sokal, R. R. 1973. *Numerical Taxonomy.* San Francisco: W. H. Freeman.
Standing, H. 1910. Note sur les ossements subfossiles provenant des fouilles d'Ampazambazimba. *Bull. Acad. Malgache [anc. ser.] 7*:61–64.
Stern, J. T., Jr., & Susman, R. L. 1983. The locomotor anatomy of *Australopithecus afarensis. Am. J. Phys. Anthropol. 60*:279–317.
Steudel, K. 1980. New estimates of early hominid body size. *Am. J. Phys. Anthropol. 52*:63–70.
Straus, W. L., Jr. 1963. The classification of *Oreopithecus.* In *Classification and Human Evolution,* ed. S. L. Washburn, pp. 146–177. Chicago: Aldine.
Susman, R. L., Stern, J. T., Jr., & Jungers, W. L. 1984. Arboreality and bipedality in the Hadar hominids. *Folia primatol. 43*:113–156.
Tattersall, I. 1973. Subfossil lemuroids and the "adaptive radiation" of the Malagasy lemurs. *Trans. N.Y. Acad. Sci. 35*:314–324.
Tattersall, I. 1982. *The Primates of Madagascar.* New York: Columbia University Press.
Villette 1911. Procès-verbal de la séance en 22 septembre 1910. *Bull. Acad. Malgache [anc. ser.] 8*:18–19.
Vuillaume-Randriamanantena, M. 1988. The taxonomic attributions of giant subfossil lemur bones from Ampasambazimba: *Archaeoindris* and *Lemuridotherium. J. Human Evol. 17*:379–391.
Walker, A. 1974. Locomotor adaptations in past and present prosimian primates. In *Primate Locomotion,* ed. F. A. Jenkins, Jr., pp. 349–381. New York: Academic Press.
Wolpoff, M. H. 1973. Posterior tooth size, body size, and diet in South African gracile australopithecines. *Am. J. Phys. Anthropol. 39*:375–394.

8

Body mass and hindlimb bone cross-sectional and articular dimensions in anthropoid primates

CHRISTOPHER RUFF

The importance of body mass as an ecological and metabolic variable has been emphasized by many authors (e.g., Fleagle 1985; Jungers 1984; Smith 1985; Steudel 1985). From the viewpoint of *mechanical* forces acting on the skeleton, body mass is also a natural variable against which to test "functional equivalence," since body weight itself constitutes a mechanical loading, and other loadings produced by muscle action are largely determined by the force required to overcome resistance to body mass inertia in movement, or to balance or prevent collapse over a supporting limb.

The significance of a mechanical approach to the study of allometric relationships between body mass and skeletal dimensions has been appreciated for some time (e.g., Schultz 1953). However, most skeletal dimensions traditionally included in allometric studies, such as long bone length or circumference, can be related to mechanical function only indirectly. Structural characteristics that more *directly* reflect mechanical loadings in vivo should provide more accurate estimates of body mass from skeletal remains, as well as aid in the reconstruction of specific locomotor and other behavioral characteristics of fossil forms.

In the present study, the relationship between body mass and two types of long bone structural features – diaphyseal cross-sectional geometry and articular surface dimensions – are examined. Both types of features should be directly related to the mechanical loads on the bones produced by gravitational and muscular forces, that is, to body mass and activity patterns. The femur and tibia were chosen for analysis,

In *Body Size in Mammalian Paleobiology: Estimation and Biological Implications,* John Damuth and Bruce J. MacFadden, eds.

Table 8.1. *Samples used in the study*

Sample	n[a]
Homo sapiens (Pecos)	119
Homo sapiens (autopsy)[b]	47, 39
Homo sapiens (Terry)[c]	60
Pan troglodytes	20
Gorilla gorilla	20
Pongo pygmaeus	20
Macaca fascicularis	20

[a]Number of individuals included in sample, except for *Homo sapiens* autopsy sample.
[b]Forty-seven femora and 39 tibiae representing 33 and 31 individuals, respectively; used in cross-sectional geometric analyses only.
[c]Used in articular analyses only.

partly because they are weight bearing in all the species examined, including bipeds as well as quadrupeds.

Methods

Samples

Five anthropoid primate species are included in the study: *Homo sapiens, Pan, Gorilla, Pongo,* and one species of macaque (*M. fascicularis*) (Table 8.1). Each nonhuman primate species is represented by a total of 20 individuals, equally divided between males and females; all are wild-shot specimens obtained from museum collections. The human samples include 119 individuals from a late prehistoric and protohistoric archaeological site in New Mexico (Ruff & Hayes 1983a) and 47 femora and 39 tibiae from a modern autopsy collection of U.S. whites (Ruff 1987; Ruff & Hayes 1988). All individuals in the samples are adult. To avoid the marked effects of age on bone structure in humans (Ruff & Hayes 1983b), the autopsy sample was limited to individuals under 60 years of age.

Cross-sectional geometry

The cross-sectional geometric properties of the femoral and tibial diaphyses examined here include bone areas and second moments of area

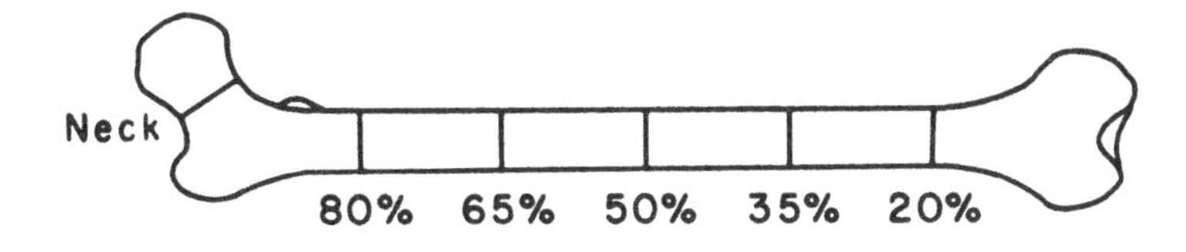

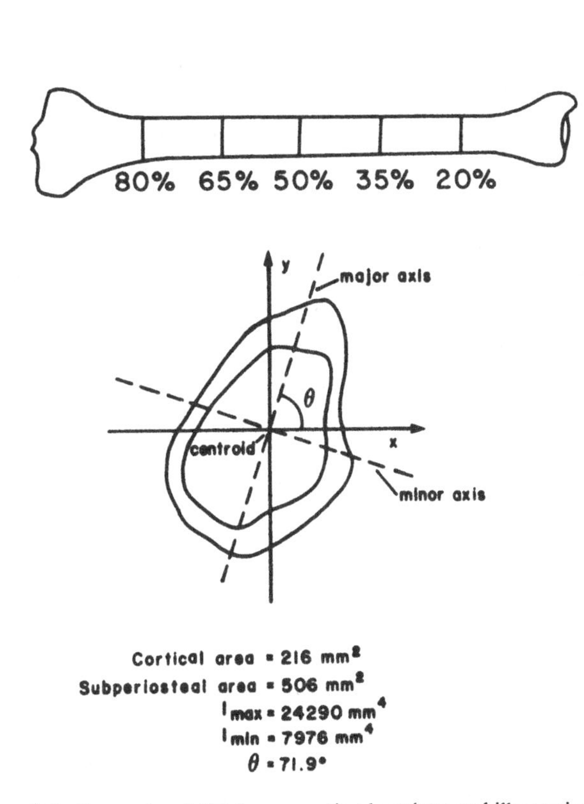

Figure 8.1. Femoral and tibial cross section locations and illustration of several cross-sectional properties calculated by SLICE (also see Ruff & Hayes 1983a). Polar second moment of area (J) = I_{max} + I_{min}. (Partly adapted from Ruff & Hayes 1983a. With the permission of the *American Journal of Physical Anthropology*.)

(SMA), also known as area moments of inertia (Figure 8.1). The biomechanical significance of these properties has been discussed at length elsewhere (Lovejoy, Burstein, & Heiple 1976; Ruff & Hayes 1983a,b). Most simply put, with long bones modeled as beams, the cross-sectional area of the cortex is proportional to axial compressive and tensile strength, whereas second moments of area are proportional to bending and torsional strength. For the purposes of this study, two cross-sectional properties were included in the analyses: cortical area (CA) and the polar second moment of area (J). The polar SMA is proportional to both torsional and average bending strength in all planes, since J equals the sum of any two SMAs measured about perpendicular bending planes.

A total of five diaphyseal cross section locations were included in the study: at 20%, 35%, 50%, 65%, and 80% of bone length (as defined in Ruff & Hayes 1983a), measured from the distal end (Figure 8.1). In addition, a superior–inferior section through the midfemoral neck was measured. All sections were taken with the bones in standardized orientation as previously described (Ruff & Hayes 1983a). Cross section contours were obtained by direct sectioning in the human material (Ruff & Hayes 1983a), and by computed tomography in the nonhuman primates (Ruff & Leo 1986). Subperiosteal and endosteal section contours were then traced on a digitizer and input to program SLICE (Nagurka & Hayes 1980), which calculates geometric section properties.

In addition to cross-sectional geometric properties, for comparative purposes bone length and the average femoral midshaft external breadth (mean of anteroposterior and mediolateral) were also measured on each specimen and entered into the analyses.

Articular surface areas

Articular surface areas of the femur and tibia were calculated from linear breadth measurements taken on the specimens with calipers (Figure 8.2). The articular surface area of the femoral head was calculated using the formula for a partial sphere, as follows: Superior–inferior (SI) and anteroposterior (AP) diameters and mediolateral depth of the femoral head articular surface (as viewed from the superior aspect perpendicular to the femoral neck axis) were measured. Using the average of the two diameters to calculate the radius of the head, r, it can be shown that the surface area is equal to $4\pi r^2 \times$ (depth/$2r$), or, simplified and in terms of the three linear measurements taken, $1.57 \times$ depth $\times$ (SI + AP). In addition to articular surface area, total volume of the femoral head was also calculated using the formula for a sphere: $4/3\pi r^3$, or $.0654 \times (SI + AP)^3$. Surface areas of the femoral condyles, tibial plateaus, and tibiotalar articulations were calculated as products of anteroposterior and mediolateral breadth measurements of each surface, assuming an approximately rectangular shape (Figure 8.2). For the femoral condyles,

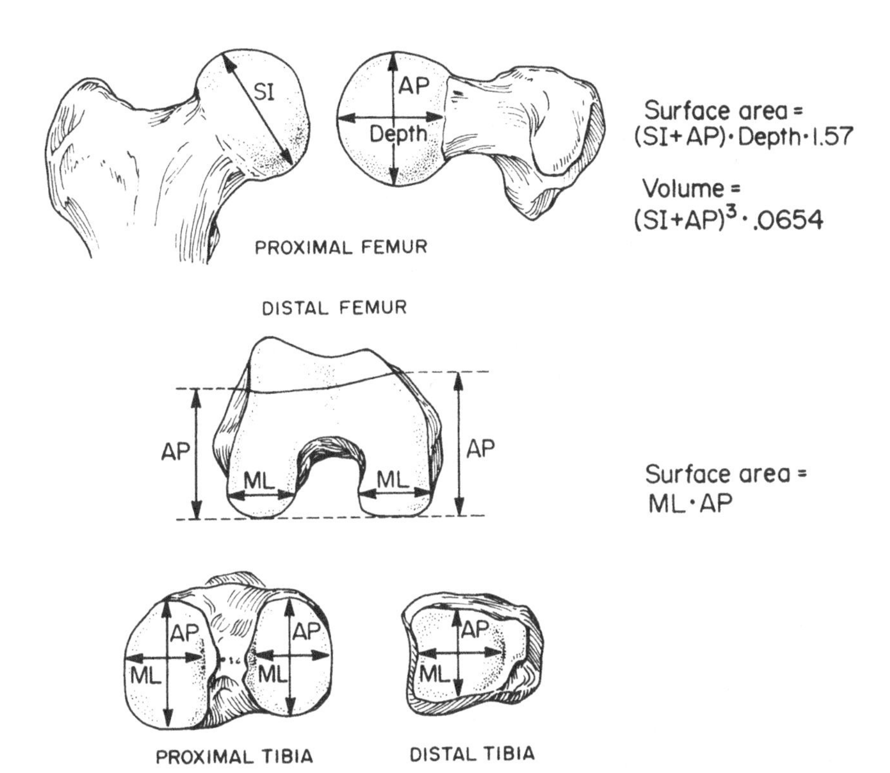

Figure 8.2. Articular dimensions of the femur and tibia measured in the study, and calculation of surface areas and femoral head volume (also see text). Note that surfaces are illustrated on human articulations for convenience, and that the lateral boundary of the patellar articular surface was drawn incorrectly in the originally published figure. (From Ruff 1988. With the permission of the *Journal of Human Evolution.*)

the region of articulation with the patella, discernible as a faintly marked ridge or relatively abrupt change in contour, was excluded from the articular surface area calculations. For these measurements, the femur was placed on an osteometric board in a standardized position with the posterior borders of the condyles on the board surface (Ruff & Hayes 1983a). The AP height of each condyle surface was then measured from the board to the edge of the patellar articular surface at its outer margin, and ML breadth measured parallel to the board just above the board surface (Figure 8.2). Tibial plateau and tibiotalar surface breadths included only the relatively flat portions of the surfaces; thus the projections of the tibial intercondylar eminences and medial malleolus were not included. The two femoral condyles and tibial plateaus were ana-

Table 8.2. *Body mass estimates (kg)*

Sample	Male	Female
Homo sapiens (Pecos)	58.9	53.8
Homo sapiens (autopsy)	80.7	65.8
Homo sapiens (Terry)	75.8	64.4
Pan troglodytes	50	40
Gorilla gorilla	164	85
Pongo pygmaeus	76	36.5
Macaca fascicularis	4.96	3.30

Note: See text and Ruff (1987) for detailed derivations of these figures.

lyzed both separately and as sums over medial and lateral articular surfaces.

Articular surface measurements on all joints were taken on the non-human primate samples. For the human samples, only measurements of the femoral head were taken. The modern human autopsy sample used in the cross-sectional geometry analyses was no longer available, and was replaced with an equivalent sample of 60 U.S. whites from the Terry Collection (Smithsonian Institution, Washington, D.C.), also approximately evenly divided between males and females and between 20 and 59 years of age.

Body weights and statistical techniques

Individual body masses were available for only the macaque sample, 13 of the orangutans, and two chimpanzees. Thus, for most of the non-human samples, mean sex-specific body mass values used in interspecies comparisons were estimated from literature values, weighted by subspecies composition of the sample (see Ruff [1987] for details). In the human samples, mean body masses were estimated from statures calculated from long bone lengths (Trotter & Gleser 1952, 1958; Genoves 1967) and published population and sex-specific weight for height tables or multiple linear regressions of weight on height and sitting height (Ruff 1987, 1988). Body masses used in the study are listed in Table 8.2.

Allometric relationships between body mass and bone structural properties were determined using least squares linear regression of $\log_{10}$ transformed data. Major axis slopes were also calculated (Kuhry & Marcus 1977) and are presented as an alternative method of scaling

Table 8.3. *Midshaft cross-sectional geometry and bone length: correlations with, and percent standard errors of estimate of body mass*

	Cortical area		Polar SMA[a]		Bone length	
Sample	*r*	%SEE	*r*	%SEE	*r*	%SEE
Femur						
Total	.994	13.8	.992	16.4	.864	69.3
Nonhuman	.996	14.8	.992	20.3	.990	18.4
Pongid	.973	15.7	.947	22.5	.960	22.1
Tibia						
Total	.978	28.9	.970	34.2	.900	93.7
Nonhuman	.992	20.6	.991	22.0	.994	22.7
Pongid	.945	23.0	.933	25.4	.948	19.3

[a]Polar second moment of area (J).

analysis. However, because correlation coefficients are usually quite high (> .95), the two methods generally yield very similar results. Reduced major axis slopes may be calculated by dividing least squares slopes by *r* (Ricker 1973). In the analyses of body mass prediction from structural properties, percent standard errors of estimate (%SEE), calculated as described in Smith (1984), are also given. %SEE may be interpreted as the ± percent deviation from the predicted value within which 68% (± 1 SD) of cases would be expected to fall (Smith 1984).

Results

Cross-sectional geometry and length

Complete tables of regression coefficients for CA and J on body mass for all 11 cross-sectional locations are given in Table 8.8 for several interspecies subgroupings. Coefficients for bone length and external breadth regressed on body mass are listed in Table 8.9. For prediction of body mass, coefficients for the reverse regressions of body mass on structural properties are given in Tables 8.10 and 8.11 for CA and J, and length and breadth, respectively.

For purposes of illustration, correlations and %SEEs of body mass using midshaft CA and J and bone length of the femur and tibia are given for the total sample, nonhuman primates, and pongids in Table 8.3. Scatterplots for the femoral properties are shown in Figures 8.3–

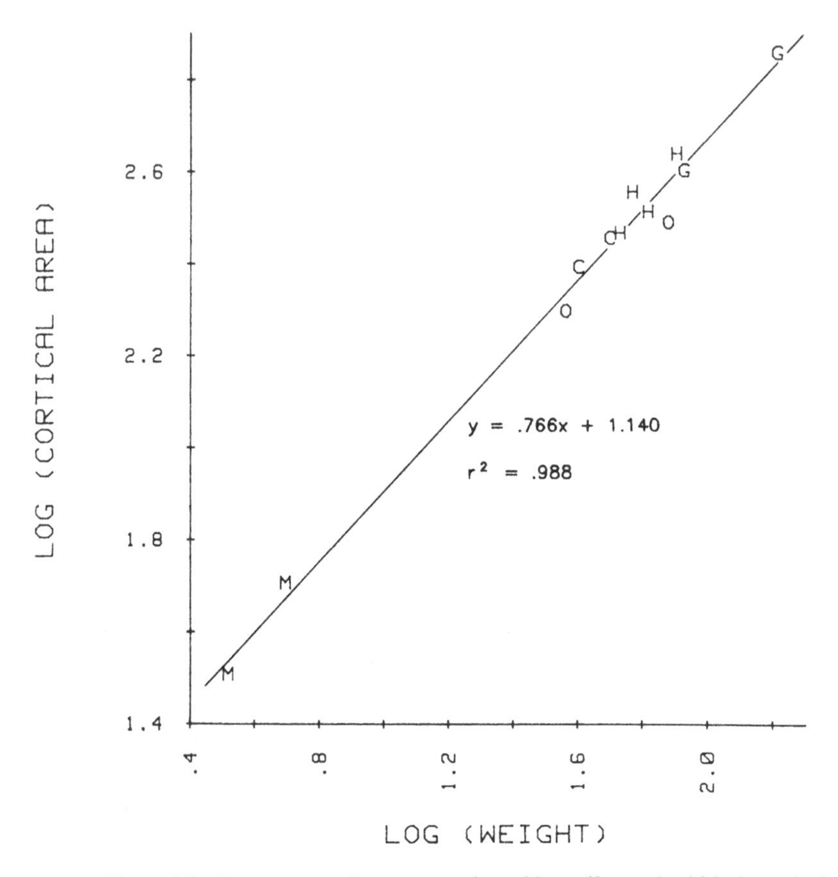

Figure 8.3. Least squares linear regression of $\log_{10}$ (femoral midshaft cortical area) on $\log_{10}$ (body mass) for sex-specific sample means. Regression line through total sample. Symbols: H, *Homo*; G, *Gorilla*; O, *Pongo*; C, *Pan*; M, *Macaca*. (From C. B. Ruff, "Structural Allometry of the Femur and Tibia in Hominoidea and *Macaca*. *Folia Primatol.* *48:* [1987] 9–49. Courtesy of S. Karger AG, Basel.)

8.5. In these plots, sex-specific means are indicated by the first letter of the species' common name – that is, H = *Homo*, G = gorilla, C = chimpanzee, O = orangutan, and M = macaque. Males are always greater than females in body weight. The two human samples are represented as (four) independent points. Least squares regression lines and coefficients are shown for the total sample, with separate lines plotted for hominoids alone or other subgroups if they are sufficiently different to be visually distinguishable on the graphs.

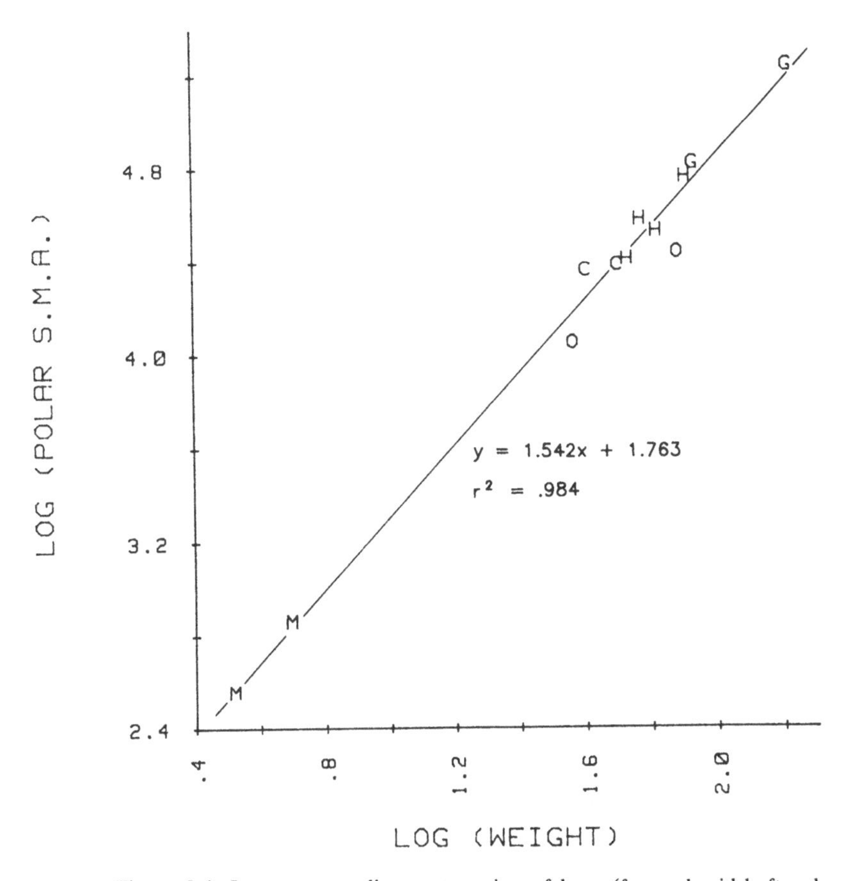

Figure 8.4. Least squares linear regression of $\log_{10}$ (femoral midshaft polar second moment of area) on $\log_{10}$ (body mass) for sex-specific sample means. Regression line through total sample. Symbols as in Figure 8.3. (From C. B. Ruff, "Structural Allometry of the Femur and Tibia in Hominoidea and *Macaca*. *Folia Primatol. 48:* [1987] 9–49. Courtesy of S. Karger AG, Basel.)

Cross-sectional bone area of the femoral diaphysis shows the tightest relationship with body mass of any of the properties examined here, regardless of which species are included in the analysis. %SEEs of body mass using this property average about 15%. The polar SMA of the femoral shaft is second best overall, with %SEEs of body mass of about 16–22%. Tibial cross-sectional properties are not as good, with %SEEs of 21–34%. Femoral and tibial lengths are relatively poorly associated with body mass across the total sample, owing mainly to the relatively long lower limbs of humans (Figure 8.5). However, even among non-

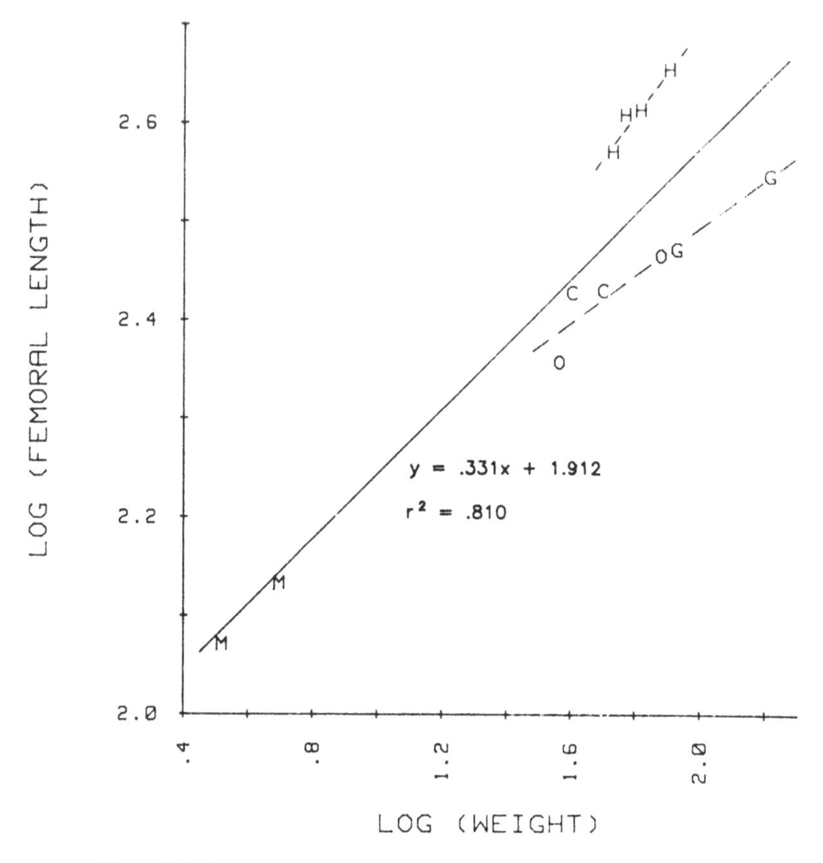

Figure 8.5. Least squares linear regression of $\log_{10}$ (femoral length) on $\log_{10}$ (body weight) for sex-specific sample means. Solid line is regression through total sample; dashed lines are regressions through humans and pongids. Symbols as in Figure 8.3. (From C. B. Ruff, "Structural Allometry of the Femur and Tibia in Hominoidea and *Macaca*. *Folia Primatol. 48:* [1987] 9–49. Courtesy of S. Karger AG, Basel.)

human primates, femoral midshaft CA is a better predictor of body mass than either of the bone lengths (Table 8.3).

In the cross-sectional analyses, the species showing the largest deviation from general scaling trends is the orangutan, which has relatively small cross-sectional dimensions for its body mass. Dropping out orangutans from species subgroupings improves predictions of body mass from femoral shaft CA or J (Table 8.10); %SEEs fall to as low as 7–9% for some sections in either the total sample or hominoids without *Pongo*.

Average femoral midshaft external breadth (Tables 8.9 and 8.11) is almost as good a predictor of body mass as femoral midshaft J, but not as good as CA. That breadth and J are closely related is not surprising, since second moments of area are largely determined by the relative outward distribution of bone in a section (i.e., external breadth), whereas cortical area should be somewhat more independent.

Slopes for CA and J on body mass, both least squares and major axis, tend to be close to isometric (.67 for CA and 1.33 for J) or slightly positively allometric, with the femur showing more positive allometry (Table 8.8; Figures 8.3, 8.4). Regression coefficients are fairly consistent, with only a few slopes or intercepts falling outside the 95% confidence limits of other sections or subgroupings. In contrast, regression coefficients for bone lengths on body mass vary widely, depending on the species included (Table 8.9; also compare Figure 8.5); this is true within nonhuman primates as well as the total sample including humans. In general, length scales negatively allometric with body mass in these species, more so in the tibia. (Note that the column headings for femoral and tibial lengths in Table 3 of Ruff [1987] were erroneously reversed; the correct coefficients are shown here in Table 8.9.) Like cross-sectional properties, external breadth is slightly positively allometric.

*Intra*species regressions were also carried out within macaques and orangutans with individual associated body weights, with detailed results presented elsewhere (Ruff 1987). In general, bone length is as good a predictor of body mass as cross-sectional properties in macaques (%SEEs of 12–14%), but not as good in orangutans (%SEEs of about 30% for length and 20% for CA and J). Macaques exhibit positive allometry in CA and J and slight negative allometry in length, whereas orangutans exhibit strong negative allometry in all three types of characteristics.

Finally, the cross-sectional data were compared to similar data reported by Biewener (1982) for a broader size range of species from mouse to horse. The results for femoral midshaft cortical area are shown in Figure 8.6, with my sex-specific means indicated by crosses and Biewener's individual data points by open circles. The allometric trends for the anthropoid primates of the present study and the broader mammalian sample are essentially indistinguishable, with regression coefficients in the two samples varying by less than 4%. Thus, these trends appear to be relatively general to mammals as a whole.

Articular dimensions

Complete listings of regression coefficients for articular dimensions on body mass and the reverse regressions of body mass on articular di-

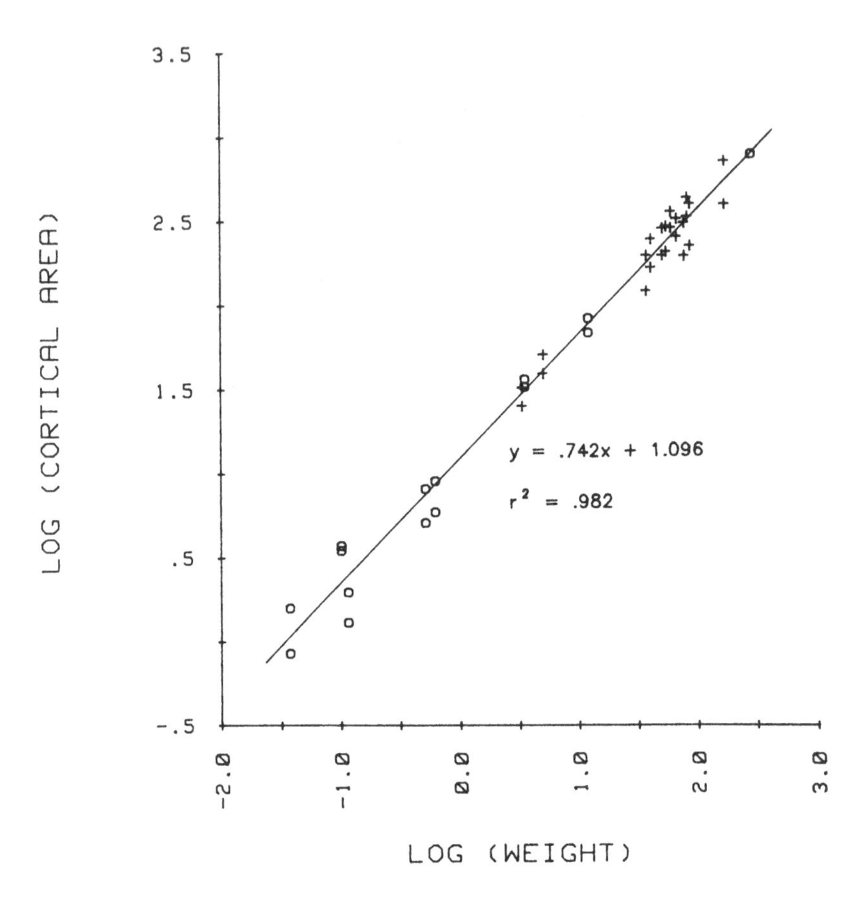

Figure 8.6. Least squares linear regression of $\log_{10}$ (cortical area) of the femoral and tibial midshafts on $\log_{10}$ (body mass) for the present study sample (+) and a sample reported by Biewener (1982) (○). The regression line is drawn through all points (sample-specific regression lines are indistinguishable). Biewener's species included here, from smallest to largest body mass, are mouse, kangaroo rat, ground squirrel, rat, guinea pig, house cat, *Colobus*, and horse. Each of his species is represented by two points, except the horse, with only the tibia. Each species in the present study sample is represented by four points (sex-specific means); for individual identification of present study sample points, see Figure 8.3. (From C. B. Ruff, "Structural Allometry of the Femur and Tibia in Hominoidea and *Macaca. Folia Primatol. 48:* [1987] 9–49. Courtesy of S. Karger AG, Basel.)

Table 8.4. *Articular dimensions: correlations with, and percent standard errors of, estimate of body mass*

	Femoral head		Femoral cond.[a]		Tibial plat.[a]		Tibiotalar		Femoral head volume	
Sample	*r*	%SEE	*r*	%SEE	*r*	%SEE	*r*	%SEE	*r*	%SEE
Total	.979	27.7	—	—	—	—	—	—	.983	25.1
Nonhuman	.997	12.2	.998	9.8	.995	15.9	.996	14.3	.998	9.0
Pongid	.988	10.2	.988	10.4	.963	18.5	.979	13.6	.995	6.5
Human	.983	3.4	—	—	—	—	—	—	.949	5.8

[a]Total combined area of medial and lateral femoral condyles, tibial plateaus.

mensions are given in Tables 8.12 and 8.13, respectively. Correlations with and %SEEs of body mass for the total sample, nonhuman primates, pongids, and humans (where applicable) are given in Table 8.4. In this table, articular dimensions around the knee have been combined into total femoral condyle and tibial plateau surface areas. Figures 8.7 and 8.8 illustrate allometric trends in femoral head articular surface area and femoral head total volume, respectively.

Correlations with and %SEEs of body mass using articular dimensions in nonhuman primates are comparable to those using diaphyseal cross-sectional area. Of the articular properties examined here, femoral head volume shows the tightest association with body mass (%SEEs, 6–9%). Tibial articulations in general are less strongly associated with body mass than with femoral dimensions, as was the case also with cross-sectional properties.

Whereas the cross-sectional diaphyseal dimensions are comparable, humans are clear outliers from nonhuman primates with respect to articular dimensions, at least of the femoral head (Figures 8.7 and 8.8). For their body mass, humans have relatively large lower limb articular dimensions. The same point has also been made by Jungers (in press). However, also of interest is the fact that humans appear to fall on a different allometric *slope* than nonhuman primates. Pongids are close to isometric in femoral head articular surface area (0.656) and femoral head volume (1.04) and tend to cluster near isometry over all of the articular dimensions examined here (Table 8.12). Including macaques increases slopes somewhat (Figures 8.7 and 8.8; Table 8.12), probably because cercopithecoids have relatively smaller lower limb articular dimensions (Jungers in press, Ruff 1988). However, neither nonhuman

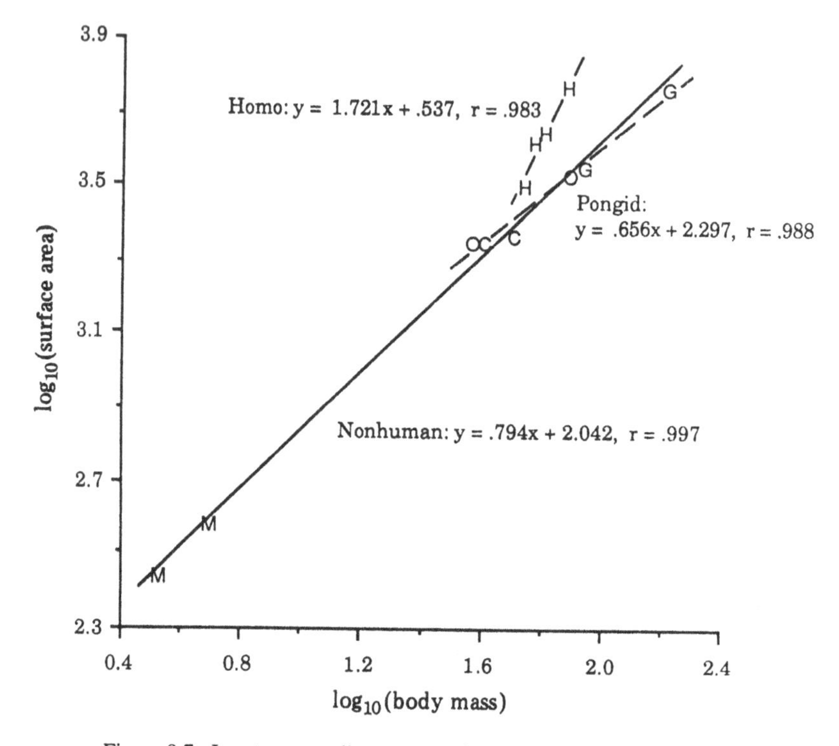

Figure 8.7. Least squares linear regression of $\log_{10}$ (femoral head surface area) on $\log_{10}$ (body mass) for sex-specific sample means. Symbols as in Figure 8.3; regression lines as in Figure 8.5.

grouping approaches the extreme positive allometry of humans, with slopes of 1.721 and 1.898 for femoral head surface area and volume, respectively. Even with only the limited number of data points available, the human and pongid slopes for articular area are barely within each other's 95% confidence intervals (0.74–2.70 and 0.51–0.80, respectively). Other intraspecies slopes for femoral head dimensions (Ruff 1988) are close to isometric (in macaques, 0.65 and 1.07 for surface area and volume, respectively) or negatively allometric (in orangutans, 0.47 and 0.84). Thus, it appears that recent human samples have not only relatively large lower limb articular dimensions, but also different scaling relationships with body mass than nonhuman primates.

Interestingly, orangutans do not show the negative deviation from general allometric trends in articular dimensions as they did for diaphyseal cross-sectional dimensions (compare Figures 8.7 and 8.8 with Figures 8.3 and 8.4). That is, despite their relatively small diaphyses,

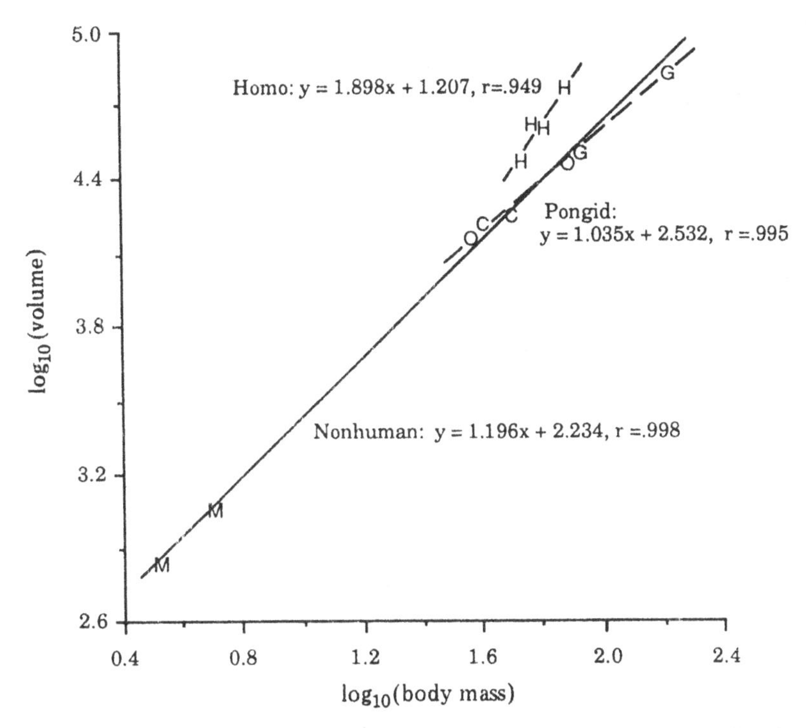

Figure 8.8. Least squares linear regression of $\log_{10}$ (femoral head volume) on $\log_{10}$ (body mass) for sex-specific means. Symbols as in Figure 8.3; regression lines as in Figure 8.5.

orangutans exhibit average-sized articular dimensions for pongids, and predictions of body mass among pongids or nonhuman primates as a whole are not improved by dropping them out.

Another interesting finding is that the medial femoral condyle shows much more positive allometry with body mass than the lateral femoral condyle, particularly among pongids (Table 8.6). Apparently, as body mass increases in these species, a greater proportion of total mass is supported by the medial condyle and relatively less by the lateral condyle.

Body mass prediction

Another measure of success in predicting body mass is the percent prediction error, %PE, calculated as [(observed mass − predicted mass)/predicted mass] × 100, first converting from log to linear scale

Table 8.5. *Percent prediction errors of body mass from cross-sectional, bone length, and articular dimensions: mean absolute values for different regression equations*

Regression equation reference group	Section and bone length dimensions[a]				
	CA	J	BD	FL	TL
Total	10	11	12	49	57
Nonhuman	10	13	14	11	15
Hominoids	8	10	12	34	36
Af. ape + *Homo*	6	7	6	35	37
Pongids	10	13	15	13	11
	Articular dimensions[b]				
	FHSA	FHV	FC	TP	TT
Total	18	17	—	—	—
Nonhuman	8	6	6	10	9
Pongids	6	5	6	11	9

Note: Mean of | [(observed wt. − predicted wt.)/(predicted wt.)] × 100 | for each regression equation. Values for each equation were calculated only for samples included in reference group (see text).
[a]CA, J, BD: cortical area, polar second moment of area, and average external breadth, femoral midshaft section; FL, TL: femoral and tibial lengths.
[b]FHSA: femoral head surface area; FHV: femoral head volume; FC, TP, TT: femoral condyle (medial + lateral), tibial plateau (medial + lateral), and tibiotalar surface areas.

(Smith 1984). %PEs of body mass for cross-sectional, bone length, and articular dimensions in several different taxonomic groupings are shown in Table 8.5. The mean absolute values of the index over each sex and species-specific data point in the group were calculated, in each case using only regression equations applicable to that species (e.g., pongid equations not applied to macaques). The "African ape + *Homo*" grouping is included for section properties since orangutans are deviant in this respect (Figures 8.3 and 8.4); "hominoids" are not included for articular properties since the Recent human samples are so deviant from pongids (Figures 8.6 and 8.7).

Femoral and tibial lengths, as expected, give very poor estimates of body mass when humans are included. Femoral midshaft cross-sectional area is a better predictor than bone length even within nonhuman groups (%PE of 10% vs. 11–15% for lengths). The best body mass predictor among nonhuman primates is femoral head volume (%PEs of 5–6%), closely followed by femoral head articular surface area and femoral

Table 8.6. *Prediction of body mass in two chimpanzees*

			Cross-sectional geometry[a]		Articular dimensions[b]	
Species	Sex	True body mass (kg)	Pred.[c]	%PE[c]	Pred.	%PE[c]
Pan t. verus	M	46.3	46.5	−0.4%	44.2	4.8%
Pan t. schweinfurthii	F	31.3	32.0	−2.2%	34.1	−8.9%

[a]Using African ape + *Homo* regression equations, average of femoral midshaft cortical area and polar second moment of area predictions.
[b]Using pongid regression equations, average of femoral head articular surface area and volume predictions.
[c]Predicted value and percent prediction error. %PE = [(observed − predicted)/predicted] × 100.

condyle surface area. Over the entire sample including humans, femoral midshaft cortical area is the best predictor (%PEs of 6–10%).

As another test of the accuracy of these regression formulae, prediction of body mass using cross-sectional and articular surface dimensions was compared in two individual chimpanzees of known body weight (Table 8.6). (The weights for these two individuals were *not* used directly in calculating mean species weights for derivation of the regression equations, so they may be taken as independent test cases.) Average body mass estimates from femoral midshaft cortical area and polar second moment of area are very close to actual values in both individuals – within 1 kg, or a 2% prediction error. Body mass estimates from femoral head articular dimensions are also good, but not quite as close, with both individuals within about 3 kg, or a 9% prediction error. Thus, in these two chimpanzees, both types of body mass estimators produce quite reasonable results, but cross-sectional dimensions appear to be somewhat more consistent predictors.

Finally, these equations were applied to estimation of body mass in some Early Miocene (18 myr) hominoids from Western Kenya. This material includes the well-known *Proconsul africanus* specimen KNM-RU 2036 (Walker & Pickford 1983), as well as several other relatively complete femora attributed to *Proconsul nyanzae* or *Proconsul africanus* from Rusinga and Mfangano Islands (Ruff et al. 1989; Walker & Teaford 1988). Cross-sectional geometric dimensions were determined through measurement of several natural breaks in the diaphyses of these specimens. Femoral head measurements were also taken on available preserved articulations (see Ruff et al. 1989 for details).

Using the hominoid-without-orangutan regression formulae for cross-sectional geometric properties, average body mass for two *P. nyanzae* individuals is estimated at 33.8 kg, whereas estimated body mass for two *P. africanus* individuals is 9.6 kg. Using the formulae for the total sample without orangutans yields average estimates of 33.0 and 8.9 kg, respectively, for these taxa. The average body mass estimate based on femoral head dimensions for five *P. nyanzae* specimens is 31.5 kg if the total sample formulae for volume and surface area are used, and 27.7 kg if the pongid formulae are used.

Thus, body mass estimates of about 9–10 kg for *P. africanus* and 28–34 kg for *P. nyanzae* are obtained from femoral cross-sectional diaphyseal and articular dimensions (the total range of individual values for *P. nyanzae* is 26–38 kg [Ruff et al. 1989]). Contrary to some interpretations based on dental remains, the new estimates support the view that *P. africanus* and *P. nyanzae* were separate species, since no living primate, or indeed land mammal, exhibits sexual dimorphism in body mass of 3:1 or more (Nowak & Paradiso 1983).[1]

Discussion and conclusions

In conclusion, it appears that structural properties of lower limb bones that directly reflect actual in vivo mechanical loadings do indeed show a very close relationship to body mass. Both cross-sectional geometric properties and articular dimensions are highly correlated with body mass. At least for cross-sectional properties, this appears to be true not only among anthropoid primates, but among mammals in general, which exhibit the same general scaling relationships.

In terms of body mass estimation, the use of either of these two types of structural properties offers major advantages over other methods. First, they are apparently proportion-independent. That is, they do not depend on assumptions regarding body proportions – for example, relative limb or skull length to body mass. Second, either method can be utilized on quite fragmentary material. All that is necessary is either a cross section from a recognizable bone region, or a reasonably complete articular surface.

These results indicate that both articular and cross-sectional dimensions of lower limb long bones provide good estimates of body mass. However, there is some indication that estimates from cross-sectional dimensions are less affected by the particular taxonomic reference group used, at least within extant anthropoid primates. For example, cercop-

Table 8.7. *Intergroup variation in least squares coefficients of regression on body mass of cross-sectional and articular dimensions*

	Cortical area				Articular surface area			
	Midfemur		Midtibia		Femoral head		Tibial plateaus	
Group	Slope	Int.[a]	Slope	Int.	Slope	Int.	Slope	Int.
Total	.766	1.14	.708	1.07	.842	2.02	—	—
Nonhuman	.756	1.14	.671	1.09	.794	2.04	.746	1.81
Pongid	.744	1.11	.652	1.12	.656	2.30	.649	1.99

Note: All regressions carried out on $\log_{10}$ transformed data.
[a] *Y* intercept.

ithecoids apparently have small lower limb articular dimensions and Recent humans have large articular dimensions relative to pongids (Figures 8.7 and 8.8; also see Jungers in press; Ruff 1988). This point is illustrated in Table 8.7, in which regression coefficients for cross-sectional area and articular surface area on body mass are compared in different subgroupings. At the same bone location (e.g., midshaft femur, tibial plateaus, etc.), slopes and intercepts are quite similar in different taxonomic subgroups for cross-sectional area, whereas intergroup differences are much greater for articular surface area coefficients (standard errors are comparable; see Tables 8.8 and 8.10). The same is true for the reverse regressions of body mass on structural properties (Tables 8.8 and 8.9). Therefore, it appears that if the taxonomic status of a specimen is uncertain, it may be better to rely on the cross-sectional estimates. However, if the specimen can be relegated to a particular group – for example, pongids – then articular dimensions, particularly those of the femoral head, appear to provide the closest estimates of body mass (Table 8.5).

Of cross-sectional properties, cortical area is a better predictor of body mass than second moments of area, for reasons discussed elsewhere (Ruff 1987). Cross-sectional properties of the femur give better body mass estimates than the tibia, very likely due to variation in the degree of mechanical support provided by the fibula in different species. Articular dimensions of the femur are also somewhat better predictors than those of the tibia on average. The positive allometry of the medial femoral condyle among anthropoids is apparently related to greater varus (bow-leggedness) of the knee in larger pongids (Ruff 1988).

Table 8.8. *Regression of cross-sectional dimensions on* log_{10} *(body mass)*

		$\log_{10}$ (CA)					$\log_{10}$ (J)				
Section	Group	*r*	Int.	Slope	SE	MA slope	*r*	Int.	Slope	SE	MA slope
Tib. 20%	Total	.992	0.985	.694	.028	.697	.981	1.363	1.520	.094	1.561
	Nonhuman	.994	0.991	.686	.031	.689	.993	1.389	1.455	.071	1.469
	Hominoid	.936	1.082	.641	.085	.668	.844	1.692	1.341	.302	1.720
	Af. ape + human	.938	1.183	.593	.090	.614	.896	2.153	1.123	.227	1.286
	Pongid	.955	1.104	.624	.097	.642	.942	1.541	1.373	.244	1.490
Tib. 35%	Total	.983	1.050	.700	.041	.708	.974	1.483	1.463	.048	1.518
	Hominoid	.873	1.095	.676	.133	.746	.796	1.771	1.306	.351	1.839
Tib. 50%	Total	.978	1.072	.708	.048	.719	.970	1.568	1.473	.117	1.537
	Nonhuman	.992	1.090	.671	.035	.675	.991	1.608	1.381	.076	1.398
	Hominoid	.834	1.156	.662	.155	.759	.777	1.851	1.318	.378	1.942
	Af. ape + human	.852	1.355	.568	.143	.624	.802	2.370	1.073	.326	1.434
	Pongid	.945	1.124	.652	.113	.677	.933	1.682	1.341	.258	1.472
Tib. 65%	Total	.977	1.087	.705	.049	.716	.975	1.657	1.499	.108	1.554
	Hominoid	.820	1.219	.633	.156	.730	.814	1.850	1.394	.352	1.911
Tib. 80%	Total	.980	1.169	.650	.042	.658	.974	1.788	1.522	.111	1.580
	Hominoid	.834	1.290	.584	.136	.654	.823	1.824	1.502	.367	2.043
Fem. 20%	Total	.992	1.187	.687	.027	.690	.997	1.825	1.590	.036	1.596
	Hominoid	.938	1.263	.645	.084	.672	.979	1.790	1.608	.217	1.660
Fem. 35%	Total	.995	1.136	.744	.024	.747	.994	1.759	1.553	.052	1.566
	Hominoid	.966	1.169	.725	.068	.743	.958	1.650	1.612	.171	1.718

Fem. 50%	Total	.994	1.140	.766	.026	.769	.992	1.763	1.542	.062	1.560
	Nonhuman	.996	1.145	.756	.029	.758	.992	1.769	1.532	.078	1.549
	Hominoid	.959	1.120	.776	.081	.802	.943	1.631	1.614	.201	1.763
	Af. ape + human	.983	1.178	.753	.058	.763	.980	1.800	1.545	.126	1.589
	Pongid	.973	1.109	.774	.092	.791	.947	1.629	1.607	.273	1.744
Fem. 65%	Total	.995	1.179	.758	.042	.761	.990	1.782	1.542	.070	1.564
	Hominoid	.966	1.056	.824	.078	.848	.933	1.580	1.652	.225	1.836
Fem. 80%	Total	.995	1.146	.772	.026	.775	.994	1.745	1.594	.058	1.609
	Nonhuman	.995	1.150	.768	.031	.771	.994	1.757	1.572	.068	1.584
	Hominoid	.967	1.025	.838	.079	.862	.950	1.708	1.614	.188	1.744
	Af. ape + human	.986	1.051	.831	.057	.841	.985	1.864	1.551	.111	1.585
	Pongid	.971	1.042	.826	.101	.847	.959	1.690	1.607	.238	1.711
Fem. Neck	Total	.976	1.087	.606	.043	.614	.986	1.624	1.459	.079	1.488
	Nonhuman	.996	1.085	.627	.024	.628	.994	1.404	1.660	.064	1.416
	Hominoid	.857	0.995	.656	.139	.733	.905	1.489	1.531	.254	1.778
	Af. ape + human	.834	0.965	.670	.181	.770	.928	1.711	1.434	.236	1.596
	Pongid	.964	1.150	.592	.081	.604	.958	1.449	1.577	.216	1.537

Note: CA, J: cortical area, polar second moment of area of femoral midshaft; Int., slope, SE: y intercept, slope, and standard error of slope, least squares linear regression; MA slope: major axis slope.

Table 8.9. *Regression of femoral and tibial lengths and average femoral midshaft external breadth on* log_{10} *(body mass)*

	log_{10} (Tibial length)					log_{10} (Femoral length)					log_{10} (Average breadth)				
Group	*r*	Int.	Slope	SE	MA slope	*r*	Int.	Slope	SE	MA slope	*r*	Int.	Slope	SE	MA slope
Total	.864	1.903	.289	.053	.296	.900	1.912	.331	.051	.338	.990	.693	.394	.017	.395
Nonhuman	.990	1.926	.238	.014	.238	.994	1.945	.282	.013	.282	.991	.696	.388	.021	.389
Hominoid	.328	2.128	.166	.169	.213	.462	2.077	.241	.163	.301	.930	.667	.408	.057	.417
Af. ape + human	.314	2.266	.100	.201	.132	.314	2.271	.146	.181	.181	.984	.728	.381	.028	.383
Pongid	.960	2.066	.162	.024	.163	.948	2.012	.240	.040	.242	.936	.668	.403	.075	.411

Note: Abbreviations as in Table 8.8.

Table 8.10. *Regression of log_{10} (body mass) on cross-sectional dimensions*

		log_{10} (CA)				log_{10} (J)			
Section	Group	r^2	Int.	Slope	%SEE	r^2	Int.	Slope	%SEE
Tib. 20%	Total	.984	−1.373	1.419	16.2	.843	−0.804	.633	26.1
	Nonhuman	.988	−1.409	1.440	17.5	.986	−0.920	.678	19.2
	Hominoid	.876	−1.252	1.366	17.6	.712	−0.376	.531	27.9
	Af. ape + human	.880	−1.534	1.484	17.6	.803	−1.177	.715	23.0
	Pongid	.912	−1.452	1.460	20.6	.887	−0.792	.646	23.6
Tib. 35%	Total	.966	−1.396	1.380	24.8	.949	−0.877	.648	31.7
	Hominoid	.762	−0.806	1.129	25.0	.634	−0.196	.485	32.0
Tib. 50%	Total	.956	−1.375	1.349	28.9	.941	−0.906	.639	34.2
	Nonhuman	.984	−1.574	1.466	20.6	.982	−1.116	.711	22.0
	Hominoid	.696	−0.665	1.052	28.8	.604	−0.128	.458	33.5
	Af. ape + human	.726	−1.226	1.276	27.7	.643	−0.765	.599	32.2
	Pongid	.893	−1.342	1.367	23.0	.870	−0.859	.650	25.4
Tib. 65%	Total	.955	−1.398	1.354	29.4	.951	−0.971	.634	30.8
	Hominoid	.672	−0.704	1.063	30.0	.663	−0.267	.475	30.6
Tib. 80%	Total	.960	−1.661	1.477	27.5	.949	−1.033	.624	31.3
	Hominoid	.696	−0.987	1.192	29.8	.677	−0.799	.567	29.8
Fem. 20%	Total	.984	−1.677	1.434	16.2	.994	−1.133	.626	9.1
	Total w/o *Pongo*	.985	−1.669	1.431	17.8	.997	−1.131	.623	7.4
	Hominoid	.880	−1.330	1.294	17.2	.958	−0.985	.595	9.9
	Af. ape + human	.861	−1.330	1.294	19.0	.968	−1.084	.613	8.5
Fem. 35%	Total	.990	−1.496	1.331	12.8	.988	−1.101	.636	13.8
	Total w/o *Pongo*	.993	−1.489	1.323	11.6	.997	−1.107	.633	7.5
	Hominoid	.933	−1.385	1.287	12.5	.918	−0.790	.569	14.1
	Af. ape + human	.949	−1.519	1.336	11.0	.970	−1.032	.618	8.5

(continued)

Table 8.10. (*cont.*)

Section	Group	$\log_{10}$ (CA)				$\log_{10}$ (J)			
		r^2	Int.	Slope	%SEE	r^2	Int.	Slope	%SEE
Fem. 50%	Total	.988	−1.453	1.291	13.8	.984	−1.099	.638	16.4
	Total w/o *Pongo*	.996	−1.454	1.284	9.1	.996	−1.114	.636	8.4
	Nonhuman	.992	−1.488	1.312	14.8	.984	−1.113	.643	20.3
	Hominoid	.920	−1.181	1.184	13.9	.889	−0.698	.551	16.5
	Af. ape + human	.966	−1.447	1.282	9.0	.960	−1.113	.643	9.6
	Pongid	.947	−1.128	1.222	15.7	.897	−0.721	.558	22.5
Fem. 65%	Total	.990	−1.521	1.306	13.4	.980	−1.100	.635	18.7
	Total w/o *Pongo*	.997	−1.526	1.300	7.4	.996	−1.125	.635	8.6
	Hominoid	.933	−1.075	1.133	12.6	.870	−0.599	.527	17.9
	Af. ape + human	.976	−1.288	1.208	7.3	.960	−1.025	.614	9.6
Fem. 80%	Total	.990	−1.451	1.281	13.4	.988	−1.060	.619	14.7
	Total w/o *Pongo*	.996	−1.452	1.274	8.5	.997	−1.107	.616	7.4
	Nonhuman	.990	−1.468	1.289	15.6	.988	−1.089	.629	17.0
	Hominoid	.935	−1.024	1.115	12.5	.902	−0.777	.559	15.5
	Af. ape + human	.972	−1.179	1.170	8.0	.970	−1.111	.626	8.4
	Pongid	.943	−1.088	1.142	16.1	.920	−0.820	.572	19.6
Fem. neck	Total	.953	−1.631	1.571	30.1	.972	−1.036	.666	22.5
	Nonhuman	.992	−1.704	1.582	14.6	.988	−1.149	.703	18.1
	Hominoid	.734	−0.634	1.120	26.6	.819	−0.471	.535	26.6
	Af. ape + human	.696	−0.442	1.037	29.4	.861	−0.771	.600	19.0
	Pongid	.929	−1.680	1.571	18.1	.918	−0.853	.634	19.7

Note: %SEE: percent standard error of estimate; other abbreviations as in Table 8.8.

Table 8.11. *Regression of log_{10} (body mass) on femoral and tibial lengths and average femoral midshaft external breadth*

	log_{10} (Femoral length)				log_{10} (Tibial length)				log_{10} (average breadth)			
Group	r^2	Int.	Slope	%SEE	r^2	Int.	Slope	%SEE	r^2	Int.	Slope	%SEE
Total	.810	−4.380	2.540	69.3	.864	−4.512	2.586	83.7	.981	−1.696	2.492	18.2
Nonhuman	.987	−6.750	3.498	18.4	.981	−7.914	4.123	22.7	.982	−1.737	2.533	22.1
Hominoid	.214	−0.421	0.888	50.2	.108	0.234	0.649	54.2	.865	−1.168	2.120	18.4
Af. ape + human	.099	0.125	0.673	55.7	.040	0.863	0.400	58.0	.969	−1.793	2.541	8.6
Pongid	.899	−7.343	3.740	22.1	.921	−11.581	5.674	19.3	.877	−1.227	2.174	24.7

Note: Abbreviations as in Table 8.10.

Table 8.12. *Regression of log_{10} (articular dimensions) on log_{10} (body mass)*

Articulation	Group	*r*	Int.	Slope	SE	MA slope
Femoral head surface area	Total	.979	2.018	0.842	.055	0.857
	Nonhuman	.997	2.042	0.794	.025	0.795
	Pongid	.988	2.297	0.656	.051	0.661
	Human	.983	0.537	1.721	.229	1.766
Femoral head volume	Total	.983	2.202	1.264	.076	1.293
	Nonhuman	.998	2.234	1.196	.028	1.199
	Pongid	.995	2.532	1.035	.052	1.041
	Human	.949	1.207	1.898	.447	2.065
Femoral condyles, medial + lateral	Nonhuman	.998	1.688	0.815	.021	0.817
	Pongid	.988	1.828	0.740	.059	0.747
Femoral condyle, medial	Nonhuman	.998	1.408	0.849	.023	0.851
	Pongid	.982	1.455	0.823	0.78	0.836
Femoral condyle, lateral	Nonhuman	.996	1.373	0.765	.027	0.767
	Pongid	.992	1.664	0.608	.039	0.611
Tibial plateaus, medial + lateral	Nonhuman	.995	1.810	0.746	.030	0.748
	Pongid	.963	1.989	0.649	.091	0.664
Tibial plateau, medial	Nonhuman	.994	1.444	0.803	.037	0.808
	Pongid	.955	1.720	0.654	.102	0.674
Tibial plateau, lateral	Nonhuman	.996	1.570	0.686	.026	0.688
	Pongid	.967	1.657	0.639	.084	0.652
Tibiotalar	Nonhuman	.996	1.417	0.717	.026	0.719
	Pongid	.979	1.658	0.587	.061	0.593

Note: Abbreviations as in Table 8.8.

The results for orangutans are interesting because they show some apparent contradictions between the two types of structural properties. There is good behavioral evidence (Ruff 1987) that orangutans load their lower limb bones relatively less than the other primates included here, and that this would explain the relatively smaller cross-sectional dimensions of their femur and tibia. However, despite this relative "unloading" of the lower limb, orangutan joint surface areas of the femur and tibia are *not* relatively smaller than those of other primates. This is probably explained by differences in other mechanical requirements of joint surfaces, specifically range of motion. For example, orangutans are capable of an extreme range of motion at their hip joint (Cant 1985), which would favor a relatively larger femoral head surface area even if

Table 8.13. *Regression of log_{10} (articular dimensions) on log_{10} (body mass)*

Articulation	Group	r^2	Int.	Slope	%SEE
Femoral head surface area	Total	.959	−2.231	1.138	27.7
	Nonhuman	.994	−2.549	1.252	12.2
	Pongid	.976	−3.372	1.487	10.2
	Human	.966	−0.240	0.561	3.4
Femoral head volume	Total	.965	−1.626	0.764	25.1
	Nonhuman	.997	−1.856	0.833	9.0
	Pongid	.990	−2.404	0.956	6.5
	Human	.900	−0.393	0.474	5.8
Femoral condyles, medial + lateral	Nonhuman	.996	−2.056	1.222	9.8
	Pongid	.975	−2.364	1.318	10.4
Femoral condyle, medial	Nonhuman	.996	−1.644	1.172	10.4
	Pongid	.965	−1.642	1.172	12.5
Femoral condyle, lateral	Nonhuman	.993	−1.770	1.297	13.7
	Pongid	.984	−2.661	1.617	8.4
Tibial plateaus, medial + lateral	Nonhuman	.990	−2.388	1.328	15.9
	Pongid	.928	−2.712	1.429	18.5
Tibial plateau, medial	Nonhuman	.987	−1.755	1.229	18.2
	Pongid	.912	−2.238	1.394	20.6
Tibial plateau, lateral	Nonhuman	.991	−2.256	1.445	14.8
	Pongid	.935	−2.310	1.464	17.4
Tibiotalar	Nonhuman	.992	−1.948	1.384	14.3
	Pongid	.959	−2.634	1.634	13.6

Note: Abbreviations as in Table 8.10.

the joint reaction force acting at any one time were relatively small. Further analyses of this kind may lead to new methods of determining relative joint-loading/joint range-of-motion criteria which would be useful in behavioral reconstruction (see Ruff 1988).

Humans differ in scaling relationships in some respects from the nonhuman anthropoid species examined here. In terms of femoral head articular dimensions, the recent human samples included here not only have relatively larger dimensions, not unexpected in a biped, but also a different scaling with body mass within species than any of the other species, specifically, extreme positive allometry. Why recent *Homo* should show this type of allometric scaling is unclear at present, but it has important implications with regard to estimating body mass in early, relatively small bodied hominids (Jungers 1988). From simple visual inspection of Figures 8.7 and 8.8, it is apparent that the human regression line, if continued, will intersect the nonhuman (or pongid) regression line rather quickly, especially for articular surface area. That is, the *expected* values for articular dimensions in small hominids may be very close to those for similarly sized nonhuman primates. In fact, at least one early hominid (A.L.288–1, "Lucy") does appear to have lower limb articular dimensions that fall closer to the living "pongid" range with respect to body mass (Jungers 1988; Ruff 1988). However, owing to the apparent extreme positive allometry of these dimensions in humans, this does not necessarily indicate a "less humanlike" support of body mass or form of locomotion (Jungers 1988).

Cross-sectional dimensions of Recent human lower limb bones, somewhat surprisingly given their bipedal form of body support, do not appear to deviate in a positive direction from general nonhuman primate allometric trends (Figures 8.3 and 8.4). However, earlier hominids do appear to deviate in the expected direction (Ruff & Walker, unpublished data). This may be explained by relatively lower activity levels in modern *Homo sapiens* compared with either earlier hominids or living nonhuman primates. This could affect cross-sectional diaphyseal and femoral neck dimensions to a greater degree than articular dimensions, an idea that is explored further elsewhere (Ruff 1988). Cross-sectional dimensions do not appear to be very good predictors of body mass in individual modern humans, unlike chimpanzees (Table 8.6; Ruff 1987). Thus, until scaling relationships in humans are better understood, it is probably inadvisable to use these regression equations for this purpose in fossil hominids.

Summary

Cross-sectional geometric and articular dimensions were measured on femora and tibiae of *Homo sapiens, Pan, Gorilla, Pongo,* and *Macaca fascicularis*. Cross-sectional geometric properties (areas, second moments of area) of five diaphyseal sites on each bone and the femoral neck were obtained by sectioning or computed tomography and tracing of images on a digitizer. Articular dimensions (articular surface areas, femoral head volume) were calculated from linear measurements taken with calipers. Only the femoral head was included in articular measurements of humans. Average species and sex-specific body weights for nonhuman primates were determined from museum records and literature values, and for the human samples from calculated stature and appropriate weight for height tables or regressions. Allometric analyses were carried out using least squares and major axis regressions of log transformed data for the total sample and for various taxonomic subgroupings.

Both cross-sectional geometric and articular dimensions are highly correlated with body mass ($r > .95$ for most comparisons; $r > .99$ for best comparisons). Percent standard errors of estimate of body mass are as low as 5–8% for the best predictors. Cortical area is a better predictor of body mass than second moments of area, and also cross-sectional and articular properties of the femur give better estimates than the tibia. Regression slopes indicate isometry or slight positive allometry for both cross-sectional and articular dimensions. Orangutans are deviant with respect to cross-sectional properties, probably reflecting lower average mechanical loadings of the lower limb. However, they show no deviation from general allometric trends in articular dimensions, apparently related to their relatively great lower limb joint excursions. Humans have relatively large femoral head articular dimensions and also appear to exhibit a different within-species allometric trend for this characteristic from that of other species, probably related to a bipedal mode of locomotion. In general, scaling relationships with body mass of cross-sectional diaphyseal dimensions appear somewhat less affected than articular dimensions by the particular taxonomic reference group selected.

These structural properties offer major advantages over other techniques of body mass estimation in that they do not rely on assumptions regarding relative body proportions and can be utilized on even quite fragmentary material. Deviations from general scaling trends appear to reflect locomotory differences between species. Utilization of the techniques on fossil material is illustrated by estimating body mass from femoral diaphyseal and articular dimensions of several specimens of the Miocene hominoids *Proconsul nyanzae* and *Proconsul africanus*.

Acknowledgments

Supported in part by Wenner-Gren Research Grant No. 4439 and NSF Research Grant No. BSN–8519749. I would like to thank the Harvard Museum of Comparative Zoology and the National Museum of Natural History, Smithsonian Institution, for loan of the nonhuman primate specimens; Richard Leakey, director, and governors of the National Museums of Kenya for permission to examine the *Proconsul* material; Dr. Steven Zucker for deriving the femoral head articular surface area formula; Dr. William Jungers for providing several manuscripts in press; and Dr. Alan Walker for assistance with the fossil hominoid material and for several helpful suggestions.

Note

1 Although the *P. africanus* KNM-RU 2036 is a subadult, its dental status indicates that it was very close to maturity at the time of death (Walker et al. 1986); it is also almost identical in size and proportions to the other *P. africanus* specimen included in the analysis (Ruff et al. 1989), which has completely closed epiphyses.

References

Biewener, A. 1982. Bone strength in small mammals and bipedal birds: do safety factors change with body size? *J. Exp. Biol. 98*:289–301.

Cant, J. G. H. 1985. Locomotor and postural behavior of orangutan (*Pongo pygmaeus*) in Borneo and Sumatra. Paper presented at annual meeting of the American Association of Physical Anthropologists, Knoxville, Tenn.

Fleagle, J. G. 1985. Size and adaptation in primates. *In Size and Scaling in Primate Biology*, ed. W. L. Jungers, pp. 1–19. New York: Plenum Press.

Genoves, S. 1967. Proportionality of the long bones and their relation to stature among Mesoamericans. *Am. J. Phys. Anthropol. 26*:67–78.

Jungers, W. L. 1984. Aspects of size and scaling in primate biology with special reference to the locomotor skeleton. *Yb. Phys. Anthropol. 27*:73–96.

Jungers, W. L. 1988. Relative joint size and hominoid locomotor adaptations with implications for the evolution of hominid bipedalism. *J. Human Evol. 17*:247–265.

Jungers, W. L. in press. Scaling of postcranial joint size in hominoid primates. In *Gravity, Posture and Locomotion in Primates*, ed. F. K. Jouffroy, M. H. Stack & C. Niemitz. Firenze: Il Sedicesimo.

Kuhry, B., & Marcus, L. F. 1977. Bivariate linear models in biometry. *Syst. Zool. 26*:201–209.

Lovejoy, C. O., Burstein, A. H., & Heiple, K. G. 1976. The biomechanical analysis of bone strength: a method and its application to platycnemia. *Am. J. Phys. Anthropol. 44*:489–506.

Nagurka, M. L., & Hayes, W. C. 1980. An interactive graphics package for calculating cross-sectional properties of complex shapes. *J. Biomech. 13*: 59–64.

Nowak, R. M., & Paradiso, J. L. 1983. *Walker's Mammals of the World,* 4th ed. Baltimore: Johns Hopkins University Press.

Ricker, W. E. 1973. Linear models in fishery research. *J. Fish. Res. Bd. Canada 30*:409–434.

Ruff, C. B. 1987. Structural allometry of the femur and tibia in Hominoidea and *Macaca. Folia Primatol. 48*:9–49.

Ruff, C. B. 1988. Hindlimb articular surface allometry in Hominoidea and *Macaca,* with comparisons to diaphyseal scaling. *J. Human Evol. 17*: 687–714.

Ruff, C. B., & Hayes, W. C. 1983a. Cross-sectional geometry of Pecos Pueblo femora and tibiae – a biomechanical investigation. I. Method and general patterns of variation. *Am. J. Phys. Anthropol. 60*:359–381.

Ruff, C. B., & Hayes, W. C. 1983b. Cross-sectional geometry of Pecos Pueblo femora and tibiae – a biomechanical investigation. II. Sex, age, and side differences. *Am. J. Phys. Anthrop. 60*:383–400.

Ruff, C. B., & Hayes, W. C. 1988. Sex differences in age-related remodeling of the femur and tibia. *J. Orthopaed. Res. 6*:886–896.

Ruff, C. B., & Leo, F. P. 1986. Use of computed tomography in skeletal structural research. *Yb. Phys. Anthropol. 29*:181–196.

Ruff, C. B., Walker, A., & Teaford, M. F. 1989. Body mass, sexual dimorphism and femoral proportions of *Proconsul* from Rusinga and Mfangano Islands, Kenya. *J. Human Evol. 18*:515–536.

Schultz, A. H. 1953. The relative thickness of the long bones and the vertebrae in primates. *Am. J. Phys. Anthropol. 11*:277–311.

Smith, R. J. 1984. Allometric scaling in comparative biology: problems of concept and method. *Am. J. Physiol. 256*:R152–R160.

Smith, R. J. 1985. The present as a key to the past. Body weight of Miocene hominoids as a test of allometric methods for paleontological inference. In *Size and Scaling in Primate Biology,* ed. W. L. Jungers, pp. 437–448. New York: Plenum Press.

Steudel, K. 1985. Allometric perspectives on fossil catarrhine morphology. In *Size and Scaling in Primate Biology,* ed. W. L. Jungers, pp. 449–475. New York: Plenum.

Trotter, M., & Gleser, G. C. 1952. Estimation of stature from long bones of American whites and Negroes. *Am. J. Phys. Anthropol. 10*:463–514.

Trotter, M., & Gleser, G. C. 1958. A re-evaluation of estimation of stature based on measurements of stature during life and of long bones after death. *Am. J. Phys. Anthropol. 16*:79–123.

Walker, A. C., & Pickford, M. 1983. New postcranial fossils of *Proconsul africanus* and *Proconsul nyanzae.* In *New Interpretations of Ape and Human Ancestry,* ed. R. L. Ciochon & R. S. Corruccini, pp. 325–351. New York: Plenum Press.

Walker, A. C., & Teaford, M. F. 1988. The Kaswanga primate site. An Early Miocene hominoid site on Rusinga Island, Kenya. *J. Human Evol. 17*: 539–544.

Walker, A. C., Teaford, M. F., & Leakey, R. E. 1986. New information concerning the R114 *Proconsul* site, Rusinga Island, Kenya. In *Primate Evolution,* ed. J. G. Else & P. C. Lee, pp. 143–149. Cambridge: Cambridge University Press.

9

Insular dwarf elephants: a case study in body mass estimation and ecological inference

V. LOUISE ROTH

In mammals, many ecological and physiological variables correlate with body mass (Blueweiss et al. 1978; Calder 1984; Peters 1983; Schmidt-Nielsen 1984). Mass data provide a common currency for making comparisons among taxa, within an ecological context: Even taxonomically disparate animals that differ in shape or constitution can be described in common units, and evaluated on a common scale. Empirically derived scaling relationships among body mass and metabolic rates, trophic strategies, and various parameters of life history (Damuth 1981, 1987; Eisenberg 1981; Peters 1983; Western 1979, 1983) allow one to consider the demands that mammals of various sizes place upon community resources, and upon one another. Estimates of body mass for fossils thus can provide insight into the dynamics of paleocommunities.

Unfortunately for specialists on the larger large mammals, however, error in mass estimates scales with an unwelcome positive allometry. In any least squares regression for which $|r| < 1$, the further the independent variable strays from the mean, the broader the confidence limits, and the wider the margin of error for the dependent variable – in this case, body mass. Inferences built upon inaccurate inferences only compound error.

One solution to this problem is to avoid the propagation of error entirely: to forego translating morphometric data into body masses, and to restrict one's questions about fossils to those that involve direct comparison of osteological measurements. (This is the approach taken in my previous studies of dwarf elephants: e.g., Roth 1982, 1984, in press.) Osteological measurements in and of themselves contain important bi-

In *Body Size in Mammalian Paleobiology: Estimation and Biological Implications,* John Damuth and Bruce J. MacFadden, eds.

ological information. They are a source of systematic characters (e.g., Bookstein et al. 1985); they reveal many aspects of functional morphology (e.g., Gregory 1929; Maynard Smith & Savage 1959; Scott 1985); and they bear directly on biomechanical hypotheses (Alexander 1977; McMahon 1973).

Yet imprecise mass estimates can yield valid inferences, provided one is aware of the major sources of variation and of their magnitudes. Imprecision of mass estimates may limit the resolution of one's comparisons, but the possibility of useful qualitative conclusions remains.

In analyses of body size and mammalian paleoecology, insular dwarfs such as the elephants *Mammuthus exilis* (from Southern California) and *Elephas falconeri* (from Malta and Sicily) are of special theoretical interest. With proper estimates of body size, it may be possible to infer what population densities occurred on the islands, and to ask how such densities compare with those of larger elephants today in crowded habitats. Given an island of a particular area and animals within a particular range of masses, one may determine the permissible size of the population, and gain insight into the selection pressures involved in the dwarfing. In addition to producing dwarf elephants, giant mice, and other animals of peculiar sizes (Foster 1965; Sondaar 1977), islands tend to be ecologically depauperate – to have lower species diversity than the mainland, and, especially, fewer species of carnivores (Patterson & Atmar 1986, Sondaar 1977; Williamson 1981). To what extent is, for example, a terrestrial-mammal fauna consisting simply of a skunk, a small fox, a shrew, a large species of mouse, and a small variety of elephant (all of which, moreover, may not have been contemporaneous) comparable to the more "balanced," diverse faunas of the Rancholabrean mainland (Cushing et al. 1984; Kurtén & Anderson 1980)? Do mainland and island communities follow the same rules of community organization?

The first portion of this paper will focus upon the methodological problem of body mass estimation in a relatively large, extinct mammal that is lacking in good living analogues. Having derived mass estimates, and enumerated and examined sources of variation and error, I will examine within those confidence limits some ecological conclusions that might be drawn.

Body mass determination for living elephants

Obtaining the body mass of a living elephant is no small task: a scale of suitable size must be made accessible to a sufficiently tractable animal. Field data are rare; data on captive exhibit animals are at times exaggerated (see Shoshani, Shoshani, & Dahlinger 1986). Alternatively, dead animals can be weighed in parts (the 76 coauthors listed for Shoshani et al.'s [1982] thorough necropsy of one individual indicate the magnitude of the task), with a suitable allowance in the estimate for fluid loss during the dissection: Sikes 1971:51 suggested 5%, and other authors have used 3–7%.

Variability of body mass

The variability of body masses determined for living elephants sets an upper limit to the precision with which mass can meaningfully be estimated from a fossil. The variation described below represents the maximum percentages I was able to obtain from a survey of the literature.

The mass of an individual elephant varies in the course of a single day. An elephant can take in 100 liters of water in a single bout, and thereby, with one drink, increase its mass by 100 kg, which is approximately 3.5% of the body mass of an average-size mature animal (Sikes 1971: 51). Stomach fill can account for up to 6% of live mass in African elephants (Buss 1961; Laws, Parker, & Johnstone 1975:172). Benedict (1936:107) reported that in one animal 12% of body mass was feces plus fluid from the intestine and body cavity; in another individual, intestinal contents accounted for 7% of body mass (which Benedict contrasts with some ruminants, in which up to 40% of total body mass is inert material or "ballast").[1] In the most extreme situations, therefore, short-term fluctuations in the mass of an elephant are unlikely to exceed 15%.

Pregnancy is an obvious source of variability in the mass of adult females over the two-year course of gestation. Laws et al. (1975:198) reported an average mass difference between pregnant and nonpregnant elephants in Uganda of 250–300 kg, and according to Hanks (1969), a 20-year old pregnant African elephant cow with a large fetus can weigh 500–750 kg (approximately 25%) more than her nonpregnant counterpart. With newborn calves weighing 120 kg and less (Flower 1943; Laws 1966; Laws & Parker 1968:330), the fetus itself constitutes a small portion of the additional mass.

Tusks, though a more permanent component of mass of any given individual, differ visibly in size between individuals of otherwise similar body proportions. As total body mass increases, tusks constitute an increasing percentage of the total (Figure 9.1), but a small one, on the order of 1%. Record-sized tusks of over 100 kg apiece have been reported, however (Sikes 1971: 112).

By far the greatest source of variation in body masses associated with

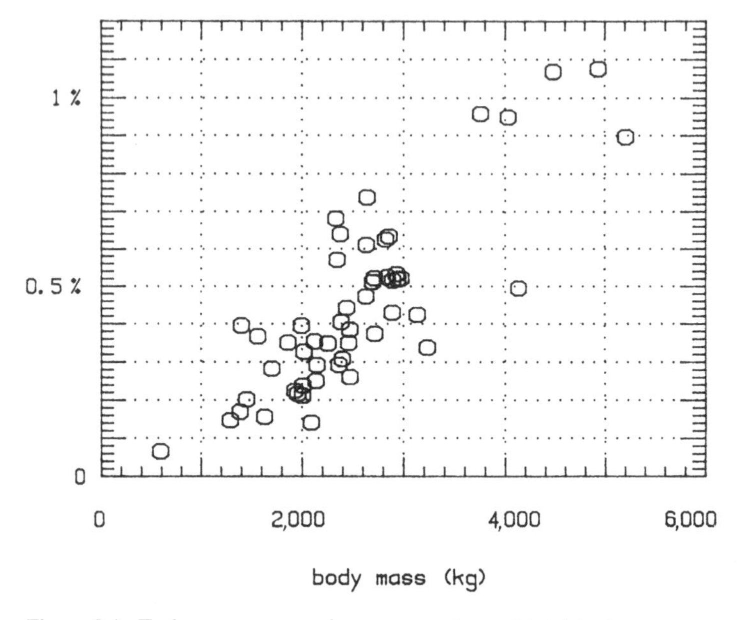

Figure 9.1. Tusk mass expressed as a percentage of total body mass among African elephants from Queen Elizabeth and Murchison Falls Parks, Uganda. (Body mass and tusk mass data from Laws 1966.)

a particular skeletal dimension are differences between individuals in body form and physical condition. Within Uganda, Laws et al. (1975:189) found Mkomasi elephants to be 9–13% heavier than Murchison Falls Park South elephants of the same shoulder height. Masses of the lighter population were taken at a favorable time of year (toward the end of the rains), and so the actual difference is probably underestimated by these figures. As Figure 9.2 illustrates, wild African and captive Asian elephants of the same shoulder height can differ in mass by as much as 100% or even 200% (that is, a factor of 2 or 3).

Methods

Choice of scaling relationships

To estimate body mass from a fossil, one needs a model: (1) a reconstruction based on a nearly complete skeleton (associated elements and well-preserved anatomy that indicate overall body proportions; e.g., Alexander 1985); (2) a modern animal (an analogue), inferred to be

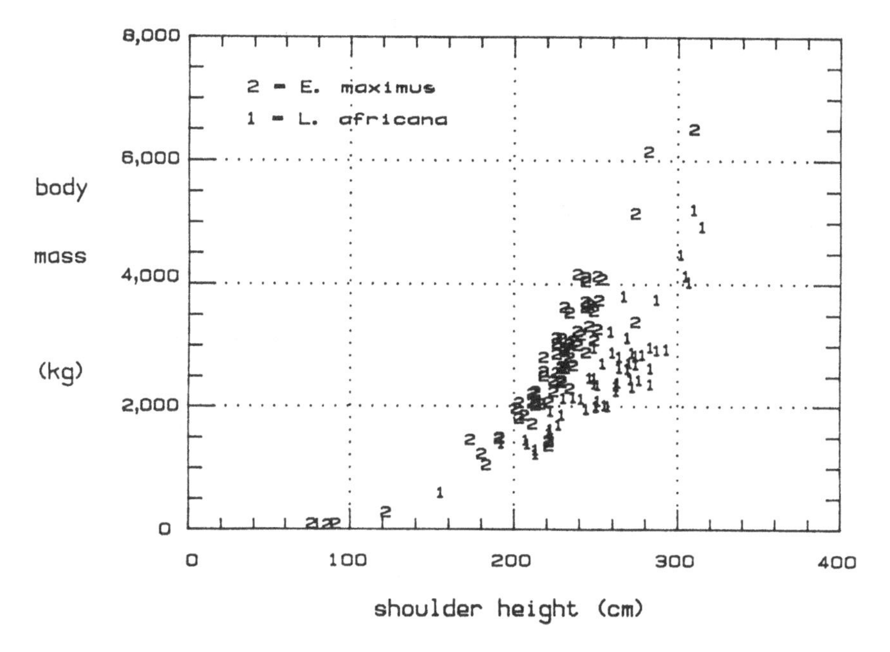

Figure 9.2. Body mass versus shoulder height for some African and Asian elephants. Data on African elephants (*L. africana*) are taken from Laws (1966), and are of wild individuals collected in Uganda. Data on Asian elephants (*E. maximus*) are mostly from Benedict (1936:22–23), and are of female captive (mostly circus) animals. The three smallest masses are from Flower (1943). Of the three heaviest individuals, the largest is "Tusko" (Lewis & Fish 1978), a captive male, and the other two are animals captured in Malaya and described by Momin Khan (1977; the masses are given for the "carcasses drained of blood," and so they may be underestimates of 3–7%). Some individuals of the same shoulder height (e.g., 280 cm) differ in mass by nearly a factor of 3.

similar in size and shape; or (3) a group of broadly analogous forms – individuals or taxa that jointly exhibit allometric trends to which the fossil is inferred to conform.

Because associated material from dwarf mammoths is uncommon, because any single reconstruction allows inferences about only a limited number of specimens, and because mature dwarf mammoths ranged widely in size, the first approach (1) is unsuitable for the present purposes. Approach (2) also presents difficulties, because in the modern fauna there are no precise analogues of the 1- to 2-m-tall elephants that inhabited continental islands during the Pleistocene. Dwarf elephants from Sicily (*Elephas falconeri*) and Santa Rosa Island, California (*Mammuthus exilis*) were smaller when they attained skeletal maturity than even the smallest mature individuals of the two extant species (Orr 1968; Roth 1987; Stock 1935). The closest living relatives of elephants – hyraxes, aardvarks, and sirenians – have distinctive body proportions and adaptations of their own, and have diverged enough to be placed in separate orders.

A desirable feature of the approach (3) used here is that once allometric relationships between bone dimensions and body mass have been discerned, they can be used to estimate body mass for fossils of a variety of different sizes. This approach – the allometric approach – requires identification of an appropriate sample of living analogues, and selection of the measurements to be taken from bones.

The model sample can be interspecific – that is, adults of various taxa from a broad range of body sizes – or ontogenetic – that is, a growth series (of elephants only).

Interspecific curves are preferable if elephants follow rules of scaling similar to other mammals. If this is the case, ungulates with bones of approximately the same dimensions as the insular dwarfs would serve as analogues for calculating body mass of the dwarfs.[2]

On the other hand, one could argue that, because elephants are so distinctive morphologically and distant genealogically from other mammals, the best analogue of a small elephant is nevertheless a small elephant, be it juvenile or adult.[3]

Both interspecific and ontogenetic models have their advantages, and both will be used here.

Static interspecific model

My choice of scaling relationships for estimating body mass was guided by the following requirements:

1. To avoid extrapolations beyond the range for which the relationship was originally calculated, the model sample had to encompass the size range of dwarf and full-sized elephants.
2. The clade defined by the model sample had to include elephants (i.e., the common ancestor of members of the sample also had to be ancestral to elephants; thus a varied selection of mammals, or a selection of ungulates and subungulates, or even a collection of mammals plus dinosaurs, is acceptable, but a collection of dinosaurs alone is not).
3. To the greatest extent possible, the variable from which body mass was to be calculated had to be directly measurable on a

fossil (i.e., not itself calculated or estimated from other measurements).

My attention focused on long bones because elephant teeth (and especially dwarf-elephant teeth) can be difficult to identify ([Roth 1982; Roth & Shoshani 1988]; their dimensions change as they continually undergo attrition and shed plates, and the particular teeth that are present depends upon the stage of ontogeny), and because of other difficulties in using teeth for mass estimates (Damuth, Fortelius, Janis, this volume).

Model interspecific samples (with the corresponding data used to generate the mass-estimation equations) were taken from the work of Alexander, Jayes, Maloiy, and Wathuta (1979) and Anderson, Hall-Martin, and Russell (1985). Alexander et al. (1979) examined the relationship between lengths of long bones and body mass in 36 species (from many orders) of quadrupedal mammals, ranging from shrews to an African elephant (*Loxodonta africana*).[4] Anderson et al. (1985) related humeral and femoral circumferences to body mass in a similar sample: 33 species, from *Microtus* to *Loxodonta* (the individual known as "Jumbo"). Both sets of curves had been generated by least squares regression on log-transformed data. Both used log (mass) as the independent variable, and so to obtain mass-estimating equations I recalculated the regressions (with raw data published by Anderson et al. 1979, or provided by Alexander, personal communication), using log (mass) as the dependent variable (see Table 9.1).

The elephant in Alexander et al.'s model sample (whose mass itself had been estimated from its height) was less than half the mass of Jumbo, the largest member of Anderson et al.'s sample. Substituting data on Jumbo (length data courtesy of R. Voss) for Alexander's elephant had a minor effect on the length–mass regression: It increased mass estimates for *E. falconeri* by approximately 1%, and for the largest mammoth by 3.2%.

The equations based upon lengths of femora and humeri were useful, whereas those for more distal bones did not closely approximate the length–mass relationship for the elephant. For the more distal bones, the elephant in Alexander et al.'s model sample fell far from the regression line (outside the 95% confidence limits, and sometimes outside the prediction limits as well) and was considerably more massive than would be predicted from the length of the bone.

All of the specimens for which I was estimating mass are larger than the geometric mean (log-average) of the model samples. If curvilinear

Table 9.1. *Equations for the estimation of body mass from humerus and femur lengths and circumferences, obtained by least-squares regression of log (mass) on log (bone measurement)*

1	Mass (kg) = $2.767 \times 10^{-5} \times$ Humerus length (mm)$^{2.675}$ 5.7 mm < Hu. L. < 830 mm
2	Mass (kg) = $1.774 \times 10^{-5} \times$ Femur length (mm)$^{2.654}$ 6.0 mm < Fe. L. < 980 mm
3	Mass (kg) = $9.448 \times 10^{-4} \times$ Humerus circumference (mm)$^{2.611}$ 4.9 mm < Hu. C. < 459 mm
4	Mass (kg) = $3.790 \times 10^{-4} \times$ Femur circumference (mm)$^{2.827}$ 5.5 mm < Fe. C. < 413 mm

Note: The sample data used in calculating these regressions were published by Anderson et al. (1985), or supplied by R. McN. Alexander (personal communication; masses from the same data set were published in Alexander et al., 1979). Beneath each equation is given the size range over which it applies.

trends were evident in the length–mass or circumference–mass relationship, it would be preferable to use regressions based solely on large members of the model samples. For Alexander et al.'s and Anderson et al.'s data, respectively, I computed separate regressions for smaller-than-average and larger-than-average members, but these regressions did not differ significantly. Moreover, the elephants in the model samples fell within 95% confidence limits and 95% prediction limits regardless of whether the regression was based upon the full sample, or upon larger-than-average members.

Ontogenetic model

In practice, bones and body mass are not commonly taken from the same individual elephants, but mass and shoulder height are (Benedict 1936; Hanks 1972; Johnson & Buss 1965; Laws et al. 1975).[5] Limb posture in elephants is columnar, and shoulder height of an osteological specimen can be approximated by summing the lengths of scapula, humerus, and ulna, with the height of the foot, and allowing 7–15 cm for the thickness of the skin.[6]

For the ontogenetic model I used regressions of mass on shoulder height that have been published for several wild populations of elephants, or that I constructed from published data. The model samples (with one exception; see "Omari" in Table 9.2) were therefore cross-

Table 9.2. *Coefficients and exponents of equations for the estimation of mass from shoulder height, derived from various populations of elephants*

Coefficient; exponent	Height range (cm)	Model sample	Reference
Loxodonta africana			
a. 1.02×10^{-4}; 3.11	79–251	Females, Zambia	Hanks[a] 1972
b. 1.267×10^{-3}; 2.631	88–270	Females, Murchison Falls Nat. Park (MFNP) South, Uganda	Laws et al. 1975
c. 5.07×10^{-4}; 2.803	85–325	Males, MFNPS, Uganda	Laws et al. 1975
d. 2.58×10^{-4}; 2.917	unstated	Females, MFNP North, Uganda	Laws et al. 1975
e. 3.06×10^{-4}; 2.890	unstated	Males, MFNPN, Uganda	Laws et al. 1975
f. 1.81×10^{-4}; 2.97	95–320	Males, Uganda	Johnson & Buss 1965
g. 8.234×10^{-4}; 2.711	117–320	Longitudinal data on captive male "Omari"	Johnson & Buss 1965
h. 2.080×10^{-4}; 2.934	89–315	MFNP & Queen Elizabeth Park, Uganda	Laws, 1966
Elephas maximus			
i. 3.071×10^{-4}; 2.917	164–279	Averages for animals grouped by age (5-yr intervals) and sex; Thailand	Tumwasorn et al. 1980
j. 4.682×10^{-5}; 3.263	76–310	Above data supplemented with 1 large captive, 3 large Malayan, and 3 newborn Indian individuals	Flower 1943; Lewis & Fish 1978; Momin Khan 1977; Tumwasorn et al. 1980
k. 3.24×10^{-5}; 3.356	173–254	Captive female circus animals	Benedict 1936
l. 2.73×10^{-5}; 3.387	76–310	Above data supplemented with 1 large captive, 3 large Malayan, and 3 newborn Indian individuals	Benedict 1936; Flower 1943; Lewis & Fish 1978; Momin Khan 1977

Note: The parameters shown here were determined by least squares regression of log (mass in kg) on log (shoulder height in cm) and subsequently converted to arithmetic (not log-transformed) coordinates. Mass (in kg) is therefore calculated as: coefficient $\times$ (shoulder ht in cm)$^{exp.}$

[a]Hanks preferred the semilogarithmic equation $\log_{10}$ (body mass [kg]) = 0.007 $\times$ (shoulder ht [cm]) + 1.73, but this formula produces inaccurately high masses for small individuals – for example, 192 kg for a 79-cm individual. Known masses for animals of this height are closer to the 81 kg predicted by Hanks's other equation.

sectional rather than ontogenetic in a strictly longitudinal sense (Cock 1966). Usually neither the raw data nor their variances were available, and so although for a single shoulder height I might calculate as many as 12 distinct mass estimates (corresponding to the 12 equations in Table 9.2, and producing the striped blocks in Figure 9.3), it was not possible to compute confidence limits. Instead, I calculated an estimate from each appropriate equation (equations in Table 9.2 were applied only to specimens within the range of heights delimited by the model sample), and used the smallest and largest mass estimates to define the zone of expectation.

Adult Columbian mammoths were often taller than the largest modern African elephants, so for some specimens of this species none of the mass estimation equations were appropriate, and extrapolation was unavoidable (see Figure 9.4).

Six museum specimens of Recent elephants were available with postcranial elements *and* recorded body masses. The accuracy of the various mass-estimation equations was tested on these specimens, which ranged from 77 to 4500 kg.

Figure 9.3 (*facing page*). Comparison of body mass estimation techniques, illustrated for three specimens. The mass estimates for the specimens – a mounted skeleton of *E. falconeri*, and humeri from two different individuals of *M. exilis* – were determined using equations in Tables 9.1 and 9.2 (Hu. L. = humerus length, Table 9.1, equation 1; Hu. C. = humerus circumference, equation 3; etc.; Sh. Ht. = shoulder height, Table 9.2). Differences between pairs of mass estimates are shown as percentages, calculated as the larger divided by the smaller estimate, minus 100%.

For *E. falconeri*, bones from the right and left sides of the animal produced slightly different estimates, and this range is shown by a thick dark bar, which is enclosed within the 95% confidence interval for the predictions. Note the great breadth of these confidence intervals; predictions are shown without their confidence intervals for the two specimens of *M. exilis*.

For shoulder heights, each vertical line within a block represents a mass estimate determined by a different equation in Table 9.2; for each specimen, only those equations with the appropriate height range were used. For UNSM #2540, shoulder height was measured directly from the mounted skeleton (with an additional 7-cm allowance for skin). For LACM #67786, three different heights were used: For the upper two blocks of estimates it was assumed that the humerus constituted 26% and 38%, respectively, of the skeletal height; the lower block shows the estimates assuming Harington et al.'s (1974) figure of 34%. The lower block for SBMNH #244 also assumes that the humerus is 34% of skeletal height, whereas for the upper block, the height was taken from an *E. maximus* specimen with a humerus of similar length.

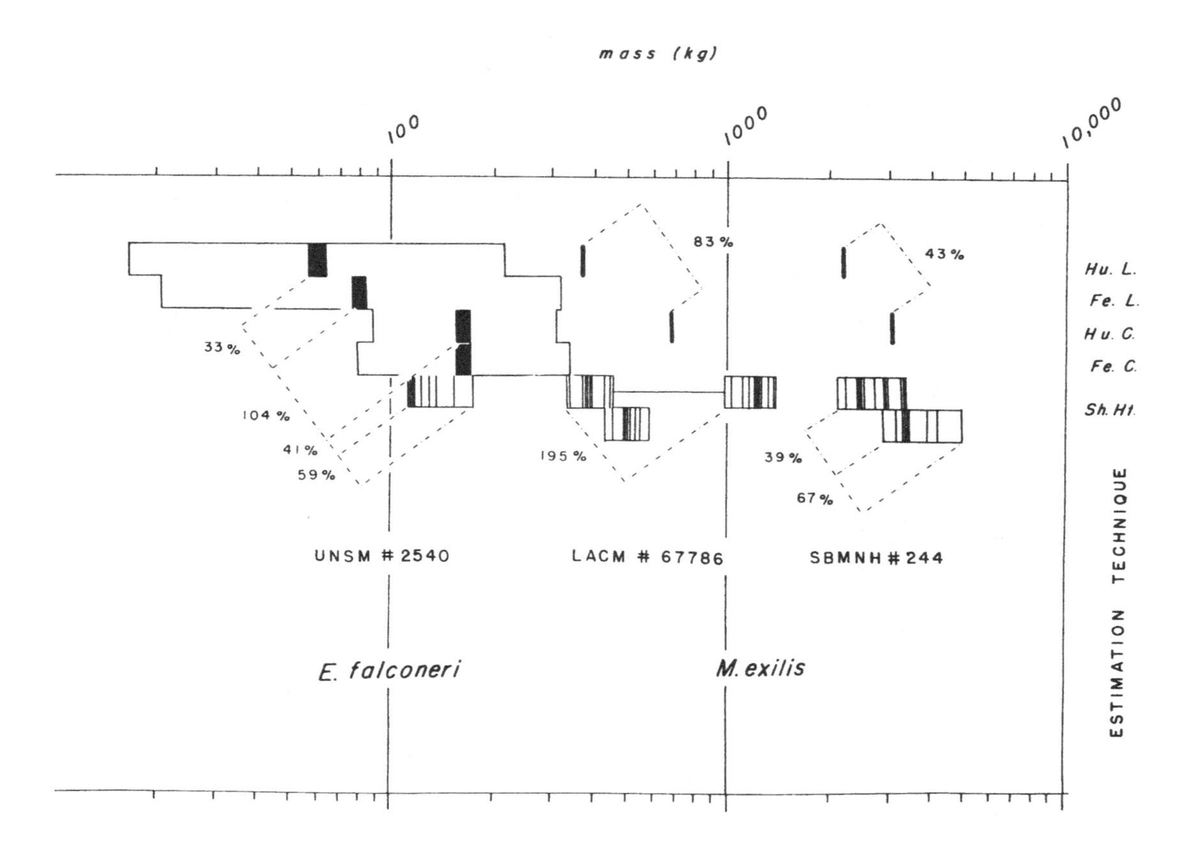
mass (kg)
100
1000
10,000
83%
43%
Hu. L.
Fe. L.
Hu. C.
Fe. C.
Sh. Ht.
33%
104%
41%
59%
195%
39%
67%
UNSM # 2540
LACM # 67786
SBMNH # 244
E. falconeri
M. exilis
ESTIMATION TECHNIQUE

Materials

For the fossils I confined my attention to bones with fully fused epiphyses. I directly measured total lengths and minimum circumferences of the long bones from one specimen of *Elephas falconeri* (UNSM #2540; see Acknowledgments for museum acronyms; circumference data courtesy of G. Corner), and obtained additional length data on this species from Ambrosetti's (1968) published account. Because Ambrosetti's data were bimodal, I used gaps in uni- (and sometimes bi-)variate plots to distinguish males and females. My division of the data did not always precisely match Ambrosetti's, but the gaps I identified divided the sample into two bell-shaped curves that closely fit normal and log-normal distributions.

For *Mammuthus exilis* I examined all of the specimens that are currently available in museum collections (SBMNH and LACM). For *M.*

Figure 9.4 (facing page). Variation in body mass estimates for *E. falconeri* (white blocks, left), *M. exilis* (shaded blocks), and *M. columbi* (white blocks, right). For *E. falconeri*, mass estimates for males and females are shown separately. Mass predictions for the average humeri (females: $n = 11$; males: $n = 9$) and femora (females: $n = 18$; males: $n = 7$) are denoted by heavy vertical lines, which are enclosed within 95% confidence intervals. Confidence intervals were calculated for the predicted value corresponding to the geometric mean of the bone measurements of the specimens in each case (Draper & Smith 1966; Sokal & Rohlf 1980). From comparisons with a small mounted specimen (UNSM #2540; see text), shoulder heights for average females and males were judged to be 100 and 130 cm, respectively, and all equations in Table 9.2 that applied to these heights were used in producing the ranges of mass estimates delimited by blocks in the figure.

M. exilis and *M. columbi* specimens are less numerous and more variable than Ambrosetti's sample of *E. falconeri* (Roth in press), and so their mass estimates are shown as ranges rather than averages; the lower and upper boundaries of the ranges are determined by the smallest and largest specimens of each species. For any single shoulder height, the equations in Table 9.2 produce a range of mass estimates. The ranges for the smallest and largest specimens are shown here as blocks, which are connected by a horizontal line.

The smallest or largest specimens of each element were not always from the same individual; thus sampling error contributes to the differences apparent between techniques. The narrow ranges of mass estimates based on femora of *M. exilis*, for example, reflect the scarcity ($n = 4$) of intact, fully fused femora. For an illustration of the differences in estimates that can be obtained by applying different techniques to a *single* specimen, see Figure 9.3. Because mammoths attained larger sizes than any extant quadruped, some specimens fall outside the size range for which the equations in Tables 9.1 and 9.2 apply. The upper limit for "appropriate" estimates is shown here with a jagged line. Mass estimates to the right of these marks are extrapolated from regressions that were calculated for smaller animals.

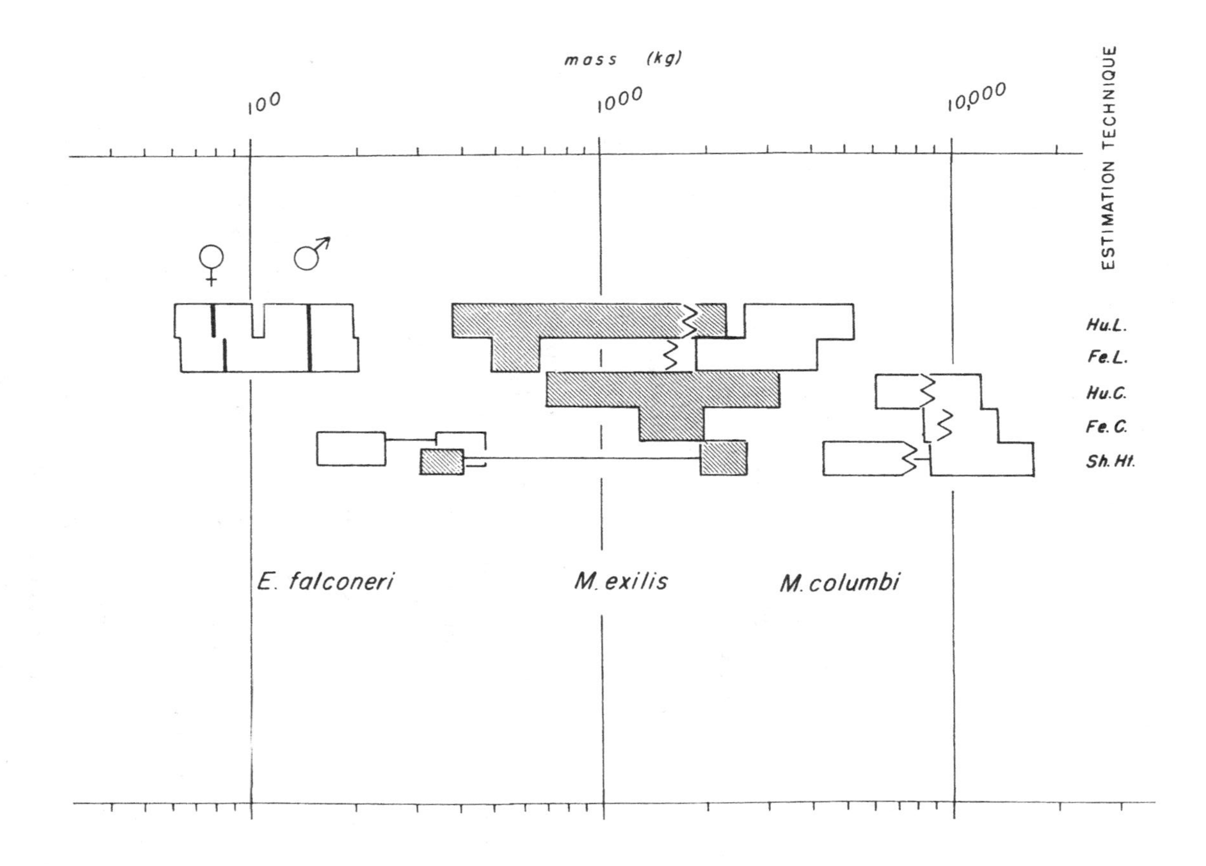

mass (kg)
100
1000
10,000
ESTIMATION TECHNIQUE
Hu.L.
Fe.L.
Hu.C.
Fe.C.
Sh.Ht.
E. falconeri
M. exilis
M. columbi

columbi I supplemented measurements I took on specimens at TMM, LACM, UNSM, UF, CMNH, and the Pratt Museum (Amherst) with data assembled by Dutrow (1980). Additional data were obtained in the collections of BMNH, YPM, AMNH, and USNM on the two extant elephant species (Asian, *Elephas maximus;* and African, *Loxodonta africana*).

Shoulder heights

Intact scapulae of *E. falconeri* are rare, and Ambrosetti (1968) published no measurements on this bone. I measured skeletal height directly on one mounted skeleton of *E. falconeri* (UNSM #2540) whose scapulae had been reconstructed; with an allowance of 7 cm for skin, the shoulder height would have been 90 cm. This specimen has humeri that are shorter, by approximately 10% and 40%, than the averages I obtained from Ambrosetti's (1968) data for females and males, respectively. More typical shoulder heights for this species, therefore, might have been 100–130 cm.

The largest (humerus SBMNH #344) and smallest (tibia SBMNH #370) known adult individuals of *M. exilis* are not represented by associated skeletons. By assuming that intermembral proportions in the fossils followed trends apparent in growth series of modern elephants (Roth 1982, 1984), and making an additional allowance for the thickness of skin (see note 6), I determined that shoulder heights for known specimens of this species ranged from 122 to 225 cm, with one very large humerus extending the range to 245 cm (Roth 1982). Using the method of Harington, Tipper, and Mott (1974), by contrast, one obtains a maximum height for *M. exilis* of 275 cm, which more closely fits the "eight or nine feet" maximum quoted by Stock (1935). The ancestors of dwarf mammoths were full-sized, and full-sized teeth are known from the islands (Cushing et al. 1984; Roth 1982) so at some point presumably the maximum height was equivalent to that of a mainland mammoth.

Harington et al. (1974) reported that humeri constitute approximately 34% (31–35%) of the shoulder heights of mammoths and elephants; my own observations of mounted skeletons confirm this figure, but broaden the range to 26–38%. On a mounted specimen of *M. exilis* (forelimb = SBMNH #192), for example, Harington's method produces a height estimate of 10 cm (6%) greater than that measured directly. If Harington's figure of 34% were used when the actual value was 30%, the result would be a 13% overestimate of height, which translates approximately to a 50% overestimate in mass. To be conservative, it is therefore desirable to employ more than one method to estimate ranges for shoulder heights (see Figure 9.4).

Results

Mass estimates vary as a result of variability between individuals, between samples, and between mass estimation equations.

Individuals not only vary in size, but also differ in their allometry from one another. If an individual differs from the "typical mammal" defined by the set of regression equations, then different measurements – the length of the femur, and the length of the humerus, for example – and their corresponding mass estimation equations can yield different estimates for the *same* animal.

Not all measurements may be available for all individuals, and such differences in sampling also contribute to differences in the distributions and ranges obtained for any given species.

Moreover, if one chooses to recognize the residual variation within the original regressions (on which the estimates are based), and calculates confidence intervals for the predictions, one defines a range, rather than a single value, of reasonable mass estimates.

In some instances, more than one mass estimation measurement (shoulder height, and length or circumference of femur or humerus) was available for a single individual, and I was able to compare estimates obtained by different methods (see Figure 9.3). The techniques were compared pairwise: Estimates obtained by one method were graphed against estimates obtained by another, and this scatterplot was compared with the line $y = x$. From this comparison the following tendencies were apparent:

1 *The 95% confidence intervals of the mass predictions obtained using different techniques always overlapped.* The intervals were, however, extremely broad. (Confidence limits for mass estimates on UNSM #2540 are illustrated in Figure 9.3.) For circumference-based masses, the factor/divisor for an estimate on a single specimen was 1.8–2.0, and the total range of masses for the specimen therefore spanned a factor of 3–4. Length-based masses (for which the confidence interval was $\times/\div$ 3.4–4.0) ranged over a factor of 12–16.

2 *Mass estimates based upon circumferences were greater than those based upon lengths.* The reason for this is apparent in Figure 9.1 of

Table 9.3. *Percent standard errors of estimates (%SEE), and percent prediction errors (%PE) for mass estimation equations in Tables 9.1 and 9.2*

Equation	%SEE	Avg. %PE	%PE of elephant in model sample	Amount of over(+)- or under(−)estimate (kg)
Table 9.1				
1 (Hum. L.)	82	53	41	(−) 724
2 (Fem. L.)	89	56	63	(−)1537
3 (Hum. C.)	34	25	30	(+)2524
4 (Fem. C.)	39	26	37	(+)3521
Table 9.2: shoulder hts.				
g	4			
h	16	11		
i	13			
j	19			
k	18	12		
l	20			

Note: An artifact of calculating these statistics is that overestimates generate lower prediction errors than underestimates; percent prediction errors for the elephant from equations 3 and 4 become equal to or greater than those from equations 1 and 2 if the observed, rather than the predicted, mass is used in the denominator. %SEE and %PE were not available for equations a–f; correlation coefficients for these equations all exceeded .965.

Anderson et al. (1985), in which the point representing the elephant is situated directly on the upper boundary of the 95% confidence limit for the regression of circumference on body mass. Elephants have limb bones that are robust in comparison to those of other mammals.

The percent prediction errors and percent standard errors of the estimate (Smith 1985; see Van Valkenburgh [this volume] for formulae) listed in Table 9.3 superficially suggest that circumferences produce more accurate mass estimates than lengths. These indices of prediction accuracy can be misleading however, because equations that underestimate body mass inevitably show larger percent errors than equations that overestimate by the same number of kilograms; the ratio is inflated by the small value of predicted mass in the denominator.

3 *For the elephants the size of or larger than a dwarf mammoth (*M. exilis, M. columbi, *and the two modern species of elephants), lengths of humeri produce higher estimates than lengths of femora, but femoral circumferences produce higher estimates than humeral circumferences.*

For *E. falconeri,* however, femoral lengths yield slightly smaller estimates than humeral lengths, and mass estimates from femoral and humeral circumferences are identical.

Intact, fully fused femora are relatively uncommon, and do not fully sample the range of sizes for *M. exilis* known from other bones. *In general, the choice of elements (femur or humerus) affected the estimate less than the choice of measurements (length or circumference).*

4 *Mass estimates derived from shoulder height* (Table 9.2) *depend in part upon the body condition and in part upon the species of the elephants in the model sample.* Laws's African population (the *1*'s in Figure 9.2) was nutritionally stressed; Benedict's Asian elephants (*2*'s) were captive. A Thai population (Tumwasorn et al. 1980) maintained in more natural circumstances than Benedict's circus sample generates predictions of mass that are lower than Benedict's, but still higher than any predicted by regressions based on African populations. Whereas the highest point on the body of an adult African elephant occurs at the shoulder, the highest point on an Asian elephant is the forehead; for their shoulder heights, therefore, adult Asian elephants are expected to be heavier than African elephants (though some of the regression lines cross, and reverse the pattern among small, young individuals).

5 *For all species except* E. falconeri, *mass estimates from shoulder heights either embrace or fall between estimates from circumferences or lengths of single bones.* One would not expect all of the techniques used here – some based on interspecific and others on intraspecific allometry – to be biased (Damuth, this volume) in the same direction, and this suggests that the true masses of the animals represented by these specimens were also intermediate (i.e., between the length- and the circumference-based estimates).

For *E. falconeri,* mass estimates from shoulder heights were high – two to three times the masses estimated from bone lengths, and similar to the estimates from circumferences (Figures 9.3 and 9.4). The true mass was still probably less than these high estimates would suggest. The largest masses are obtained using the height–mass equations based upon African elephants. Equations for Asian samples (which as members of the same genus may be better analogues) predict lower masses, that, for UNSM #2540, are intermediate between length- and circumference-based estimates (Figure 9.3). My subjective impression of UNSM #2540 is of an animal closer in size and build to a warthog, Grant's gazelle, or impala (60–90 kg), as suggested by bone lengths, than to the wildebeest or oryx (150–180 kg) indicated by bone circumferences. (Body masses of these ungulates were obtained from Alexander et al. 1979 and Anderson et al. 1985.) Height-based regressions could overestimate masses if the proportionately large head of a juvenile rendered it more massive than a dwarf of the same height.

6 Comparison of mass estimates based on skeletal dimensions with known body masses for the same individuals (of *E. maximus* and *L. africana*) revealed the following: *(a) For the three specimens smaller than 2000 kg, bone lengths consistently provided the best mass estimates; estimates based on shoulder height either embraced or slightly overestimated, and circumferences consistently overestimated, the actual body mass of these animals. (b) For the three specimens larger than 2500 kg, conversely, shoulder heights and bone circumferences yielded accurate and similar estimates, whereas bone lengths underestimated actual body mass.*

In sum, estimates of mass can be obtained for dwarf elephants from lengths and circumferences of humeri and femora, and these estimates are consistent with masses based upon shoulder heights. Although 95% confidence intervals for single estimates are too broad to be informative, use of more than one mass estimation technique allows one to define a range that is likely to encompass the actual mass of the animal. For small elephants (less than approximately 2000 kg), bone lengths appear to yield the most accurate mass estimates, whereas larger masses are better approximated by using mass estimation equations based on bone circumferences. The differences in the estimates derived by using different techniques range from 40% to 275% at most, and thus they approximate the range of masses that is actually observed among elephants of a given height. Because two elephants of identical shoulder height can differ in body mass by a factor of 2, however, it will not be possible to estimate masses for fossils with greater resolution.

Nevertheless, given sufficiently conservative assumptions, useful hypotheses can be constructed and tested. I will present below a simple but plausible predictive model for the dwarfing of elephants on islands. Even mass estimates with broad ranges can yield useful qualitative conclusions.

Inferences

Damuth (1981, 1987) observed an inverse relationship between body mass and population density among mammalian herbivores. Island populations were excluded from the analysis, and in Damuth's regressions, mass was the independent and density the dependent variable. An implication is that for mammalian herbivores, population densities are constrained by body size.

On islands, large herbivores characteristically are smaller than their relatives on the mainland (Foster 1965). Large predators are absent, and without this selective constraint, adult body size of herbivores may vary more freely than on the mainland. Insular populations are constrained, however, by geography, and high densities unrelieved by emigration tend to remain high. These observations raise the possibility that the causal relationship described above is reversed: For large herbivores on islands, could high population density be an agent in dwarfism?

The most frequently cited cause of dwarfism in large mammalian herbivores on islands is resource limitation: An isolated population attains high density, which, unrelieved by emigration, leads to overcrowding, habitat destruction, stunting, and ultimately natural selection for individuals that fare well on a low plane of nutrition (e.g., among others, Case 1978; Lawlor 1982; Lomolino 1985; Roth 1982). To explain the extreme cases of size reduction observed for Pleistocene elephants, conditions much like those in many dense and stressed East African elephant habitats today have been invoked (Johnson 1972, 1980; Roth 1987). Although the scenario seems appropriate, the analogy with modern populations reasonable, and the verbal argument logical, a question arises: From the magnitude of the population densities observed among highly stressed modern populations of elephants, could one predict a body size reduction of the appropriate magnitude?

Model for dwarfism

1. We assume that the use of energy resources (per unit area) by the population is proportional to population density times the metabolic rate of an average adult individual. Metabolic rate depends upon body mass, and is determined from a regression equation based upon Kleiber's data for mammals (Kleiber 1961:205).
2. Our expectation of average body-size reduction will be expressed in terms of a reduction in energy use by the population.[7] The energy-use reduction factor is predicted by calculating the ratio of energy use by modern populations of elephants at (a)

very high density, and (b) a density sustained comfortably in unstressed populations. Body size (and metabolic rate) is held constant here, and so we simply use the ratio of the two densities. However, because the best high- and low-density figures available in the literature pertain to different species (and therefore different sizes) of modern elephants, a correction term, k, must be applied: $k \times$ (published density for *E. maximus*) = (corrected density for *L. africana*), where k = (body mass of *E. maximus*/body mass of *L. africana*)$^{0.75}$. The exponent (0.75) is derived from Damuth's (1981) regression of population density on body mass.

3 We then compute the corresponding reduction in energy use brought about by the reduction of large Pleistocene elephants to the body size of the dwarfed insular forms. For this, we hold density constant, and calculate metabolic rate from the following equation:

$$\text{Metabolic rate (kcal/day)} = 65.9\ (\text{body mass [kg]})^{0.768}$$

(This equation differs from Kleiber's [1961: 206] because it incorporates data from an elephant. The elephant datum was among those Kleiber excluded "because conditions [for measuring metabolic rate] were not comparable." I judged it preferable to include the elephant data rather than extrapolate a regression that ended at 600 kg. It is worth noting that Nagy's [1987] exponent for field metabolic rates of eutherian herbivores was 0.77.) Body mass for the fossils is determined from their linear dimensions, using the mass estimation equations of Table 9.1 or 9.2.

The comparison of predicted and actual reduction in resource use is in fact a comparison of ratios: of high and low densities (in 2) and high and low metabolic rates (in 3). Each formula cited in points 1–3 above is obtained from a regression with its own intrinsic variability, and this variability must be incorporated into confidence limits.

To account for variability in the estimates, we begin by considering two empirical distributions of bone lengths – one for full-sized elephants, and one for dwarfs. From these, we construct distributions of body masses, using a mass estimation equation from Table 9.1, to which we add a random normal variate of mean = 0 and SD determined by the standard error of the prediction. (As discussed in this paper earlier, the standard errors of predictions allow for even greater variability in masses than is observed among individual elephants, and are therefore conservative.) Masses are then converted to metabolic rates using the equation given in point 3 above. A second random normal variate is added to each predicted value of metabolic rate to account for its standard error. For each of the two populations of bones, we now have a distribution of metabolic rates. We take the average metabolic rate for each population, and then form their ratio. Repeating this process (by computer) 10,000 times generates a distribution of ratios, whose expected value we take as its mean, and whose 95% confidence limits we determine by removing the smallest 250 and largest 250 ratios from the tails. The mean of a/b, however, $\neq$ the reciprocal of the mean of b/a. For comparison, therefore, a distribution of the reciprocal ratios is also generated.

Example

I chose for this analysis *E. falconeri* because it appears to be less variable (Roth 1987) and less time-transgressive than available samples of *M. exilis* (cf. Ambrosetti 1968; Orr 1968). The ultimate ancestor of *E. falconeri* on the mainland was full-sized *E. namadicus* (or "*E. antiquus,*" as European specimens of this species are known [Kurtén 1968; Maglio 1973]). I also considered an intermediate size-class of elephants, for convenience chosen to be equivalent in size to the fossil Maltese form *E. mnaidriensis. Elephas mnaidriensis* and *E. melitensis* (which is smaller than *E. mnaidriensis,* larger than *E. falconeri,* and also from Malta) have long been assumed to be evolutionary intermediates, descended from *E. namadicus* and ancestral to the tiny *E. falconeri* (Adams 1874; Busk 1867), but recent dates and stratigraphic correlations suggest that these forms are actually younger than *E. falconeri* (Belluomini & Bada 1985; Burgio & Cani 1988). I will not, therefore, use species designations when I refer to putative evolutionary intermediates of intermediate size.

Table 9.5 lists the expected ratios of average metabolic rates (calculated, ultimately, from lengths of femora) for all pairs of size classes. The ratio of energy use between high and comfortable densities (Table 9.4) falls within the 95% confidence limits of two of the ratios: that corresponding to the body size reduction from the mainland form to an intermediate-sized form on the island, and that for the transition from the intermediate to *E. falconeri.* These observations do not change if a lower peak density is employed, or if a smaller average mass for African

Table 9.4. *Ratio of energy use for populations of differing densities*

The numerator is energy use of the high-density population; the denominator is energy use of the comfortable-density population; the ratio = E. Assuming all of the following:

1. A high density (for *L. africana*) of 4.5/km^2 (Laws et al. 1975)
2. A comfortable density (for *E. maximus*) of 1.0/km^2 (Eisenberg & Seidensticker 1976)
3. Average body masses of 2860 and 1810 kg for *L. africana* and *E. maximus*, respectively (Damuth 1987; Eisenberg & Seidensticker 1976)
4. An exponent of 0.75 with a 95% confidence interval of 0.70–0.80 for mass correction of the densities (Damuth 1981)

Then:

$$E = 6.3 \pm 0.2$$

If a high density of 3.9 (Laws et al. 1975) is used instead, $E = 5.5 \pm 0.1$.
If a mass of 3000 kg is used for *L. africana*, $E = 6.6 \pm 0.2$.
If a mass of 2500 kg is used for *L. africana*, $E = 5.7 \pm 0.1$.
If a mass of 2500 kg and a high density of 3.9 are used, $E = 4.97 \pm 0.08$.

Table 9.5. *Ratios of metabolic rates of large and small elephants*

Pairs of size classes	R^a	$\bar{R}^b$	$L_1{}^c$	$L_2{}^c$
E. namadicus / Intermediate	1.98	2	1.35	3
E. namadicus / *E. falconeri*	14.38	14.5	9.9	21
Intermediate / *E. falconeri*	7.23	7	5	11

Note: Metabolic rates were estimated from femur lengths given by (or estimated from data in) Adams (1874), Ambrosetti (1968), Busk (1867), and Kurtén (1968).
[a] R = ratio of average metabolic rates (ignoring variance in the mass estimation and metabolic-rate estimation regressions).
[b] $\bar{R}$ = Average ratio of average metabolic rates (incorporating the variance in the regressions). The extent to which the average ratio and the reciprocal of the average of its reciprocal agree determines the precision with which it (and L_1 and L_2) is reported.
[c] L_1, L_2 = 95% confidence limits for $\bar{R}$.

elephants is assumed when k, the correction factor for the densities, is calculated. The conclusion is in this sense robust.

The difference in body size between *E. namadicus* and *E. falconeri* is, in contrast, too great. If the simplifying assumptions employed in this model are valid,[8] and if (as this model assumes) we imagine body size reduction to be a compensatory reduction of energy use associated with high population densities, then *E. falconeri* must have arisen through something other than a single such adjustment. One possibility involves a more iterative process: overcrowding, followed by dwarfing, followed by still further increases in density and further dwarfing. Alternatively, a declining resource base (see note 8) could produce further dwarfing by a similar process but without a secondary increase in density, because as resources decline, the body size for which a specified population density is comfortable also declines.

Summary

This paper examines methodological issues relating to body mass estimation in a large extinct mammal for which there is no precise living analogue. Body masses for elephants smaller than two metric tons appear to be best approximated by allometric equations based on bone lengths, whereas for larger animals, masses are better estimated from bone circumferences. Confidence limits for body mass estimates on Pleistocene insular dwarf elephants are inevitably broad; the reasons for this include the fact that even masses measured directly on living elephants vary greatly with any given skeletal dimension. Within the envelope of variation in body mass estimates, some conclusions can be drawn, however.

We address the following question: Assuming that body size reduction among large herbivores on islands is associated with high population densities, and that it produces a compensatory reduction of energy resource use, can one account for the magnitude of size reduction in insular Pleistocene elephants simply by assuming that they attained population densities analogous to those sustained in isolated, overcrowded modern populations of African elephants?

Mass estimates derived from limb bone lengths of fossil elephants are used to estimate ratios of energy resource use by populations (of a given density) of animals of different sizes. Variation in the estimates is incorporated into confidence limits, and the ratios compared with the ratio of energy use by populations of differing densities. Despite the variability of the estimates, and as simplistic as this model is, it allows us to conclude (for example) that the evolution of the diminutive body size of *Elephas falconeri* (from the Pleistocene of Sicily and Malta) either (1) accompanied a decrease in the resource base on the islands, or (2) involved an intermediate-sized ancestor (smaller than the mainland form *E. na-*

madicus), and at least two successive episodes of overcrowding and dwarfing associated with those high population densities (or both).

Notes

1 A necropsied female *E. maximus* of approximately 2100–2500 kg reportedly had maximum capacities of the stomach and intestinal tract of 76.6 and 616.76 1, respectively (Shoshani et al. 1980, 1982). The capacities of these organs were obtained after they had been removed from the body cavity, and expansion was unconstrained by other tissues, so the approximate 30% of body mass that these volumes total may be excessive.

2 Because body size and adaptations for feeding and locomotion are strongly interrelated (Eisenberg 1981, this volume), a dwarf elephant may have less in common ecologically with its full-sized kin than with (for example) a peccary or a tapir, which, though genealogically more remote, are closer in size. Indeed, Sondaar (1977) suggested that in conjunction with dwarfism, elephants, hippos, and ruminants on Mediterranean islands underwent shifts in diet and/or locomotion. It could additionally be argued that juvenile full-sized elephants are unsuitable analogues of dwarfed adults, because juveniles are subject to different functional demands and morphological constraints. For example, juvenile elephants, like the juveniles and unlike the adults of any species, are not equipped to defend themselves or flee from predators (Graham 1976); yet they must be tall enough to nurse, and fast enough to keep up with the herd (Pennycuick 1975), and their limb proportions presumably reflect these requirements. Young elephants, unlike adult dwarfs, are morphologically constrained to be intermediates on a growth trajectory that produces graviportal adults.

3 A juvenile elephant is living proof that its particular combination of body proportions is functional. Large predators were absent from Sicily and the California Channel Islands, so selection pressures on mainland and island (including, conceivably, those that maintain differences between ontogenetic and adult interspecific allometry; see note 2) differed. For intermembral proportions (Roth 1984) and for the shapes of individual bones (Roth 1982), ontogenetic and interspecific allometry in elephants are similar in some, although not all, respects.

4 Alexander et al. also regressed diameters on mass, but I did not attempt to use diameters in mass estimation because the measurement is difficult to define consistently and precisely. On an elephant limb bone, the midshaft diameter in the sagittal plane (the measurement used by Alexander et al.) is influenced by the degree of torsion in the bone, and by strongly developed muscle insertions. The shapes of the bones are complex, and vary along the length of the shaft (e.g., a femur is anteroposteriorly compressed proximally, but rounds out to become nearly cylindrical distally), and measurements oriented with respect to a body axis ("sagittal plane"), and positioned distally a particular fraction of the bone's length, do not necessarily specify anatomically homologous points. I found "minimum circumference," the measurement used by Anderson et al. (1985), to be less ambiguous.

5 A body of data on elephants exists that permits mass estimates to be made for a particular specimen from heart weight, mass of hindlimb, foot circumference, age, dental stage, etc. Such *estimates* could be related to bony dimensions, and the relationship used in turn to predict body masses from bony dimensions of new specimens. However, in several-step inferences, error can become compounded and inaccuracies are magnified. For the ontogenetic models I employed, I used only mass data measured directly on living or fully necropsied individuals.

6 The thickness of the skin on the dorsum of one 227-cm tall (reclining) female Asian elephant was reported to be 3.2 cm (Shoshani et al. 1982). If an additional amount of twice that is allowed for the sole of the foot and its associated connective tissue, the total skin thickness is approximately 10 cm. This amount was added to the skeletal heights of individuals between 150 and 300 cm tall to obtain shoulder heights; for larger individuals I used 15 cm (corresponding to the "10 inches" used by Dutrow [1980] for full-sized mammoths), and for smaller ones I used 7 cm.

7 Although no particular selective mechanism is explicitly invoked in this model, this argument (like that of Wassersug et al. 1979) might be construed as involving group selection. A discussion phrased in terms of individual selection will be presented elsewhere (Roth & Mercer, unpublished).

8 Among these assumptions is that changes in body size do not materially affect the resource base of the population. Small ungulates in fact tend to feed more selectively than do full-sized elephants (though, as cecal digesters, elephants are likely to tolerate a high-fiber diet at a smaller body size than could a ruminant; Janis 1976), high densities of elephants can destroy habitats (Laws, Parker, & Johnstone 1975), and small animals cannot knock down or reach foliage in the upper branches of taller trees. These factors act to decrease the resource base, thereby reducing the tolerable population density and intensifying the forces leading to size reduction.

Acknowledgments

For assistance, and for access to (the often ungainly) specimens in their care, I thank the staff and curators of the vertebrate paleontology sections of the Santa Barbara Museum of Natural History (SBMNH), Los Angeles County Museum (LACM), University of Texas at Austin (TMM), University of Nebraska (UNSM), Florida State Museum (UF), Cleveland Museum of Natural History (CMNH), and Pratt Museum, Amherst College; and the mammals divisions of the U.S. National Museum (USNM), Yale Peabody Museum (YPM), American Museum of Natural History (AMNH), and British Museum (Natural History) (BMNH). Kim Bryan (BMNH), Robert Voss (AMNH), and R. George Corner (UNSM) supplied some crucial measurements, and R. McNeill Alexander shared his unpublished data, all on very short notice. John

Mercer translated the algorithm that generated the results in Table 9.5 into FORTRAN, and served frequently and graciously as a sounding board. Donald Burdick and Lee Altenberg supplied valuable statistical advice, and Enzo Burgio, Jeheskel Shoshani, and Joseph Dudley provided some useful references. Discussions with and comments by John Damuth have been both instructive and stimulating. This work was supported by grants from the (U.S.) National Science Foundation (BSR-8516818) and the Duke University Research Council.

References

Adams, A. L. 1874. On the dentition and osteology of the Maltese fossil elephants, being a description of remains discovered by the author in Malta between the years 1860 and 1866. *Trans. Zool. Soc. Lond. 9*:1–122.

Alexander, R. McN. 1977. Allometry of the limbs of antelopes (Bovidae). *J. Zool. Lond. 183*:125–146.

Alexander, R. McN. 1985. Mechanics of posture and gait of some large dinosaurs. *Zool. J. Linn. Soc. 83*:1–25.

Alexander, R. McN., Jayes, A. S., Maloiy, G. M. O., & Wathuta, E. M. 1979. Allometry of the limb bones of mammals from shrews (*Sorex*) to elephant *(Loxodonta). J. Zool. Lond. 189*:305–314.

Ambrosetti, P. 1968. The Pleistocene dwarf elephants of Spinagallo (Siracusa, south-eastern Sicily). *Geol. Romana 7*:277–398.

Anderson, J. F., Hall-Martin, A., & Russell, D. A. 1985. Long-bone circumference and weight in mammals, birds and dinosaurs. *J. Zool. Lond. 207*: 53–61.

Belluomini, G., & J. L. Bada 1985. Isoleucine epimerization ages of the dwarf elephants of Sicily. *Geology 13*:451–452.

Benedict, F. G. 1936. *The Physiology of the Elephant.* Washington, D.C.: Carnegie Institution.

Blueweiss, L., Fox, H., Kudzma, V., Nakashima, D., Peters, R., & Sams, S. 1978. Relationships between body size and some life history parameters. *Oecologia (Berlin) 37*:257–272.

Bookstein, F., Chernoff, B., Elder, R., Humphries, J., Smith, G., & Strauss, R. 1985. *Morphometrics in Evolutionary Biology.* Philadelphia: Academy of Natural Sciences, Special Publ. No. 15.

Burgio, E., & Cani, M. 1988. Considerazioni sulla successione filogenetica degli elefanti fossili Siciliani. In *Abstracts of the International Conference on Early Man in Island Environments (Oliena, Sardinia, 25 Sept.–2 Oct. 1988).* Nuoro: Comitato Corbeddu.

Busk, G. 1867. Description of the remains of three extinct species of elephant, collected by Capt. Spratt, C. B., R. N., in the ossiferous cavern of Zebbug, in the island of Malta. *Trans. Zool. Soc. Lond. 6*:227–306.

Buss, I. O. 1961. Some observations on food habits and behavior of the African elephant. *J. Wildlife Mgmt. 25*:131–148.

Calder, W. A. 1984. *Size, Function, and Life History*. Cambridge, Mass.: Harvard University Press.

Case, T. J. 1978. A general explanation for insular body size trends in terrestrial vertebrates. *Ecology 59*:1–18.

Cock, A. G. 1966. Genetical aspects of metrical growth and form in animals. *Q. Rev. Biol. 41*:131–190.

Cushing, J., Daily, M., Noble, E., Roth, V. L., & Wenner, A. 1984. Fossil mammoths from Santa Cruz Island, California. *Q. Res. 21*:376–384.

Damuth, J. 1981. Population density and body size in mammals. *Nature 290*: 699–700.

Damuth, J. 1987. Interspecific allometry of population density in mammals and other animals: the independence of body mass and population energy-use. *Biol. J. Linn. Soc. 31*:193–246.

Draper, N. R., & Smith, H. 1966. *Applied Regression Analysis*. New York: John Wiley.

Dutrow, B. L. 1980. Metric analysis of a Late Pleistocene mammoth assemblage, Hot Springs, South Dakota. Master's thesis, Southern Methodist University, Dallas.

Eisenberg, J. F. 1981. *The Mammalian Radiations*. Chicago: University of Chicago Press.

Eisenberg, J. F., & Seidensticker, J. 1976. Ungulates in southern Asia: a consideration of biomass estimates for selected habitats. *Biol. Conserv. 10*: 293–308.

Flower, S. S. 1943. Notes on age at sexual maturity, gestation period and growth of the Indian elephant, *Elephas maximus*. *Proc. Zool. Soc. Lond. [A]: 113*:21–27.

Foster, J. B. 1965. The evolution of the mammals of the Queen Charlotte Islands, British Columbia. *Occas. Pap. British Columbia Provincial Mus. 14*:1–130.

Graham, R. W. 1976. Pleistocene and Holocene mammals, taphonomy, and paleoecology of the Friesenhahn Cave local fauna, Bexar County, Texas. Ph.D. dissertation, University of Texas, Austin.

Gregory, W. K. 1929. Mechanics of locomotion in the evolution of limb structure as bearing on the form and habits of the titanotheres and the related odd-toed ungulates. In *The Titanotheres of Ancient Wyoming, Dakota, and Nebraska*. Washington, D.C.: U.S.G.P.O. Monogr. No. 55.

Hanks, J. 1969. Growth in weight of the female African elephant in Zambia. *East Afr. Wildlife J. 7*:7–10.

Hanks, J. 1972. Growth of the African elephant *(Loxodonta africana)*. *East Afr. Wildlife J. 10*:251–272.

Harington, C. R., Tipper, H. W., & Mott, R. J. 1974. Mammoth from Babine Lake, British Columbia. *Can. J. Earth Sci. 11*:285–303.

Janis, C. 1976. The evolutionary strategy of the Equidae and the origins of rumen and cecal digestion. *Evolution 30*:757–774.

Johnson, D. L. 1972. Landscape evolution on San Miguel Island, California. Ph.D. dissertation, University of Kansas, Lawrence. Ann Arbor: University Microfilms.

Johnson, D. L. 1980. Episodic vegetation stripping, soil erosion, and landscape modification in prehistoric and Recent historic time, San Miguel Island, California. In *The California Islands: Proceedings of a Multidisciplinary Symposium,* ed. D. M. Power. Santa Barbara: Santa Barbara Museum of Natural History.

Johnson, O. W., & Buss, I. O. 1965. Molariform teeth of male African elephants in relation to age, body dimensions and growth. *J. Mammal. 46*:373–384.

Kleiber, M. 1961. *The Fire of Life.* New York: John Wiley.

Kurtén, B. 1968. *Pleistocene Mammals of Europe.* London: Weidenfeld and Nicolson.

Kurtén, B., & Anderson, E. 1980. *Pleistocene Mammals of North America.* New York: Columbia University Press.

Lawlor, T. E. 1982. The evolution of body size in mammals: evidence from insular populations in Mexico. *Am. Nat. 119*:54–72.

Laws, R. M. 1966. Age criteria for the African elephant *Loxodonta a. africana. East Afr. Wildlife J. 4*:1–37.

Laws, R. M., & Parker, I. S. C. 1968. Recent studies on elephant populations in East Africa. In *Comparative Nutrition of Wild Animals,* ed. M. A. Crawford, pp. 319–359. Symp. Zool. Soc. Lond. No. 21.

Laws, R. M., Parker, I. S. C., & Johnstone, R. C. B. 1975. *Elephants and their Habitats: The Ecology of Elephants in North Bunyoro, Uganda.* Oxford: Clarendon Press.

Lewis, G., & Fish, G. 1978. *I Love Rogues.* Seattle: Superior Publishing.

Lomolino, M. V. 1985. Body size of mammals on islands: the island rule reexamined. *Am. Nat. 125*:310–316.

Maglio, V. J. 1973. Origin and evolution of the Elephantidae. *Trans. Am. Philos. Soc. 63* (3):3–149.

Maynard Smith, J., & Savage, R. J. G. 1959. The mechanics of mammalian jaws. *School Sci. Rev. 40*:289–301.

McMahon, T. 1973. Size and shape in biology. *Science 179*:1202–1204.

Momin Khan, M. K. 1977. Aging of elephants: estimation by foot size in combination with tooth wear and body dimensions. *Malayan Nature J. 30*: 15–23.

Nagy, K. A. 1987. Field metabolic rate and food requirement scaling in mammals and birds. *Ecol. Monogr. 57*:111–128.

Orr, P. C. 1968. *Prehistory of Santa Rosa Island.* Santa Barbara: Santa Barbara Museum of Natural History.

Patterson, B. D., & Atmar, W. 1986. Nested subsets and the structure of insular mammalian faunas and archipelagos. *Biol. J. Linn. Soc. 28*:65–82.

Pennycuick, C. J. 1975. On the running of the gnu (*Connochaetes taurinus*) and other animals. *J. Exp. Biol. 63*:775–799.

Peters, R. H. 1983. *The Ecological Implications of Body Size.* Cambridge: Cambridge University Press.

Roth, V. L. 1982. Dwarf mammoths from the Santa Barbara, California Channel Islands: size, shape, development, and evolution. Ph.D. dissertation, Yale University, New Haven. Ann Arbor: University Microfilms.

Roth, V. L. 1984. How elephants grow: heterochrony and the calibration of developmental stages in some living and fossil species. *J. Vertebr. Paleontol. 4*:126–145.

Roth, V. L. in press. Dwarfism and variability in the Santa Rosa Island mammoth: an interspecific comparison of limb bone sizes and shapes in elephants. In *Proceedings of the Third California Islands Symposium,* ed. F. G. Hochberg. Santa Barbara: Santa Barbara Museum of Natural History.

Roth, V. L., & Mercer, J. M. n.d. A model of natural selection for dwarfism in insular mammals. Unpublished manuscript.

Roth, V. L., & Shoshani, J. 1988. Dental identification and age determination in *Elephas maximus. J. Zool., Lond. 214*:567–588.

Schmidt-Nielsen, K. 1984. *Scaling: Why Is Animal Size So Important?* Cambridge: Cambridge University Press.

Scott, K. M. 1985. Allometric trends and locomotor adaptations in the Bovidae. *Bull. Am. Mus. Nat. Hist. 179* (article 2):197–288.

Shoshani, J., et al. [76 coauthors]. 1980. An abstract on the dissection of a female Asian elephant *Elephas maximus maximus* (Linnaeus, 1758). *Elephant 1* (4):44–46.

Shoshani, J., et al. [76 coauthors]. 1982. On the dissection of a female Asian elephant *Elephas maximus maximus* (Linnaeus, 1758) and data from other elephants. *Elephant 2* (1):3–93.

Shoshani, S. L., Shoshani, J., & Dahlinger, F. C. 1986. Jumbo: origin of the word and history of the elephant. *Elephant 2* (2):86–122.

Sikes, S. K. 1971. *The Natural History of the African Elephant.* London: Weidenfeld & Nicolson.

Smith, R. J. 1985. The present as a key to the past: body weight of Miocene hominoids as a test of allometric methods for paleontological inference. In *Size and Scaling in Primate Biology,* ed. W. L. Jungers. New York: Plenum Press.

Sokal, R. G., & Rohlf, F. J. 1980. *Biometry,* 2nd ed. New York: W. H. Freeman.

Sondaar, P. Y. 1977. Insularity and its effect on mammal evolution. In *Major Patterns in Vertebrate Evolution,* ed. M. K. Hecht, P. C. Goody, & B. M. Hecht. New York: Plenum Press.

Stock, C. 1935. Exiled elephants of the Channel Islands, California. *Sci. Mo. 41*:205–214.

Tumwasorn, S., Udsanakornkul, S., Leenanuruksa, D., Kaeophrommarn, C., & Pichaicharnnarong, A. 1980. Some weight and body measure estimates of Asiatic elephants *(Elephas maximus). Thai J. Agric. Sci. 13*:179–185.

Wassersug, R. J., Yang, H., Sepkoski, J. J., Jr., & Raup, D. M. 1979. The evolution of body size on islands: a computer simulation. *Am. Nat. 114*: 287–295.

Western, D. 1979. Size, life history, and ecology in mammals. *Afr. J. Ecol. 17*:185–204.

Western, D. 1983. Production, reproduction, and size in mammals. *Oecologia (Berlin) 59*:269–271.

Williamson, M. 1981. *Island Populations.* Oxford: Oxford University Press.

10

Skeletal and dental predictors of body mass in carnivores

BLAIRE VAN VALKENBURGH

In this chapter, I explore the accuracy of four measures, head–body length, skull length, occiput-to-orbit length, and lower first molar length as predictors of body weight for fossil mammalian predators. Measurements of the skull and dentition were chosen over those of the postcranial skeleton because fossil mammals are often known only from these elements. Head–body length was included because it was thought to provide a better predictor of body mass than the dimensions of single skeletal elements, and could be used when an entire skeleton is available.

Materials and methods

The study sample consists of 72 species representing seven families of mammals: Dasyuridae (n = 2), Ursidae (n = 8); Procyonidae (n = 5), Mustelidae (n = 19), Viverridae (n = 4), Hyaenidae (n = 4), and Felidae (n = 16) (Table 10.1). Average body mass and head–body lengths were estimated for males and females of each species from the literature (Table 10.1). If separate mass values for the sexes were not found, the estimate for the species was used. In most cases the estimates shown in Table 10.1 represent the average of the midpoints of published ranges of mass and lengths; thus they are not sample means. In a few cases, ranges were not available, and therefore, the estimate is the average of the available data. Cranial and dental measurements were made on one male and one female of each species whenever possible. Owing to a lack of cranial material, five species are represented by only

In *Body Size in Mammalian Paleobiology: Estimation and Biological Implications,* John Damuth and Bruce J. MacFadden, eds.

Table 10.1. *List of species measured, data used in the analyses (log_{10} of raw values), and sources for body mass and head–body length data*

Species	Sex	BM	HBL	SKL	OOL	M_1L	Sources
Canidae							
Canis lupus	F	1.613	3.122	2.318	2.097	1.415	1, 2
	M	1.648	3.139	2.356	2.146	1.446	
C. latrans	F	1.114	2.940	2.225	1.991	1.320	1, 3
	M	1.114	2.940	2.272	2.053	1.362	
C. mesomelas	F	0.903	2.813	2.146	1.940	1.207	4, 5
	M	0.903	2.813	2.152	1.940	1.253	
Speothos venaticus	F	0.954	2.822	2.134	1.968	1.161	1
	M	0.954	2.822	2.114	1.940	1.124	
Vulpes vulpes	F	0.863	2.775	2.117	1.875	1.190	1
	M	0.863	2.775	2.093	1.863	1.146	
V. velox	F	0.415	2.644	1.987	1.799	1.086	1, 3
	M	0.415	2.644	2.004	1.806	1.097	
Urocyon cinereoargenteus	F	0.643	2.782	2.076	1.886	1.068	1, 3
	M	0.643	2.782	2.013	1.826	1.072	
Dusicyon culpaeus	M	1.079	2.903	2.201	1.968	1.236	18
	M	1.079	2.903	2.173	1.944	1.220	
Cerdocyon thous	F	0.845	2.813	2.111	1.892	1.134	6, 7
	M	0.845	2.813	2.143	1.914	1.152	
Alopex lagopus	F	0.760	2.754	2.090	1.886	1.104	1
	M	0.760	2.754	2.093	1.908	1.137	
Nyctereutes procyonoides	F	0.845	2.724	2.041	1.857	1.093	1
	M	0.845	2.724	2.033	1.845	1.121	
Otocyon megalotis	F	0.574	2.716	2.093	1.881	0.863	1, 4
	M	0.574	2.716	2.072	1.851	0.833	
Chrysocyon brachyurus	M	1.362	3.097	2.336	2.117	1.354	1
Lycaon pictus	F	1.297	2.948	2.250	2.041	1.387	3, 5
	M	1.326	2.948	2.286	2.079	1.407	
Ursidae							
Selenarctos thibetanus	F	2.000	3.190	2.400	2.270	1.328	1, 8
	M	2.000	3.190	2.479	2.338	1.336	
Tremarctos ornatus	F	2.130	3.217	2.279	2.152	1.241	1
	M	2.130	3.217	2.362	2.207	1.299	
Ursus americanus	F	2.155	3.228	2.389	2.199	1.272	1, 3
	M	2.155	3.228	2.393	2.204	1.270	
U. arctos	F	2.322	3.275	2.502	2.314	1.362	3
	M	2.505	3.311	2.599	2.427	1.449	
Thalarctos maritimus	F	2.545	3.336	2.519	2.338	1.326	1, 3
	M	2.598	3.336	2.602	2.442	1.316	
Melursus ursinus	F	2.009	3.204	2.450	2.279	1.188	1, 3
	M	2.009	3,204	2.470	2.312	1.236	
Helarctos malayanus	F	1.678	3.051	2.352	2.201	1.225	1, 8
	M	1.797	3.076	2.386	2.243	1.238	
Ailuropoda melanoleuca	M	2.070	3.130	2.455	2.286	1.522	1

Table 10.1. (*cont.*)

Species	Sex	BM	HBL	SKL	OOL	M_1L	Sources
Procyonidae							
Bassariscus astutus	F	0	2.556	1.838	1.705	0.851	1, 3
	M	0	2.556	1.867	1.751	0.839	
Procyon lotor	F	0.978	2.734	1.987	1.820	0.987	1, 3, 9
	M	0.978	2.734	2.033	1.863	0.968	
Potos flavus	F	0.362	2.667	1.914	1.808	0.716	1, 10
	M	0.362	2.667	1.906	1.812	0.708	
Nasua nasua	F	0.954	2.756	2.025	1.792	0.813	3, 11
	M	0.954	2.756	2.072	1.826	0.886	
Ailurus fulgens	F	0.574	2.757	2.013	1.845	1.068	1
	M	0.574	2.757	2.009	1.886	1.045	
Mustelidae							
Mustela vison	F	−0.081	2.568	1.735	1.659	0.820	1, 3
	M	0.009	2.599	1.782	1.703	0.881	
M. nigripes	F	−0.041	2.610	1.829	1.732	0.940	1
	M	−0.041	2.610	1.823	1.751	0.929	
M. frenata	F	−0.854	2.332	1.599	1.509	0.724	3
	M	−0.569	2.393	1.663	1.581	0.771	
Gulo gulo	F	1.230	2.889	2.121	2.000	1.294	1, 3, 9
	M	1.312	2.889	2.155	2.041	1.367	
Pteroneura brasiliensis	F	1.255	3.097	2.121	2.049	1.283	1
	M	1.255	3.097	2.185	2.140	1.272	
Martes americana	F	−0.114	2.620	1.827	1.726	0.886	1, 3
	M	0.079	2.650	1.870	1.767	0.949	
M. pennanti	F	0.653	2.749	1.987	1.857	1.064	1, 3
	M	0.732	2.749	2.053	1.944	1.121	
Spilogale interrupta	F	−0.337	2.362	1.698	1.601	0.863	1, 3
	M	−0.137	2.362	1.730	1.642	0.898	
Ictonyx striatus	F	0.114	2.525	1.745	1.650	0.833	1, 4
	M	0.114	2.525	1.745	1.645	0.833	
Mephitis mephitis	F	0.491	2.556	1.806	1.690	0.987	1, 3
	M	0.491	2.556	1.826	1.699	0.996	
Taxidea taxus	F	0.881	2.732	2.104	2.000	1.143	1, 3
	M	0.881	2.732	2.104	1.996	1.104	
Lutra canadensis	F	0.929	2.839	2.033	1.978	1.114	1, 3
	M	0.929	2.839	2.053	1.954	1.167	
Conepatus mesoleucas	F	0.362	2.607	1.844	1.715	0.949	1, 3
	M	0.362	2.607	1.871	1.718	0.949	
Enhydra lutris	F	1.505	2.987	2.143	2.090	1.201	1, 3
	M	1.505	2.987	2.155	2.083	1.193	
Aonyx capensis	F	1.267	2.989	2.041	1.968	1.201	1
Eira barbera	F	0.623	2.806	2.000	1.886	0.914	1, 10
	M	0.623	2.806	2.057	1.929	0.959	
Galictis vittata	F	0.415	2.709	1.908	1.826	1.000	1, 10
	M	0.415	2.709	1.919	1.833	1.025	
Meles meles	F	1.114	2.836	2.064	1.914	1.179	1
	M	1.114	2.836	2.083	1.929	1.188	

Table 10.1. (*cont.*)

Species	Sex	BM	HBL	SKL	OOL	M_1L	Sources
Mellivora capensis	F	1.021	2.839	2.124	2.053	1.111	3
	M	1.021	2.839	2.149	2.041	1.130	
Viverridae							
Civettictis civetta	F	1.079	2.881	2.185	2.033	1.170	1, 4
	M	1.079	2.881	2.155	2.000	1.167	
Mungos mungo	F	0.279	2.574	1.863	1.771	0.672	1, 4
	M	0.279	2.574	1.857	1.748	0.681	
Nandinia binotata	F	0.279	2.708	1.978	1.892	0.886	1, 4
	M	0.279	2.708	1.954	1.813	0.908	
Arctictis binturong	F	1.041	2.897	2.161	2.009	0.978	1
	M	1.041	2.897	2.140	1.968	0.940	
Hyaenidae							
Hyaena brunnea	F	1.607	3.023	2.342	2.173	1.373	1
	M	1.607	3.023	2.369	2.207	1.407	
Proteles cristatus	F	0.987	2.832	2.158	1.987	.	1, 4
	M	0.987	2.832	2.140	1.954	.	
Crocuta crocuta	F	1.740	3.079	2.378	2.220	1.420	12
	M	1.690	3.041	2.344	2.199	1.389	
Hyaena hyaena	F	1.462	3.021	2.297	2.121	1.301	1, 4
	M	1.462	3.021	2.324	2.161	1.316	
Felidae							
Felis yagouaroundi	F	0.869	2.785	1.996	1.908	0.940	1, 3
F. wiedii	F	0.431	2.813	1.968	1.863	0.940	1, 13, 14
	M	0.431	2.778	2.009	1.903	0.940	
F. temmincki	F	1.176	2.903	2.130	2.000	1.158	1, 3, 13
	M	1.176	2.903	2.061	1.924	1.076	
F. viverrina	F	1.079	2.954	2.124	2.004	1.045	14
	M	1.079	2.954	2.100	1.996	1.037	
F. serval	F	1.025	2.919	2.021	1.908	0.978	4
	M	1.025	2.919	2.079	1.949	0.987	
F. caracal	F	1.217	2.851	2.025	1.929	1.114	4, 13
	M	1.217	2.851	2.064	1.968	1.076	
F. aurata	F	1.025	2.912	2.134	2.004	1.041	4
Lynx rufus	F	0.978	3.013	2.053	1.924	1.045	1, 3
	M	1.033	2.919	2.049	1.924	1.057	
Puma concolor	F	1.740	3.146	2.248	2.114	1.199	1, 3, 13
	M	1.833	3.121	2.279	2.161	1.260	
Panthera onca	F	1.653	3.114	2.283	2.149	1.318	1, 3, 10, 13
	M	1.740	3.114	2.362	2.225	1.318	
P. leo	F	2.179	3.253	2.410	2.238	1.403	1, 5
	M	2.236	3.276	2.473	2.305	1.431	
P. tigris	F	2.121	3.230	2.410	2.243	1.380	1, 13
	M	2.279	3.380	2.435	2.279	1.461	
P. pardus	F	1.613	3.158	2.279	2.130	1.230	1, 5, 13
	M	1.658	3.158	2.314	2.161	1.272	

Table 10.1. (*cont.*)

Species	Sex	BM	HBL	SKL	OOL	M_1L	Sources
Neofelis nebulosa	M	1.342	2.924	2.223	2.072	1.204	1, 13, 16
	M	1.342	2.924	2.212	2.053	1.190	
Acinonyx jubatus	F	1.776	3.225	2.140	2.013	1.207	1, 5
	M	1.787	3.255	2.212	2.068	1.248	
Uncia uncia	F	1.505	3.057	2.176	2.041	1.230	1, 13, 15
	M	1.505	3.057	2.199	2.053	1.233	
Marsupialia – Dasyuridae							
Sarcophilus harrisii	F	0.690	2.821	2.068	1.903	0.991	1
	M	0.886	2.821	2.107	1.924	1.017	
Thylacinus cynocephalus	F	1.398	3.061	2.310	2.076	1.083	17
	M	1.398	3.061	2.378	2.161	1.086	

Abbreviations: BM, log body mass; HBL, log head–body length; SKL, log skull length; OOL, log occiput-to-orbit length; M_1L, log M_1 length.
Sources: (1) Walker 1964; (2) Mech 1966; (3) Burt & Grossenheider, 1976; (4) Kingdon 1977; (5) Schaller 1972; (6) Nowak & Paradiso 1983; (7) Schaller 1983; (8) Prater 1965; (9) Grinnell, Dixon, & Linsdale 1937; (10) Eisenberg, O'Connell, & August 1979; (11) Leopold 1959; (12) Kruuk 1972; (13) Guggisberg 1975; (14) Lekagul & McNeeley 1977; (15) Blomqvist 1978; (16) Davis 1962; (17) Moeller 1970; (18) T. Fuller, pers. comm.

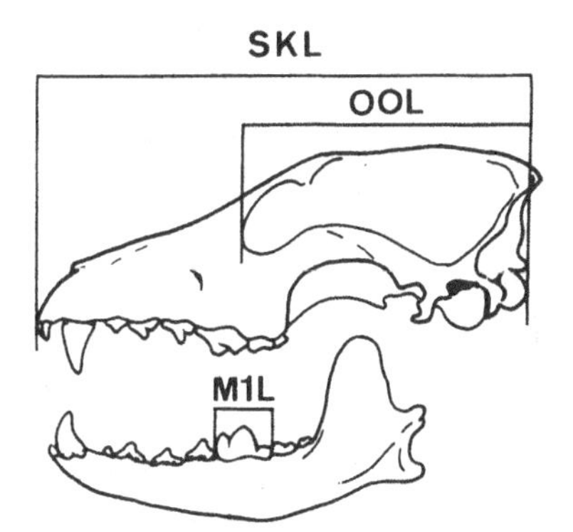

Figure 10.1. Cranial and dental measurements used in this study: SKL, total skull length from posterior surface of occipital condyles to anterior of premaxilla; OOL, occiput-to-orbit length, measured from occipital condyles to anterior edge of orbit M_1L, total length of lower first molar, measured at alveolar margin. The skull shown is that of a domestic dog, *Canis familiaris*.

one sex. All skulls are part of the Mammalogy collection of the United States National Museum.

The measurements include total skull length (SKL), occiput-to-orbit length (OOL), and lower first molar length (M_1L) (see Figure 10.1 for details). Although the results for occiput-to-orbit length are similar to those for skull length, the former is included because complete fossil skulls are relatively rare. The first lower molar tooth was chosen because of its functional importance. This tooth, along with the upper fourth premolar, is typically well-developed in most carnivorans. Together, they form the major meat-slicing tool of the carnivore tooth row (Ewer 1973; Van Valkenburgh 1989) and therefore might be expected to scale in a predictable way with body size. By contrast, the other cheek teeth, canines and incisors, tend to vary more in function and relative development among species, reflecting both dietary and behavioral differences (Van Valkenburgh 1989; Van Valkenburgh & Ruff 1987).

Least squares regression of $\log_{10}$ transformed data is used to model the association between body mass and skeletal measures. The significance of differences in the regressions was tested by analysis of covariance using the ANCOVA program of the software package SYSTAT (Wilkinson 1986). Because the correlation coefficient (r) is often a poor indicator of the predictive power of the independent variable (Smith 1981), two additional parameters that reflect residual variation were calculated. These are the percent prediction error (%PE) and the percent standard error of the estimate (%SEE). Following Smith (1981, 1984), %PE was calculated for each case as:

$$[(\text{Observed} - \text{Predicted}) / \text{Predicted}] \times 100$$

Thus, for each case, %PE indicates the percent difference between the actual weight and that predicted by the regression. In addition, the mean of the absolute values of the %PEs for a given regression provides a comparative index of predictive accuracy among regressions.

The second parameter, %SEE, also reflects the overall ability of the independent variable to predict the dependent variable. As described by Smith (1984), %SEE is calculated as follows: 2 is added to the (log) standard error of the estimate, and the antilog is taken. The result is equal to 100 plus the %SEE. For example, given a log standard error of the estimate of .219, the %SEE would be calculated as the antilog of 2.219 (=166) less 100, or ±66%. Assuming a normal distribution, 68% of the actual values would be expected to fall within +66% and −66% of the predicted value.

These two parameters, as well as slopes, y-intercepts, standard errors, and the correlation coefficient (r), are provided for each regression (Tables 10.2 and 10.3). The regressions were performed for the entire sample and for subsets of the sample. The subsets include four carnivore families (Canidae, Ursidae, Mustelidae, and Felidae) and three size groups (< 10

Table 10.2. *Regression of log_{10} (mass) on log_{10} (skeletal measures)*

Measure	Group	Slope	Int.	*r*	SE	%SEE	%PE
HBL[a]	Total	2.88	−7.24	.96	.185	53	36
	Canidae	2.30	−5.58	.96	.093	24	17
	Ursidae	2.98	−7.43	.95	.083	21	15
	Mustelidae	2.81	−7.08	.92	.238	73	48
	Felidae	2.72	−6.83	.92	.188	54	36
SKL[b]	Total	3.13	−5.59	.95	.220	66	47
	Canidae	2.86	−5.21	.93	.117	31	21
	Ursidae	2.02	−2.80	.70	.193	56	39
	Mustelidae	3.39	−6.03	.95	.199	58	40
	Felidae	3.11	−5.38	.92	.196	57	38
OOL[c]	Total	3.44	−5.74	.95	.207	61	42
	Canidae	3.08	−5.03	.94	.114	30	22
	Ursidae	1.98	−2.38	.64	.207	61	42
	Mustelidae	3.29	−5.53	.93	.220	66	47
	Felidae	3.54	−5.86	.92	.196	57	37
M_1L[d]	Total	2.97	−2.27	.83	.377	138	97
	Canidae	1.82	−1.22	.87	.158	44	27
	Ursidae	0.49	1.26	.43	.250	78	46
	Mustelidae	3.48	−3.04	.93	.220	66	45
	Felidae	3.05	−2.15	.95	.613	41	28

Note: Sample sizes for the family regressions are as follows: Canidae, 27; Ursidae, 15; Mustelidae, 37; Felidae, 3. Total, using the total sample of 63 species.
Abbreviations: HBL, head–body length; SKL, skull length; OOL, occiput–orbit length; M_1L, lower first-molar length; int., *y*-intercept; *r*, correlation coefficient; %SEE, percent standard error of estimate; %PE, mean absolute value of percent prediction error.
[a]Slopes do not differ significantly among families, but *y*-intercepts differ significantly between felids and canids ($p < .05$). Elevations of the mustelid and ursid regressions were not compared because the lack of overlap in body mass values with other families makes such comparisons of dubious value.
[b]Felid and mustelid curves differ from that of ursids in slope ($p < .10$); canid and felid regressions differ significantly in elevation ($p < .01$).
[c]No significant differences among families.
[d]Canid and ursid regressions differ significantly in slope from felid and mustelid regressions ($p < .01$).

kg, 10–100 kg, and > 100 kg). Separate regressions are not shown for the remaining carnivore families because sample sizes were too small. Finally, the regressions are used to predict the body mass of six Oligocene mammals, all members of extinct families of predators. They include three sabertooth cats (Nimravidae; *Hoplophoneus primaevus, H. occidentalis, Dinictis felina*), two bear-dogs (Amphicyonidae; *Daphoenus vetus, D. hartshornianus*), and two creodonts (Creodonta; *Hyaenodon*

Table 10.3. *Regression of log_{10} (mass) against log_{10} cranial and dental measures for size categories*

	Size group	Slope	Int.	*r*	SE	%SEE	%PE
HBL	>100 kg	2.46	−5.78	.83	.111	29	18
	10–100	2.05	−4.77	.84	.143	39	28
	6–10	0.90	−1.68	.35	.136	36	26
	<6	2.13	−5.39	.76	.212	63	43
SKL	>100 kg	1.56	−1.61	.66	.149	41	26
	10–100	1.71	−2.42	.69	.190	55	38
	6–10	1.01	−1.28	.36	.134	36	25
	<6	2.55	−4.56	.81	.193	56	39
OOL	>100 kg	1.51	−1.25	.61	.155	43	28
	10–100	2.30	−3.37	.79	.161	45	30
	6–10	0.80	−0.70	.30	.137	37	26
	<6	2.70	−4.55	.77	.212	63	41
M_1L	>100 kg	0.57	1.45	.27	.190	55	36
	10–100	1.19	−0.09	.58	.215	64	44
	6–10	0.36	0.43	.29	.137	37	26
	<6	1.21	−0.93	.42	.299	99	55

Note: Abbreviations as in Table 10.2. Sample sizes for the size groups: >100 kg, 17; 10–100 kg, 56; 6–10 kg, 36; <6 kg, 30.

horridus, H. crucians). All are represented by complete or nearly complete skeletons from North American deposits of Orellan age (Clark, Beerbower, & Kietzke 1967; Scott & Jepsen 1941).

Results

Head–body length

The regression of log body mass against log head–body length (HBL) for the total sample resulted in a correlation coefficient of .96 and an average percent prediction error of 36 (Figure 10.2 and Table 10.2). The %SEE is 53, indicating that 68% of the actual mass values lie within ±53% of their predicted values. If body mass increased as the cube of a linear measurement, the slope of the regression line would be 3, and the relationship between the two measures would be isometric and follow the expectation of geometric similarity. However, the slope of the line, 2.88, is significantly less than 3 ($p < .10$, Student's *t*, 2-tailed), indicating that mass is negatively allometric with respect to head–body length for the sampled carni-

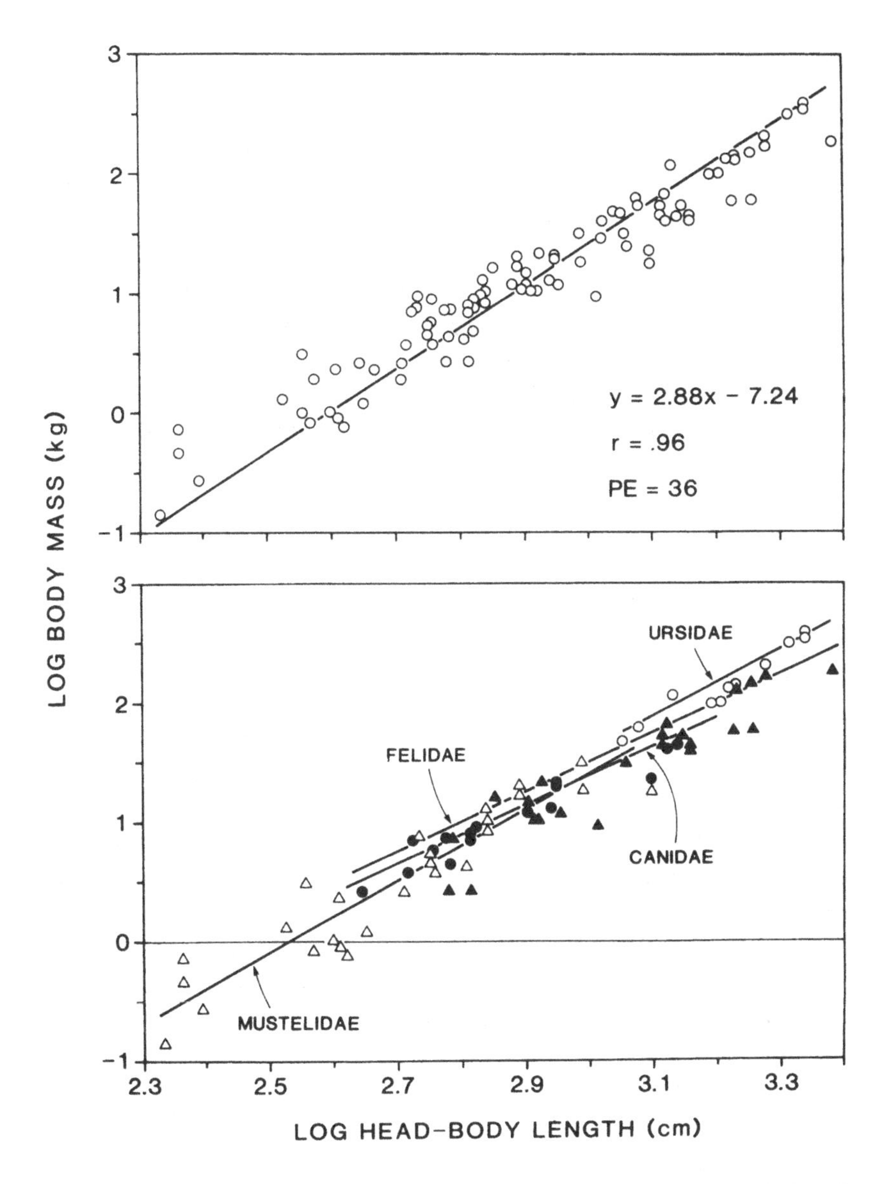

Figure 10.2. Plot of $\log_{10}$ body mass against $\log_{10}$ head–body length for all individuals (top) and for four families (bottom). Symbols for family plot: Ursidae, open circles; Felidae, solid triangles; Canidae, solid circles; Mustelidae, open triangles. See Table 10.1 for species list. Regression statistics for all lines are in Table 10.2.

vores. Thus, larger carnivores are about 40% lighter than would be expected if they were geometrically similar to small species.

When body mass is regressed against head–body length for each of the four families, prediction errors drop markedly for canids and ursids (%PE = 17 and 15, respectively) but not for mustelids and felids (%PE = 48 and 36, respectively) (Table 10.2). None of the slopes of the separate regressions differ significantly and all are less than 3, but the curve for felids has a significantly greater elevation than that of canids ($p < .05$). Thus, felids tend to be heavier than canids of similar body length.

The regressions by size category provide some improvement over the total sample regression (Table 10.3). With the exception of the smallest size group, the %PE ranges from 18 to 26, and the %SEE ranges from 29 to 36.

Skull length

The correlation coefficient of skull length and body mass is high (.95), but skull length does not appear to predict body mass as well as does head–body length (%PE = 47, %SEE = 66; Table 10.2 and Figure 10.3). However, as with regressions of body mass on HBL, separate regressions for the families are better (%PE = 21 to 40). Similarly, skull length predicts body mass more accurately within the size groups than across the total sample (%PE = 25 to 39, Table 10.3). Across the total sample, body mass is positively allometric relative to skull length (slope = 3.13, Table 10.2), indicating that larger species tend to have relatively shorter skulls. However, there are differences among the families in slope; body mass increases with skull size faster in felids and mustelids than in ursids ($p < .10$). Although the slopes of the canid and felid regressions are not significantly different, the two differ in elevation ($p < .01$; Figure 10.3). Felids have relatively more massive, robust bodies than canids of similar skull length.

Occiput-to-orbit length

The results for occiput-to-orbit length (Figure 10.4) generally follow those for skull length. Prediction errors for both the total, and family and size group regressions are similar to their skull length counterparts (Tables 10.2 and 10.3). Likewise, the allometric pattern is similar to that for skull length; mass is positively allometric with respect to occiput-to-orbit length, but in contrast to the SKL regressions, there are no

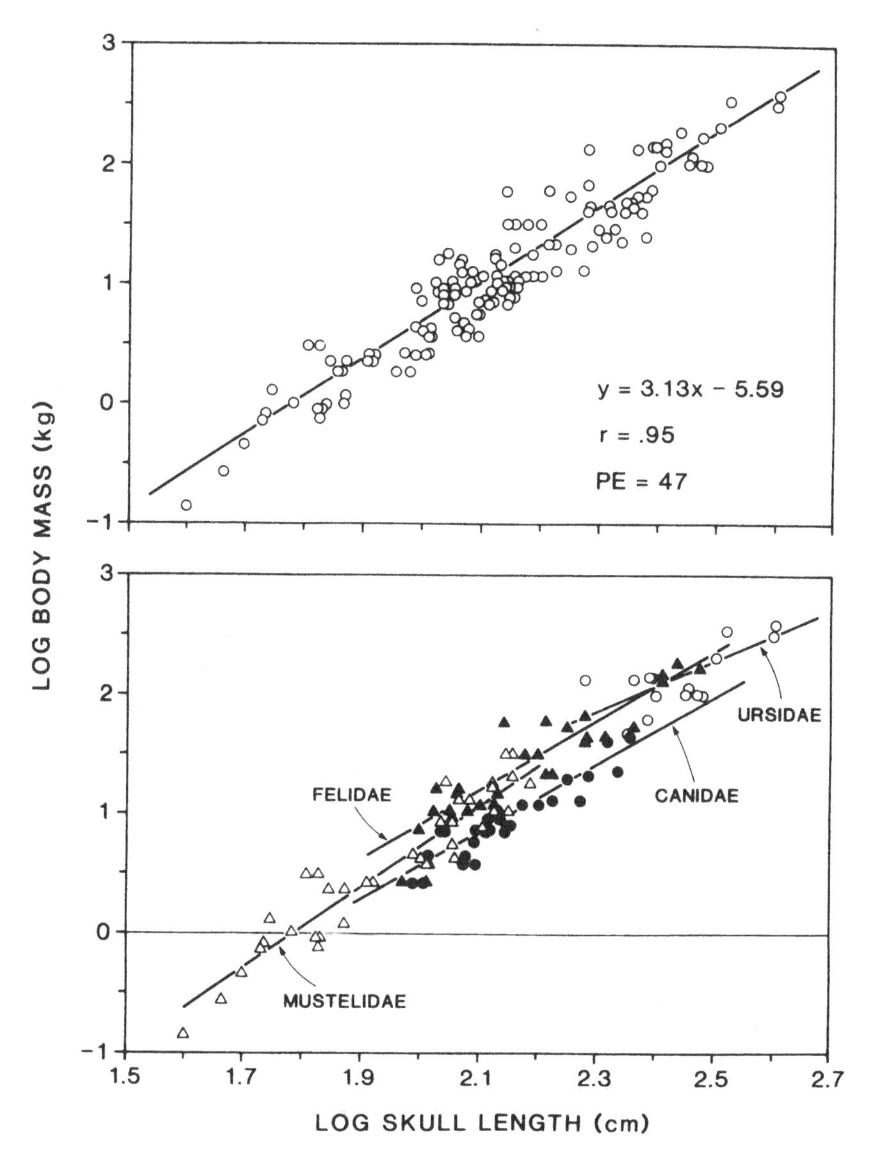

Figure 10.3. Plot of $\log_{10}$ body mass against $\log_{10}$ skull length for all individuals (top) and for four families (bottom). For symbols, species list, and regression statisics, see Figure 10.2.

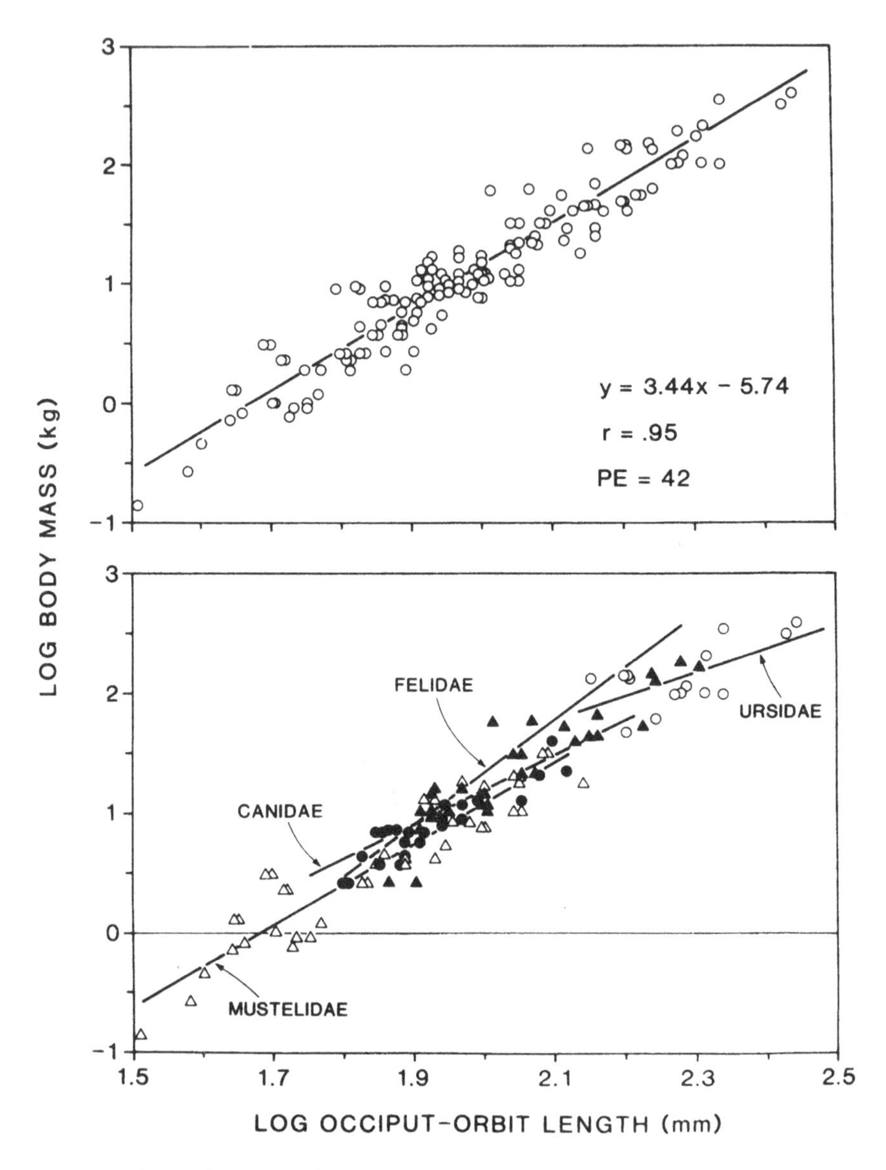

Figure 10.4. Plot of $\log_{10}$ body mass against $\log_{10}$ occiput-to-orbit length for all individuals (top) and for four families (bottom). For symbols, species list, and regression statistics, see Figure 10.2.

significant differences among the four families. Apparently, the disparities in scaling of skull size between felids and mustelids, on the one hand, and ursids on the other, reflect differences in the relative length of the skull anterior to the orbit (i.e., snout length).

M_1 length

Of the four skeletal and dental measures, lower M_1 length (Figure 10.5) is the poorest predictor of body weight when the entire sample is considered. The regression for the entire sample has a %SEE of 138 and an average %PE of 97. However, the errors are much smaller for the family regressions, with values similar to those of the skull and occiput-to-orbit length regressions. Interestingly, the prediction error for felids is smaller for M_1 length than for the skull or body length measures (%PE = 28 as opposed to 36–38; Table 10.2).

Among the size groups, prediction errors are about 10–20% greater for M_1 length than for the other measures in all but the 6- to 10-kg size group (Table 10.3). In this group, prediction errors are similar for the four skeletal and dental measures.

The allometry of body mass and tooth size varies considerably among the families. For the entire sample, mass increases nearly as the cube of M_1 length (slope = 2.97) and thus follows the expectation of geometric similarity. However, canids and ursids differ significantly ($p < .01$) from felids and mustelids in the relationship of body to tooth size. Whereas mass scales negatively with tooth length in canids and ursids, the opposite is true of mustelids and felids (Figure 10.5; Table 10.2). Thus, for a given M_1 length, canids and ursids are less massive than felids and mustelids.

Prediction of weight of extinct predators

The mass values for six Oligocene predators, all of which belong to extinct orders or families of carnivorous mammals, were estimated by using the total sample regression equations for mass against HBL and SKL (Tables 10.2 and 10.4). In three of the six species (*Hoplophoneus occidentalis, Dinictis felina, Daphoenus vetus*), the mass estimates based on HBL and SKL are relatively similar. However, in both *Hyaenodon* and the smaller *Hoplophoneus* species, the two estimates differ by as much as a factor of 4.5, with skull length producing a much higher estimate than head–body length.

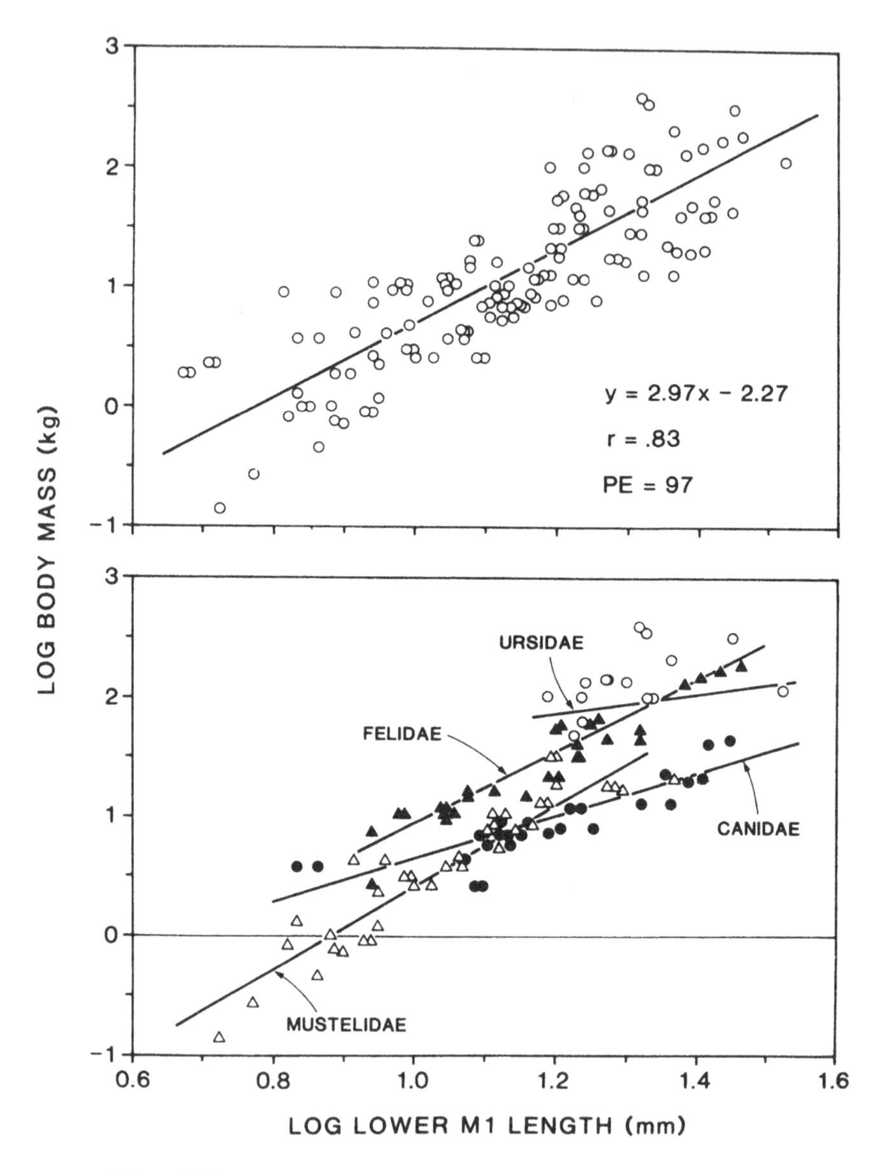

Figure 10.5. Plot of $\log_{10}$ body mass against $\log_{10}$ lower M_1 length for all individuals (top) and for four families (bottom). For symbols, species list, and regression statistics, see Figure 10.2.

Discussion

Comparison of the regressions

Although the regressions based on the total sample tend to have higher correlation coefficients than those based on the family or size subsets, their predictive accuracy tends to be relatively poor (Tables 10.2 and 10.3). As discussed in detail by Smith (1984), the correlation coefficient is not always a good indicator of the strength of a bivariate relationship when the range of values included in a regression is either relatively large or small. When the range of values is great and the slope far from zero, r values tend to be high, even if residuals are fairly large; similarly, when the range is small and slopes are lowered, r values tend to be low. In this paper, the span of body sizes is quite large, ranging from a minimum of 0.85 kg (*Mustela nivalis*) to a maximum of 396 kg (*Thalarctos maritimus*), about a 400-fold difference. Thus, it is not surprising that correlation coefficients tend to be large for regressions based on the total sample, and larger than those based on size-restricted subsets of the sample (Tables 10.2 and 10.3). Consequently, as pointed out by Smith (1981, 1984), the percent prediction error and standard error of the estimate are much better indicators of the correspondence between independent and dependent variables.

Based on percent prediction error (%PE) and percent standard error (%SEE), head–body length appears to be the best predictor of body mass when the total sample of 72 species is considered, followed by the two estimates of skull size, SKL and OOL, and then lower molar length (M_1L) (Table 10.2). The greater accuracy of head–body length relative to the other measures is not unexpected; it is a more comprehensive estimate of body size than the length of the skull or a tooth. Apparently, among carnivore species the relationship of body length to mass tends to be more similar than that of skull or tooth size to mass.

On average, the regressions for families are better predictors of body mass than those based on the entire sample, indicating that body mass and the skeletal measures scale somewhat differently among the families. For canids and ursids, head–body length is the best predictor, followed by skull size (SKL and OOL) and lower molar length (M_1L). However, in felids, M_1 length produces the most accurate predictions, whereas in mustelids, skull length is the best predictor. The relatively close association between body size and tooth length in felids may reflect their extreme dental specialization; cats have a reduced dentition with no postcarnassial molars and the M_1 is the largest lower cheek tooth, functioning primarily to slice meat (Ewer 1973; Van Valkenburgh 1989). Given its critical importance, M_1 length might be more closely tied to body mass in felids than in other, less dentally specialized predators.

Body mass predictions based on the size group regressions are more accurate than those produced by the total sample in all but the smallest (<6 kg) size group; prediction errors are one-half to two-thirds as large (Tables 10.2 and 10.3). Among the two largest size groups, head–body length is the best predictor, followed by the skull size measures (SKL, OOL) and tooth length (M_1L). The two groups of smaller species do not show this pattern; instead, prediction errors vary little among the four measures, except for M_1L which appears to have relatively high prediction errors for the <6-kg group.

The greater accuracy of the size group regressions relative to the total sample suggests that body size influences the scaling relationship between mass and the studied skeletal measures. The slopes of the size regressions are frequently much less than those for the total sample or family regressions and all fall below 3, the value expected for geometric similarity. Thus, within a given size class, mass tends to rise relatively slowly with increase in length of the measured characters.

Sources of error and variation

As discussed above, much of the inaccuracy of the total sample regressions can be explained as a result of differences in scaling among families or size groups. The total sample includes a wide variety of locomotor types, such as terrestrial cursors (canids), swimmers (otters), climbers (cats and procyonids), and diggers (badgers). Each of these locomotor types is associated with a suite of morphological adaptations, such as long limbs for running or short limbs for digging (Hildebrand 1982; Taylor 1989; Van Valkenburgh 1987). Consequently, one would not expect the relationship between body length (or other linear dimensions) and mass to be the same or very similar in all these species. For example, felids tend to be heavier than canids for a given head–body or skull length (Figure 10.2 and 10.3). No doubt, the greater robustness of felids reflects their climbing and prey-killing behaviors, whereas the slender build of canids reflects their cursorial abilities. In addition, the relatively short skull of felids is ideal for their typical killing behavior; it improves the mechanical advantage of jaw muscles for the single, forceful bite that is used to subdue prey (Emerson & Radinsky 1980; Ewer 1973; Radinsky 1981a,b).

The influence of within-group diversity on prediction errors is apparent in the families as well as across the whole sample. The canid regressions for mass against skull or body length resulted in much smaller prediction errors than those of the equivalent felid and mustelid regressions (Table 10.2). This suggests that the diversity of body shapes, and perhaps of locomotor types, is lower among canids than felids or mustelids. Mustelids include swimmers, climbers, and diggers; felids include runners and climbers; canids are almost all entirely terrestrial (Ewer 1973; Hildebrand 1952, 1954; Van Valkenburgh 1987).

In addition to the variance within the data set due to species differences in mass/length scaling, there are two other primary sources of variance, one due to methodological problems that can be overcome, and the other more complex and less tractable. Prediction errors could probably be lowered if body mass and length data were compared for the same individual, or averages of skeletal measures and mass based on large samples were used. In fact, the lowest prediction errors were usually associated with regressions that compared average values of body mass and head–body length (Table 10.2). Ideally, each species should be represented by matched weights and skeletal measurements for a sample of individuals rather than species mean weights and skeletal measurements for one or two individuals as was done here. Unfortunately, that sort of data is relatively rare and needs to be collected by field workers.

The second problem is the magnitude of intraspecific variability in body mass. Ranges of recorded mass values for a species are often quite large; of the 72 species, the mean ratio between maximum and minimum values was 2.6 (Figure 10.6). If the range midpoint is taken as the mean, then the average range spans ± 44% of the mean value. This is not likely to result from poor weight estimation in the field. Rather, it reflects a real phenomenon: Body mass is a highly variable characteristic of a species, probably more so than skeletal and dental measures. Individual size will vary daily in response to ingested food and water, and over longer time spans, owing to pregnancy, age, and season (e.g., pre- and post-hibernation). Given the wide range of body mass values recorded for most living species, we should not feel dismayed by the large prediction errors for estimates based on regressions of skeletal measures.

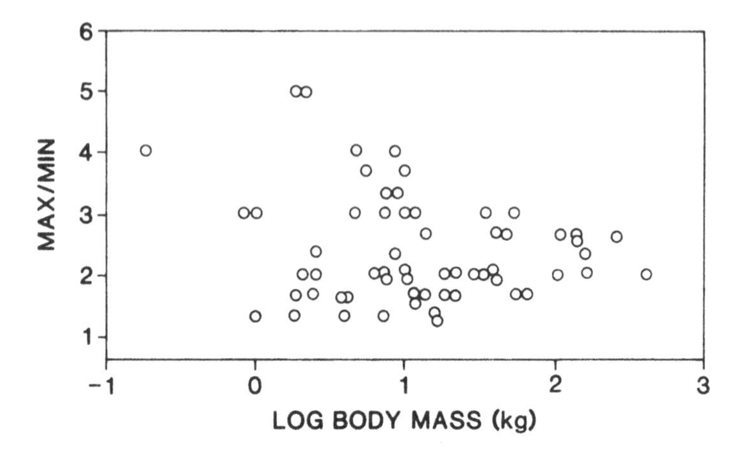

Figure 10.6. Plot of the ratio of maximum to minimum recorded mass for many species listed in Table 10.1 against body size. Species were excluded from the plot if data on maximum and minimum sizes were not available.

Interestingly, the span of recorded body mass values is particularly large for many small species, such as the skunks where the mean ratio between maximum and minimum mass values was 6.1 (Figure 10.6). Moreover, a comparison of prediction errors among the size groups shows that the smallest group always had the greatest error values (Table 10.3). The greater variability in recorded mass for small species might result from measurement error if the weighing instrument provided data at a relative coarse scale (e.g., to the nearest half kilogram). Estimation errors for small species result in very large percent prediction errors since a difference of 1 kg is a much larger percentage of a small than a large mass. Alternatively, it might reflect a real phenomenon that has ecological and evolutionary implications. For example, because of their relatively greater metabolic needs, small species need to ingest relatively larger quantities of food and water per day than do large species (Kleiber 1961). Thus, individual body mass undergoes proportionately greater fluctuations in small than large species.

Prediction of body mass for extinct predators

In general, I would recommend that, whenever possible, the mass of fossil carnivores be estimated from head–body length using the regression equation for its family. If, as is usually true, head–body length is not available, skull or occiput-to-orbit length is the next choice, and

Table 10.4. *Predicted weights for six Oligocene predators based on the total sample regressions for log_{10} body mass against log_{10} HBL and SKL*

Species	Wt. (kg) based on:	
	HBL	SKL
Hoplophoneus occidentalis	66	69
Hoplophoneus primaevus	13	19
Dinictis felina	20	17
Daphoenus vetus	36	33
Hyaenodon horridus	31	131
Hyaenodon crucians	9	25

Note: Abbreviations as in Table 10.2.
Source: Data for fossil species from Van Valkenburgh (1987).

finally M_1 length. In the case of felids, M_1 length may actually be the best choice if the cat appears similar to living felids in relative tooth size.

In three of the six Oligocene species examined, the body mass estimates predicted by skull length and head–body length differ little (*Hoplophoneus occidentalis, Dinictis felina, Daphoenus vetus*) (Table 10.4). However, in the remaining three, the estimate based on skull length exceeds that based on body length by at least a factor of 1.5. This lack of congruence between the two estimates suggests that the ratio of skull length to body length in these species differs greatly from most living carnivores. This appears to be particularly true for the large creodont, *Hyaenodon horridus*. The relatively large head of this species as compared to a modern carnivore, such as the dog, is obvious (Figure 10.7). Although the proportions of the smaller creodont (*H. crucians*) and sabertooth (*H. primaevus*) are not so extreme, they differ enough to produce marked inconsistencies in the weight estimates. In all three cases, it seems that skull length overestimates and head–body length may underestimate mass, although not to an equivalent degree. Given this, a better estimate of mass must lie somewhere between the two extremes, and thus my choice, for lack of a better alternative, would be to use an average of the two. Even if the estimate is twice or half the actual mean weight, it can still easily lie within the body size range of the species.

The Oligocene example shows that there are no easy rules for applying these regressions to fossil species. It would be foolish to use the same

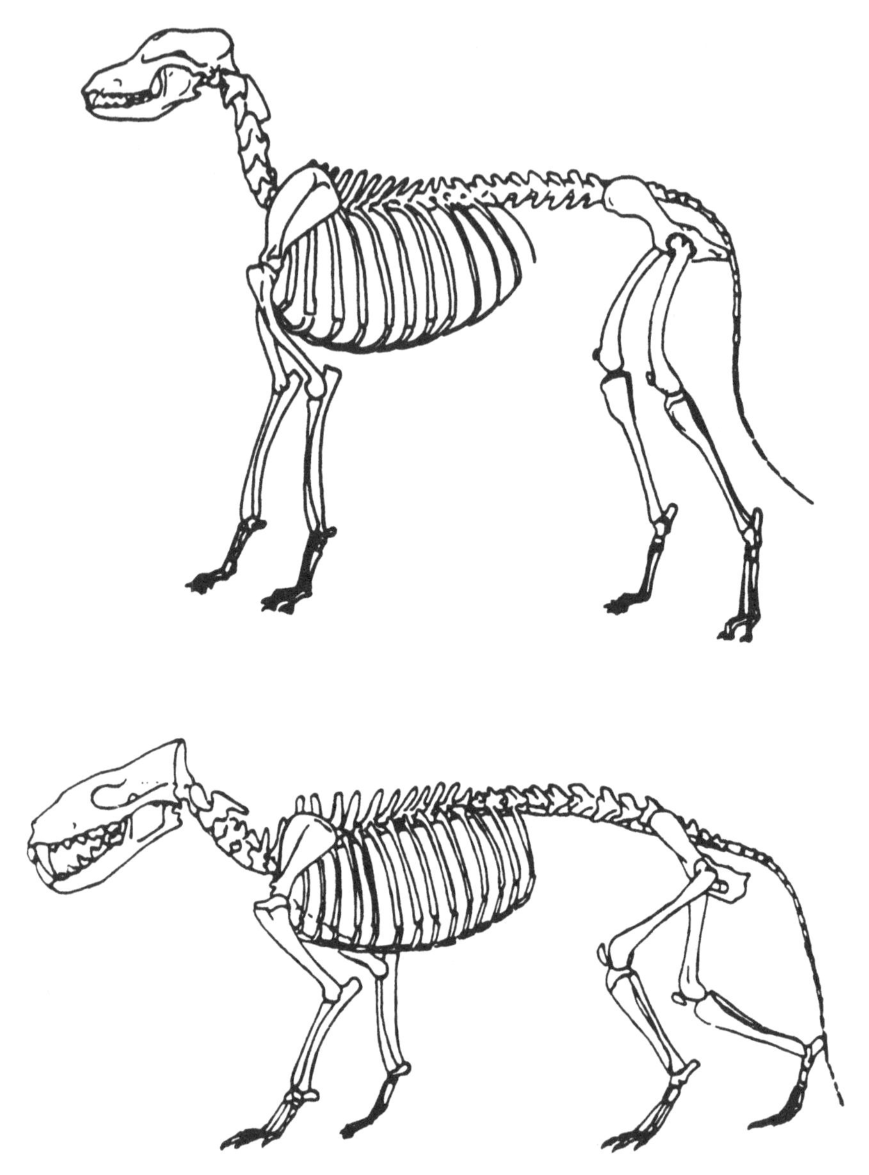

Figure 10.7. The skeleton of a domestic dog (top) and the Oligocene creodont, *Hyaenodon horridus* (bottom), drawn to similar head–body length to illustrate the relatively large skull of the creodont. (Dog from Ellenberger, Dittrich, & Baum 1956; creodont, after Scott & Jepsen 1941.)

measure or regression equation to estimate mass for all species. By comparing the results from different equations, important scaling differences between extinct and living species are highlighted. Although it would be pleasing to have a single, accurate regression that could be mindlessly applied to fossil species, it would hardly be realistic. Species vary in size and allometric patterns, and, as shown above, even the means provided for living species are probably poor estimates. We have to assume that mass estimates for extinct animals, as well as living ones, are a rough approximation of the actual mean. Nevertheless, these estimates can be used with confidence to explore paleobiological questions of ecological and evolutionary interest.

The sorts of paleobiological questions for which these equations are appropriate are those of relatively broad scope, such as the evolution of scaling differences over a fairly long timespan, such as relative brain volume, tooth size, or limb length within or across families (e.g., Jerison 1973; Radinsky 1978; Scott 1987). The differences must be great enough to exceed the error margins of the mass estimates. It would not, for example, be wise to attempt to document Hutchinsonian size-ratios (Hutchinson 1959) between similar species in fossil faunas based on estimated body mass. Indeed, given the variability of body size within species, it may be equally unwise to do the same sort of analysis on living communities based on species' mean weights. However, mass estimates produced by these equations can be used to look at general patterns of body size distribution within fossil and living communities (e.g., Andrews, Lord, & Nesbit-Evans 1979; Damuth 1982; Van Couvering 1980; Van Valkenburgh 1985, 1988, 1989) or examine associations between environment and body size (e.g., Janis 1982; Vrba 1980). For example, several studies have shown that savannah environments tend to include more species of relatively large size than do forest habitats (Coe 1980; Western 1980). Thus, in addition to sedimentological and paleobotanical data, the spectrum of body sizes within a fossil community can provide evidence of vegetation type. Mass estimates based on these regressions could also be used to study overall trends in the evolution of body size in carnivores such as has been done for other groups (e.g., MacFadden 1986; Martin 1986), and thus investigate the applicability of Cope's rule to the history of a clade (Stanley 1972).

Because body size is such an important ecological characteristic of carnivores (Gittleman 1985, 1986), future work should attempt to refine the predictions of body mass for fossil species. This could be done by comparing skeletal measurements and mass within the same individual whenever possible, and enlarging the sample sizes for each species. Moreover, the association between body mass and other skeletal measurements, such as the cross-sectional geometry of long bones and the area of articular surfaces (cf. Jungers, Ruff, this volume) should be explored. Cross-sectional properties of long bones reflect in vivo mechanical stresses due to loading (Currey 1984) and hence may be more closely correlated with body mass than are linear measurements of the skeleton. In fact, prediction errors based on the cross-sectional geometry of long bones of primates are lower than those of this study (%PE <15; Ruff, this volume). Clearly, this new and exciting approach needs to be applied to more mammals, especially those with diverse locomotor habits such as carnivores.

Despite the ease with which an animal's mass can be measured in the field, body size is a difficult parameter to estimate from bones and teeth. The problem of body size estimation is likely to be solved by a multifaceted approach that uses a variety of skeletal measures, both linear and cross-sectional, to tease apart the relationship between locomotor behavior and the scaling of mass. Undoubtedly, the investigation will produce many interesting asides concerning the evolution of body shape and size in mammals.

Acknowledgments

For assistance in collecting the data, I thank L. M. Perkins. For advice and critical review of the manuscript, I thank J. Damuth and R. K. Wayne. For access to specimens in their care, I thank the curators of the Division of Mammals, U.S. National Museum, Washington, D. C. The work was supported in part by a graduate fellowship from the American Association of University Women.

References

Andrews, P. J., Lord, J. M., & Nesbit-Evans, E. M. 1979. Patterns of ecological diversity in fossil and modern mammalian faunas. *Biol. J. Linn. Soc. 11*:177–205.

Blomqvist, L. 1978. Distribution and status of the snow leopard. In *International Pedigree Book of Leopards,* Panthera uncia, Vol. 1, ed. L. Blomqvist, pp. 6–21. Finland: Helsinki Zoo.

Burt, W. H., & Grossenheider, R. P. 1976. *A Field Guide to the Mammals,* 3rd ed. Boston: Houghton Mifflin.

Clark, J., Beerbower, J. R., & Kietzke, K. K. 1967. Oligocene sedimentation,

stratigraphy, paleoecology and paleoclimatology in the Big Badlands of South Dakota. *Fieldiana: Geol. Mem.* 5:1–158.
Coe, M. 1980. The role of modern ecological studies in the reconstruction of paleoenvironments in sub-Saharan Africa. In *Fossils in the Making,* ed. A. K. Behrensmeyer & A. P. Hill, pp. 55–71. Chicago: University of Chicago.
Currey, J. 1984. *The Mechanical Adaptations of Bones.* Princeton, N.J.: Princeton University Press.
Damuth, J. 1982. Analysis of the preservation of community structure in assemblages of fossil mammals. *Paleobiology* 8:434–446.
Davis, D. D. 1962. Mammals of the lowland rain-forest of North Borneo. *Bull. Singapore Nat. Hist. Mus.* 31:1–129.
Eisenberg, J. F., O'Connell, M. A., & August, P. V. 1979. Density, productivity and distribution of mammals in two Venezuelan habitats. In *Vertebrate Ecology in the Northern Neotropics,* ed. J. F. Eisenberg, pp. 187–207. Washington, D.C.: Smithsonian Institution Press.
Ellenberger, W., Dittrich, H., & Baum, H. 1956. *An Atlas of Animal Anatomy for Artists.* New York: Dover Publications.
Emerson, S., & Radinsky, L. 1980. Functional analysis of sabertooth cranial morphology. *Paleobiology* 6:295–312.
Ewer, R. F. 1973. *The Carnivores.* Ithaca: Cornell University Press.
Gittleman, J. L. 1985. Carnivore body size: ecological and taxonomic correlates. *Oecologia* 67:540–554.
Gittleman, J. L. 1986. Carnivore life history patterns: allometric, phylogenetic, and ecologic associations. *Am. Nat.* 127:744–771.
Grinnell, J., Dixon, J. S., & Linsdale, J. M. 1937. *Fur-bearing Mammals of California.* Berkeley, Calif.: University of California Press.
Guggisberg, C. A. W. 1975. *Wild Cats of the World.* London: David and Charles.
Hildebrand, M. 1952. An analysis of body proportions in the Canidae. *Am. J. Anat.* 99:217–256.
Hildebrand, M. 1954. Comparative morphology of the body skeleton in Recent Canidae. *Univ. Calif. Publ. Zool.* 52:399–470.
Hildebrand, M. 1982. *Analysis of Vertebrate Structure,* 2nd ed. New York: John Wiley.
Hutchinson, G. E. 1959. Homage to Santa Rosalia, or why are there so many kinds of animals? *Am. Nat.* 43:145–159.
Janis, C. M. 1982. Evolution of horns in ungulates: ecology and paleoecology. *Biol. Rev.* 57:261–318.
Jerison, H. 1973. *Evolution of the Brain and Intelligence.* New York: Academic Press.
Kingdon, J. 1977. *East African Mammals,* Vol. 3, Part A (Carnivores). London: Academic Press.
Kleiber, M. 1961. *The Fire of Life: An Introduction to Animal Energetics.* New York: Wiley.
Kruuk, H. 1972. *The Spotted Hyaena.* Chicago: University of Chicago Press.
Lekagul, B., & McNeeley, J. A. 1977. *Mammals of Thailand.* Bangkok: Bangkok Association for Conservation of Wildlife.

Leopold, A. S. 1959. *Wildlife of Mexico.* Berkeley: University of California Press.

MacFadden. B. J. 1986. Fossil horses from "Eohippus" (*Hyracotherium*) to *Equus,* Cope's law, and the evolution of body size. *Paleobiology 12*: 355–369.

Martin, R. A. 1986. Energy, ecology, and and cotton rat evolution. *Paleobiology 12*:370–382.

Mech, L. D. 1966. *The Wolves of Isle Royale.* Fauna of the National Parks of the United States, Fauna Series No. 7, pp. 1–210. Washington, D.C.: U.S. Dept. of the Interior.

Moeller, H. 1970. Vergleichende untersuchungen zum evolutionsgrad der gehirne grober raubbeutler (*Thylacinus, Sarcophilus* und *Dasyurus*) I. Hirnegewicht II. Hirnform unf Furchenbild. *Z. Zool. Syst. Evolutionsforsch. 8*:69–80.

Nowak, R. M., & Paradiso, J. L. 1983. *Walker's Mammals of the World,* 4th ed. Baltimore: Johns Hopkins Press.

Prater, S. H. 1965. *The Book of Indian Mammals,* 2nd ed. Bombay: Bombay Natural History Society.

Radinsky, L. 1978. Evolution of brain size in carnivores and ungulates. *Am. Nat. 112*:815–831.

Radinsky, L. 1981a. Evolution of skull shape in carnivores. 1. Representative modern carnivores. *Biol. J. Linn. Soc. 15*:369–388.

Radinsky, L. 1981b. Evolution of skull shape in carnivores. 2. Additional modern carnivores. *Biol. J. Linn. Soc. 16*:337–355.

Schaller, G. B. 1972. *The Serengeti Lion.* Chicago: University of Chicago Press.

Schaller, G. B. 1983. Mammals and their biomass on a Brazilian ranch. *Arquiv. Zool. 31*:1–36.

Scott, K. M. 1987. Allometry, weight prediction and variation in the skeleton of living and fossil equids. *J. Vertebr. Paleontol. 7*:27A (abstracts).

Scott, W. B., & Jepsen, G. 1941. The mammalian fauna of the White River Oligocene. *Trans. Am. Philos. Soc. 28*:747–980.

Smith, R. J. 1981. Interpretation of correlations in intraspecific and interspecific allometry. *Growth 45*:291–297.

Smith, R. J. 1984. Allometric scaling in comparative biology: problems of concept and method. *Am. J. Physiol. 246*:R152–R160.

Stanley, S. M. 1972. An explanation for Cope's rule. *Evolution 27*:1–26.

Taylor, M. E. 1989. Locomotory adaptations of carnivores. In *Carnivore Behavior, Ecology, and Evolution,* ed. J. L. Gittleman, pp. 382–409. Ithaca, N.Y.: Cornell University Press.

Van Couvering, J. A. H. 1980. Community evolution in East Africa during the Late Cenozoic. In *Fossils in the Making,* ed. A. K. Behrensmeyer & A. P. Hill, pp. 272–298. Chicago: University of Chicago Press.

Van Valkenburgh, B. 1985. Locomotor diversity in past and present guilds of large predatory mammals. *Paleobiology 11*:406–428.

Van Valkenburgh, B. 1987. Skeletal indicators of locomotor behavior in living and extinct carnivores. *J. Vertebr. Paleontol. 7*:162–182.

Van Valkenburgh, B. 1988. Trophic diversity in past and present guilds of large predatory mammals. *Paleobiology 14*:155–173.
Van Valkenburgh, B. 1989. Carnivore dental adaptations and diet: a study of trophic diversity within guilds. In *Carnivore Behavior, Ecology, and Evolution,* ed. J. L. Gittleman, pp. 410–436. Ithaca, N.Y.: Cornell University Press.
Van Valkenburgh, B., & Ruff, C. B. 1987. Canine tooth strength and killing behaviour in large carnivores. *J. Zool. 212*:379–397.
Vrba, E. S. 1980. The significance of bovid remains as indicators of environment and predation patterns. In *Fossils in the Making,* ed. A. K. Behrensmeyer & A. P. Hill, pp. 41–54. Chicago: University of Chicago Press.
Walker, E. P. 1964. *Mammals of the World.* Baltimore: Johns Hopkins Press.
Western, D. 1980. Linking the ecology of past and present mammal communities. In *Fossils in the Making,* ed. A. K. Behrensmeyer & A. P. Hill, pp. 41–54. Chicago: University of Chicago Press.
Wilkinson, L. 1986. *SYSTAT: The System for Statistics.* Evanston, Ill.: SYSTAT.

11

Problems with using fossil teeth to estimate body sizes of extinct mammals

MIKAEL FORTELIUS

It is always possible to get an approximate idea of body size from tooth size within any reasonably homogeneous group, such as the mammals. In the case of morphologically similar, closely related species, quite precise relative sizes can also be determined, no matter how imprecise the estimates of absolute size. For many purposes this has been, and still is, enough. This is worth remembering.

However, the search for simple explanations has brought body size into the focus of theoretical interest in many branches of biology (e.g., Eisenberg 1981; Peters 1983; Schmidt-Nielsen 1984; Maiorana, this volume). Application of body size theory in paleobiology obviously requires some consistent and reasonably accurate means of producing body sizes from fossils. The only adequate method available at present seems to be estimation using allometric equations, and often the only parts of the animal that are preserved in sufficient numbers to be useful are teeth. Much of body size estimation is therefore necessarily based on allometric studies of teeth.

In this paper I shall concentrate on some of the considerable problems involved in allometric estimation generally, and particularly in body size estimation from cheek teeth. I shall do so by first discussing general problems and then presenting and discussing real (if faulty) data for recent ungulates. Some relevant aspects, such as the issue of probabilism and determinism, have been much discussed already and will not be mentioned here (for a useful collection of papers, see Saarinen [1982]).

In *Body Size in Mammalian Paleobiology: Estimation and Biological Implications,* John Damuth and Bruce J. MacFadden, eds.

My emphasis is on the biological relationships rather than on methodology.

Theory: hope and despair

Allometry and estimation

A very useful application of allometric relationships has been to treat them as "criteria of subtraction" (Gould 1975), as null hypotheses about what organisms of any group are expected to be like on average. Species that behave unexpectedly are singled out as worthy of further study. As far as body size estimation goes, two aspects seem particularly relevant. First, allometric relationships, conceived of as hypotheses, are propositions with some domain, some set of circumstances to which they are applicable (Van Valen 1976, 1982). Different samples of living species give different allometric equations, and the trick is to combine fossils and equations correctly. The most optimistic theoretical justification for body size estimation is the assumption that groups of organisms exist which satisfy different sets of functional similarity criteria. Allometric equations derived from living representatives of such groups, if they exist, should represent true scaling (*sensu* Prothero 1986), and should be reliably applicable to fossil representatives of the same groups.

Secondly, real biological variation (scatter about the regression line not resulting from measurement error) can be studied if the data are good enough. Thus there is hope that allometric techniques might in fact also be helpful in identifying the hypothetical functionally similar groups. I shall return to this idea below. In the following I will deliberately ignore the immense difficulties of inadequate data and/or incorrect technical procedures, and concentrate only on the conceptual problems. For recent discussions of methods relevant in this context I refer to Prothero (1986), Smith, German, and Jungers (1986), several papers in Jungers (1985), and to Jungers (this volume).

Using allometric equations to estimate unknowns thus assumes (1) that empirically established relations are more than descriptive, and (2) that their domains can be determined with sufficient accuracy. In one way or another, these assumptions have been the subject of most discussions of body size estimation published in recent years. Schmidt-Nielsen (1984), for example, was very skeptical about the use of allometric equations for extrapolation beyond the range covered by the source data. However, the difference between extrapolating beyond a range and interpolating within it may be less crucial than determining the domain of the equation, which appears to be the real issue here. In many cases we may be assuming a domain that is too wide. For any set of similarity criteria our data represent only a small and possibly biased sample of the species that have actually, or could have potentially, satisfied those criteria. Often we do not even have data for all such species that exist now.

Using allometric equations to estimate body masses also means performing an averaging operation, that is, eliminating variation due to anything other than the relationship that is being used. An ecological model of a Miocene fauna based on an estimated body-size distribution will inevitably be a trivialized version of the real thing, populated only by species of average body mass (relative to whatever independent variable was used). We may even trivialize the present in this way, if we exclude outliers to produce neater equations (which may well be the correct technical procedure in most cases). There is an essentialist (or idealist or whatever) element in this which I find vaguely disturbing; we should not miss the music of ecology (Simberloff 1982:85).

Furthermore, as with all actualistic procedures applicability decreases as one encounters cases less and less similar to Recent ones. There is a kind of functional "pull of the Recent" involved: One may never know of interesting systematic differences between fossil and living species eliminated from the data by the assumption that they do not exist. These are real problems, but they can be tackled, as shown by Damuth (this volume). Indeed they should, since we hope to study the past, not a warped reflection of the present!

We cannot, of course, use the same allometric relationship to estimate something and to identify deviant species, but we can use several different ones and compare the results. If the cross sections of weight-bearing limb bones are the best estimators of body mass, as seems to be the case (Ruff; Jungers, this volume), then these should be used whenever available. It should be possible in principle to construct proportion diagrams of different species of mammals (such as log ratio diagrams, based on body parts or perhaps even on derived mass estimates). These could then be used for comparison and for reference in the numerous cases when only teeth are available. It might also be possible to use relationships that manifestly are *not* highly dependent on body mass to identify functional groups independently of size. Thus there is potential use for both "good" and "bad" equations in the sense of Smith (1985).

The idea of laws that define the construction of animals is, of course, an old one. Cuvier had high hopes about this:

> In a word, the shape of the tooth implies the shape of the condyle, that of the scapula, that of the nails, just as the equation of a curve implies all its properties; and just as by taking a property separately as the basis of a particular equation, one would find again and again both the ordinary equation and all the other properties, similarly the nails, the scapula, the condyle, the femur and all the other bones taken separately, give the tooth, or give each other; by beginning with any of them, someone with a rational knowledge of the laws of organic economy could reconstruct the whole animal. (Cuvier 1834:181, translated by Buffetaut 1987:56)

Later authors have generally been less optimistic, but Rensch, for example, did express a similar idea:

> Just as a physicist is able to make predictions on the basis of known laws, we can *predict* what the proportions of the organs would be, for example, in a rodent (or would have had to have been in a fossil rodent, of which only fragments are known), which is *smaller* than any of the forms known so far. In view of the fact that we are not dealing with laws, but rather with rules for complicated living beings, such predictions are not absolutely certain but will be true with a probability of about 70–85%. (Rensch 1952:391, original emphasis; my clumsy translation does poor justice to the elegance of the original.)

However, although the idea is not new, we may be in a somewhat better position to apply it to quantitative data than was possible previously.

Unexplained, purely empirical relationships can be used for estimation and have been so used. It is, however, extremely difficult to justify this practice theoretically, since there is no way to determine their domains. Without explanatory models allometric relationships thus tend to remain rather isolated and barren. Sometimes explanatory models may be derived from first principles (MacMahon 1973), but more commonly they are "mixed" (Prothero 1986), including empirical components. Whether systems of only empirical relationships should be called models is perhaps a moot point; certainly they are not true explanations. Such combinations may be very useful heuristically, however, in relating numbers to biological reality. Indeed, even a partly qualitative model is a great deal better than no model at all. The problem of "metabolic" scaling of mammalian postcanine teeth illustrates this point.

The biology of tooth size–body size relationships

As noted by at least Van Valen (1960), Pilbeam and Gould (1974; Gould 1975), and a great many authors since, tooth size must in some way be related to food processing needs and ultimately to metabolic rate. Van Valen (1960) considered crown height and longevity, a subject that I have discussed elsewhere (Fortelius 1985, 1987). Pilbeam and Gould (1974) were concerned with short-term requirements and occlusal area, an issue obviously related to tooth size–body size relationships. According to these authors, occlusal area should scale directly with metabolic rate as body mass raised to approximately 0.75. Since isometric scaling of an area to a volume (mass) gives an exponent of 0.67, the hypothesis of "metabolic scaling" implies positive allometry of tooth size to body size. This is now well known to conflict with empirical data, which strongly suggest overall approximate isometry (e.g., Creighton 1980; Fortelius 1985; Gingerich, Smith, & Rosenberg 1982; Kay 1975, 1978; see also the other dental papers in this volume).

The solution of this paradox is potentially simple, although good data are still lacking (Fortelius 1985). The solution, however, is strictly applicable only to general relationships, perhaps at the level of superfamily and above. Lower-level relationships are more complicated, and, unfortunately, often of greater interest to paleobiologists. I will treat the general case first, and discuss problems arising from special cases in the next sections.

Animals chew food of a certain volume at a certain rate. For theoretical purposes it is probably justifiable to think of food processing as a process whereby each chew results in the final comminution of a volume of food. In reality, of course, all the food is circulated in the mouth and chewed repeatedly. If all else is assumed equal, this theoretical volume processed per chew may reasonably be assumed isometric to mouth volume and the "effective volume" between tooth rows. These may be difficult to define in practice, but should be more or less isometric to tooth "volume" (area raised to 3/2). Heart rate and breath rate, and a host of other physiological rates, are proportional to body mass raised to an exponent close to -0.25 (e.g., Peters 1983). If chewing rate is also assumed to scale as body mass to -0.25 and processed volume for one chew as body mass to 1.00, then volume processed per unit time will scale as body mass to 1.00 − 0.25 = 0.75, with metabolic rate, if tooth size is isometric to body size. That is, isometry will in fact represent true metabolic scaling.

How realistic are the assumptions above? I know of no useful data to assess the validity of the assumption of isometry of volume processed per chew to tooth size, but both White (1959) and Wolpoff (1985) took it for granted. It is certainly more plausible intuitively than the suggestion that processing is a direct function of area, which apparently ignores the presence of food altogether. It must be possible to approach this question experimentally, either as such or indirectly via predictions that follow from the hypothesis. For example, if the mouth behaves as an isometric system, then particle size should do so as well. Hyraxes should ingest and swallow particles about ten times smaller in diameter than should rhinoceroses eating the same ideal (nongrainy) food. (Exact testing is, of course, impossible, since all else is never equal.)

For chewing rate the situation is not much better. The study of Hendrichs (1965) is available, but the data do not appear to be very accurate (Fortelius 1985). Allometric equations calculated from his data for chewing cycle duration (the reciprocal of chewing rate), give major axis slopes of 0.13–0.16 for different categories (Fortelius 1985). Rather inadequate data collected by myself give slopes of 0.19–0.23 (Fortelius 1985). More accurate data are needed for any degree of confidence. (It is true but perhaps unimportant that my data give 95% confidence limits including the predicted value of 0.25 in all cases.)

Nevertheless, it is reassuring that occlusal stress must at least be nearly independent of body size. This is because chewing muscle mass is nearly isometric to body mass (at least for primates [Cachel 1984]). Therefore both muscular cross sections and occlusal areas will scale approximately as body mass to 0.67. Because muscular force is approximately proportional to cross section, occlusal stress (force/area) will then be a dimensionless constant, at least in principle (but see Smith et al. [1986] and Prothero [1986] for methodological difficulties). All else being equal, geometrically similar teeth will be functionally similar regardless of size. (The main problem here is food, which comes in discrete sizes.)

The model above is unnecessarily simplified, indeed unrealistic. Animals do not spend all their time chewing at an even rate. It is thus perfectly possible that the proportion of time spent feeding has a component dependent on body mass. This could easily be compensated for by a reciprocally scaled chewing rate, to produce a long-term realized chewing rate proportional to body mass to −0.25. If chewing rate scales as body mass to about −0.20, as seems to be a reasonable guess at the moment (see above), then we should expect either the proportion of total time spent chewing to scale as body mass to about −0.05, or volume comminuted per chew to scale as body mass to about 0.95. Larger animals should either spend a somewhat smaller proportion of their time feeding, which is contradictory to the anecdotal data known to me (Janis 1976; McKay 1973), and might be worth investigating. It may be more likely, however, that tooth size is in fact slightly negatively allometric to body size (see example below and Table 11.1), resulting in a slight negative allometry of the volume comminuted per chew.

But this is still unnecessarily simple. We may envision a direct compensatory relationship between long-term chewing rate and volume of food processed per chew (influenced by differences in food properties, tooth shape, and relative tooth size). Real metabolic rates also vary, as do energy content of foods. There is in fact considerable regularity in this (McNab 1980, 1986), but insufficient for accurate predictions for individual species, and certainly so for fossil ones. As shown by many captive and domestic mammals, the proximate regulatory feedback systems, such as between blood-sugar level and feeding behavior, are also perfectly capable of making individuals function, even when the diet is altered radically. In view of how chaotic things might potentially appear and still form part of a basically simple system, it is somewhat reassuring, at least from the point of view of body size estimation, that relatively simple and strong relationships can, in fact, be demonstrated.

There are no evident contradictions in this "model" of the observed tooth size–body size relationship, and it is even possible to generate testable predictions from it, at least in principle. In its weakest form, it simply states that animals function, which is true but uninteresting. Successively stronger versions are more interesting, but may be too simplified to be of much use. This is an empirical question, and speculation in the absence of better data is futile. The relationship itself is, of course, still unexplained. Any explanation of the scaling of metabolic rate, or some other basic variable, will automatically extend to this particular case, too.

One should not be too greedy. Saying that tooth size is isometric to body size within most taxa at, perhaps, superfamily level or higher in *living mammals* is simply stating an empirical fact (and also ignores, for convenience, the variation observed). Saying that this should *always* theoretically be the case is setting a frightening number of physiological and behavioral factors equal, factors that cannot be determined for fossil

Table 11.1. *Correlation and major axis regression of dental variables on body mass for various groups of ungulates (and condylobasal skull length on body mass and dental variables on skull length for a sample of bilophodont and bunodont species including* Tapirus indicus, Tapirus *sp. (South American),* Sus scrofa, Potamochoerus porcus, *and* Tayassu tajacu

Group/variables	*N*	*r*	*b*	95%CL (*b*)	*a*
Nonselenodonts					
BM:TRUA	20	.979	0.649	0.584–0.719	0.115
TRLA	20	.977	0.618	0.553–0.688	0.087
M^2A	20	.973	0.668	0.592–0.750	−0.699
TRUL	20	.982	0.308	0.279–0.338	0.500
TRLL	20	.971	0.308	0.271–0.347	0.491
M^2L	20	.972	0.324	0.286–0.364	−0.302
Selenodonts					
BM:TRUA	23	.968	0.580	0.513–0.651	0.281
TRLA	22	.964	0.569	0.498–0.645	0.183
M^2A	22	.969	0.600	0.523–0.682	−0.505
TRUL	24	.966	0.289	0.255–0.323	0.512
TRLL	22	.962	0.269	0.234–0.305	0.617
M^2L	22	.958	0.306	0.264–0.350	−0.264
Bovids					
BM:TRUA	13	.983	0.623	0.552–0.699	0.029
TRLA	13	.973	0.618	0.525–0.718	−0.073
M^2A	12	.980	0.649	0.561–0.746	−0.769
TRUL	13	.976	0.299	0.258–0.340	0.452
TRLL	13	.973	0.299	0.253–0.346	0.466
M^2L	12	.959	0.324	0.258–0.392	−0.342
Rhinos + hyraxes					
BM:TRUA	8	.998	0.654	0.614–0.696	0.162
TRLA	8	.998	0.623	0.580–0.666	0.047
M^2A	8	.996	0.681	0.624–0.741	−0.648
Bilophobunodonts					
BM:TRUA	5	.989	0.556	0.409–0.725	0.606
TRLA	5	.990	0.427	0.320–0.543	1.150
M^2A	5	.985	0.516	0.338–0.698	0.075
CBL	5	.983	0.226	0.148–0.307	1.368
CBL:TRUA	5	.996	2.442	2.090–2.910	−2.716
TRLA	5	.998	1.872	1.665–2.122	−1.394
M^2A	5	.999	2.267	2.126–2.424	−3.009

Note: See Table 11.2 for species included.
Abbreviations: *N*, sample size; *r*, correlation coefficient; *b*, major axis slope; CL (*b*), confidence limits for the slope; *a*, *y*-intercept. Other abbreviations as in Figures 11.1–11.3.

taxa. There is probably no meaningful parameter that one might term "functional tooth size" and that would relate to metabolism in a simple manner. Nevertheless, empirical data suggest that teeth will probably give useful if crude estimates in most cases. But unusual animals do occur, and for them estimates based on the average relationship may be wildly incorrect.

In the following sections I shall discuss practical aspects of body size estimation from teeth, given that internally consistent, empirical, and probabilistic relationships is the best basis available.

The use of dental morphology

To produce a body mass to accompany a fossil tooth, one will not usually want the general mammalian relationship as a basis for estimation, since this would give a very approximate estimate at best. It is accordingly necessary to decide on some suitable reference population of living species on which to base the estimate – that is, determine an equation within the domain of which the tooth lies. The only clue may be the shape of the tooth itself.

There are two main ways in which dental morphology is important. A trivial one is that the ratio of crown base area (measured) to effective occlusal area (implied in theoretical analysis) is different in different morphological groups. The other is that animals with similarly shaped teeth tend to eat similar foods, and so should have similar relations between their teeth and bodies. For the purpose of fueling the fire of life all flesh is definitely not grass: Mammalian meat contains about 30 times as much energy per unit mass as do vegetables such as lettuce and celery (Anonymous 1985). Much smaller differences may conceivably affect the relationship enough to cause serious errors in the estimates.

An attractive thought that is partly false is that tooth size, chewing rate, and metabolic rate are all closely determined by body size, for reasons that are not well understood (Fortelius 1988). Tooth shape, on the other hand, is independent of body size. Dietary adaptation will therefore be reflected in tooth shape but not in the other variables.

If this were really true, it would make things very much easier, but of course it is not. For one thing it is impossible to make a sharp distinction between tooth size and tooth shape. Furthermore, nothing is strictly body-size dependent, as metabolic rates, digestive physiology, and other factors are also influenced by adaptation. Even so, it may be true enough to be helpful, and I shall argue below that tooth length measures primarily size whereas tooth width includes a strong shape component. This gives some practical meaning to an otherwise meaningless abstraction. Thus there is some theoretical justification for the common practice of interpreting occlusal morphology, or relative tooth width, in terms of food properties.

Two kinds of food properties may be distinguished: nutritive and mechanical. The former will affect the amount to be comminuted per chew (mainly tooth area), the latter how comminution will be achieved (mainly occlusal topography). (For a somewhat different and more detailed classification see Lucas, Corlett, & Luke 1986.) Both will influence tooth shape, conceivably in sometimes conflicting ways (see, for example, the discussion of anisodonty in Fortelius 1985). But we are still far from understanding how most foods are actually comminuted, and it also seems unlikely that occlusal morphology will usually reflect the average food of an animal; especially demanding constituents will probably have a disproportionate influence.

However, the theoretical argument of Lucas et al. (1986) that the chewing of large volumes requires above all a widened tooth row is both plausible and encouraging. It can obviously be utilized to support the idea that length is much less affected by adaptation than is width, and so can be used as a relatively pure size measure. The data presented later in this paper also support this hypothesis, as do the corresponding data of Damuth (this volume) and Janis (this volume).

If this seems vague or farfetched, remember that the original assumption was that dietary adaptation is reflected only in dental morphology, but not in physiology or behavior. Since this is documented to be untrue to some unknown degree, our theoretical foundations appear rather shaky. What about empirical knowledge?

Some groups are much less problematical than others. Bovids and cervids, and indeed the whole of the selenodont artiodactyls, are probably best treated by comparison with the relevant recent taxon. Shared dental morphology and almost certainly digestive physiology make these groups very homogeneous. Differences exist, of course, but correlations between body mass and dental measurements are high for living species (Table 11.1; see also Janis, this volume; Kay 1975; Lucas 1980). Other large homogeneous groups might be found among the rodents. Carnivores, in the sense of specialized terrestrial meat eaters, may also be a relatively homogeneous group, although reduction of the dentition anterior or posterior to the carnassials in different groups causes complications. Carnivores also eat food of very high energy content and spend much less time chewing than do herbivores, which could mean that tooth size–body size relationships are more relaxed. The results of Creighton (1980) and Van Valkenburgh (this volume) lend some support to this idea.

Rhinocerotids and tapirids would probably be easy if more species survived; as it is, estimates are unreliable (Fortelius 1985; Prothero & Sereno 1982). The same is true of most ungulate families except bovids and cervids. The "archaic" ungulates, in particular, constitute a real problem, since *no* living species with a similar dental morphology exist. What should be done about pantodonts, dinocerates, embrithopods, most tapiroids and proboscideans, and so on? And what about the South American extinct orders? For these cases even a limited functional understanding would be helpful.

In the next section I will outline some relationships that may assist in the selection of suitable groups for comparison. There are probably good reasons not to extrapolate between ecologically, behaviorally, and/or physiologically very different groups, even if occlusal morphology is similar (e.g., kangaroos, manatees, and tapirs). On the other hand, there are no compelling reasons why the groups should be monophyletic, or even consist of closely related taxa. I concentrate on ungulates which is the group most familiar to me.

Last resorts

An obvious source of trouble is anisodonty: Should one use upper or lower teeth when comparing forms with very different upper and lower teeth? As will be seen below, the answer is not simple, for anisodonty affects plots regardless of whether upper or lower teeth are compared. However, anisodonty, expressed quantitatively as the ratio of upper to lower widths of occluding, serially homologous teeth, also discriminates surprisingly well between dental morphological categories (Fortelius 1985). Presumably this is because it reflects, in a very direct way, the degree of lateral movement during the power stroke, which in turn must be a very fundamental variable of chewing mechanics. There is much overlap between adjacent categories, and morphology itself is, of course, a much more sensitive tool for assigning teeth to morphological groups. Nevertheless, it may be helpful to study this index in the case of forms with no good living analogues, in order to decide which of the living groups would be most suitable for comparison. The hope embodied in this suggestion is that we would be comparing similar mouths even if teeth differ somewhat.

For example, there is a tendency to describe the teeth of certain early ungulates (many "condylarths," and, among others, the equid *Hyracotherium*) as bunodont, but these forms actually have indices of anisodonty of about 1.5, whereas truly bunodont forms (pigs, gomphotheres, hominoids, bears) have indices of about 1.0–1.3. This suggests that such early, morphologically primitive ungulates might compare better with living selenodont artiodactyls which have indices of about 1.4–1.6. Simple selenodont teeth can be derived from such primitive ungulate teeth by increasing cusp height and very little else.

We are very short of bilophodont large ungulates today; there are only five species of tapir, four of which are closely related. In contrast, a great number of fossil bilophodont species are known. What would be the best group(s) to include in a reference population, in addition to tapirs, to increase size range and sample size? The Perissodactyla might be a natural group to choose. But living horses are very specialized plagiolophodonts, and rhinoceroses are equally specialized ectolophodonts, with the high anisodonty that goes with a dominant ectoloph cutting edge (for a discussion of these and other terms, see Fortelius 1985). If we decide to pay any attention at all to occlusal morphology, we should probably avoid using these taxa.

The living peccaries are small bilophodont forms and might be suitable. Bilophodont and bunodont forms have similar anisodonty values, and intermediate morphologies as well as pure states are seen in listriodont and other pigs. This suggests that both morphologies function in the same type of mouth, including, of course, jaw design, musculature etc. Perhaps pigs, excluding the obviously specialized *Hylochoerus* and *Phacochoerus* (e.g., Kingdon 1979), would also be a good choice? An obvious complication here is that tapirs are hindgut fermenters, whereas pigs have a limited foregut fermentation.

What about trilophodont tapiroids, dilambdodonts such as brontotheres and chalicotheres, or brachydont horses? Nothing very suitable is available here. Perhaps a combination of hyraxes, tapirs, and rhinoceroses would be the best approximation? Even for rhinocerotoids the five living species of rhinoceros are not very useful since the correlation between tooth size and body mass is rather low, so something like the above might do better. On the other hand, the high correlation resulting from an extended size range may give false confidence. In cases such as these, nondental estimates are badly needed.

The Proboscidea is another difficult case. Only two living species exist, and both have the specialized elephantid dentition. How should one estimate the body sizes of zygolophodont and bunodont proboscideans, or of such beasts as stegodonts? The answer may be that we should avoid dental estimates altogether, or, perhaps, use some very generalized "hyrax-to-elephant" relationship, with duly widened confidence limits for the estimates (see Roth, this volume). Extrapolating from a bunodont–bilophodont (peccary–pig–tapir) reference population might work for the "mastodons," but the difference in size is worrying, particularly since even the much smaller hippopotamus plots far off the peccary–pig–tapir line (as shown below).

Dental wear must eventually provide useful information about diet, despite the complications that have dashed the originally high hopes. Wear should be studied at many scales (both micro and macro), and studies should include effects resulting from dental morphology and structure as well as from foods. Unusual wear patterns might help to identify opportunistic species that have abandoned their inherited diet (to which they are morphologically adapted). However, the fact remains that we are still very much at the mercy of empirical relations which we do not understand at all well. We should compare like with like, but we do not really know which kind of similarity is most important.

Example: the living nonselenodont ungulates

In this section I shall present an example and discuss it in relation to the theoretical arguments outlined above.

Figures 11.1 and 11.2 show bivariate plots of log-transformed data (various tooth size measures against body mass) for a sample of ungulates. Plotted are various bunodont and lophodont species, but the selenodonts were omitted since they would have obscured the patterns (for the relevant parameters, see Table 11.1). All lines shown are major axes. The data base is the same as in Fortelius (1985), with most species represented by one to three individuals and body masses taken from the literature (see Table 11.2). Needless to say, the interpretations must be very tentative. It is reassuring, however, that both Damuth and Janis (this volume) have obtained a highly similar pattern from at least predominantly independent data.

An obvious result already referred to above is that tooth row length is a better predictor of body mass than is tooth row area. This is valid for both upper and lower tooth rows (cf. Figures 11.1A,B and 11.2A,B), and, indeed also, for single teeth (Figures 11.1C, and 11.2C). To me at least, this suggests that length comes close to being a true size variable, whereas width includes a major component of shape and therefore dietary adaptation.

Another result is that different groups show consistent differences. The line for the Bovidae, shown in all plots, always has a lower intercept and a slightly lower slope than the line for the sample shown. For upper tooth measurements which include width, rhinoceroses and hyraxes plot above the general line. The line for these highly anisodont forms is shown in Figure 11.1A and C.

All these relationships are nearly isometric (Table 11.1), but there are also remarkable regularities here. First, the best estimate is below isometry in all cases except for M^2A in rhinoceroses + hyraxes, although significantly so only for the sample of selenodont artiodactyls and the special case of bilophodont and bunodont forms discussed below. Such weak negative allometry would be consistent with a chewing rate scaling at an exponent somewhat higher than -0.25, as discussed in the previous section. Secondly, upper tooth-row area always has a higher slope than lower tooth-row area, perhaps because the most anisodont forms also happen to be the largest ones. This difference is not observed for tooth row length – another point in favor of using length measurements for body size estimation.

An intriguing but admittedly uncertain result is that the bilophodont–bunodont group which was singled out on a priori grounds in the preceding section indeed does show a very good fit to a line which, for dental area, has a much lower slope than the more general lines. This is the dashed line in Figure 11.1A–C, based on *Tapirus indicus, Tapirus* sp. (South American), *Sus scrofa, Potamochoerus porcus*, and *Tayassu tajacu*. (This is true: I wrote the theoretical part of this paper before I made any plots, and this rather striking result was a surprise to me.)

Figure 11.1 (*facing page*). Log-transformed bivariate plot of dental area measurements (mm^2) against body mass (BM, g) for living, nonselenodont ungulates (for species see Table 11.2). All lines are major axes (for parameters see Table 11.1). The heavy line is for all the data points shown; the lower thin line is for bovids (not shown); the upper thin line is for hyraxes and rhinos only; the dashed line is for tapirs, pigs (excluding the specialized *Phacochoerus* and *Hylochoerus*), and the peccary *Tayassu tajacu*. Symbols: E = equid, Hi = hippopotamus, Hy = hyracoid, R = rhinocerotid, S = suid, Tp = tapirid, Ts = tayassuid. (A) TRUA = length of upper postcanine tooth row × width of M^2. (B) TRLA = length of lower postcanine tooth row × width of M_2. (C) M^2A = length × width of M^2.

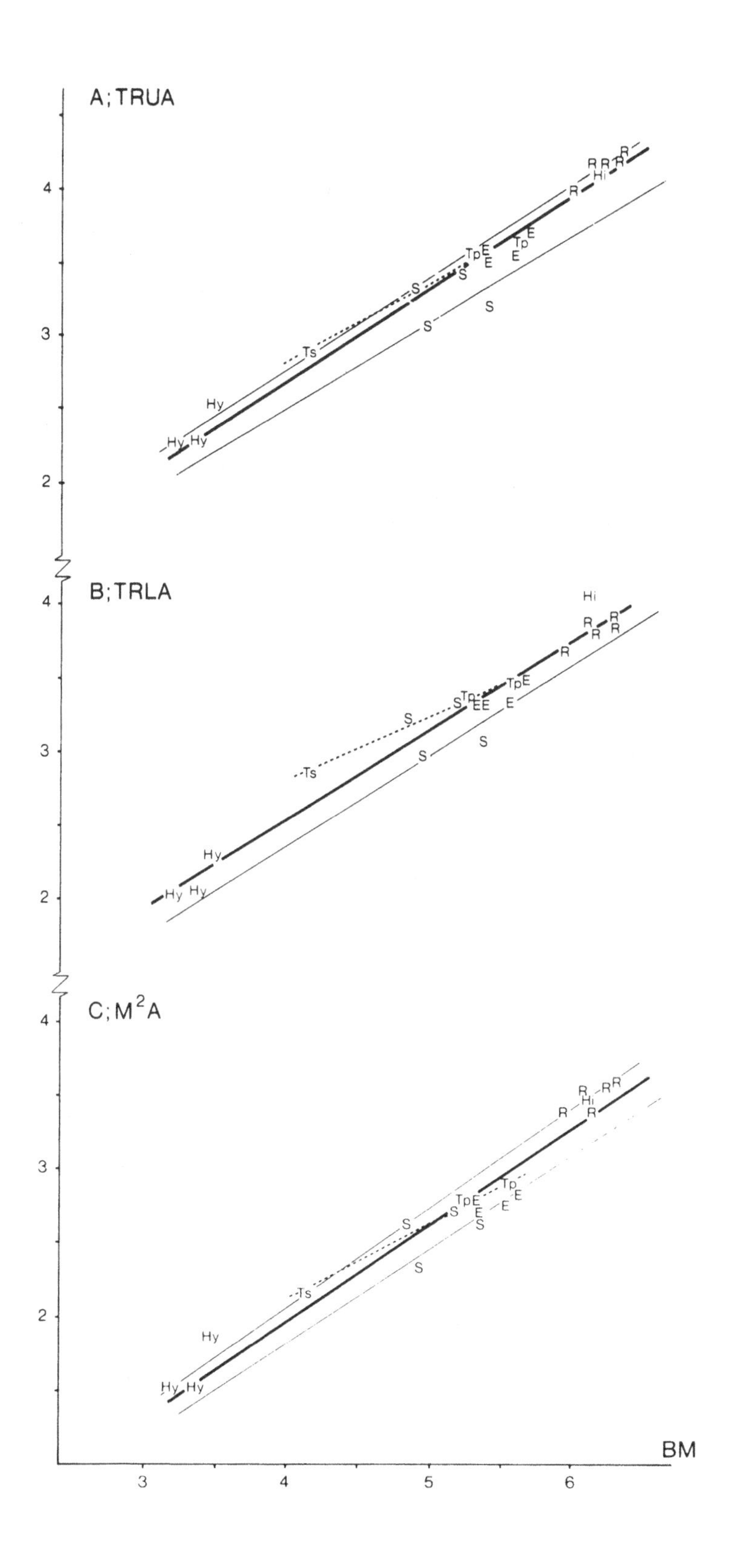
A; TRUA
B; TRLA
C; M^{2}A
BM
Hy
Ts
S
Tp
E
R
Hi
2
3
4
5
6

The hippopotamus, however, always plots with the rhinoceroses, high above the bilophodont–bunodont line. This is contrary to my expectations. The pattern may well be spurious; at least the domain of this relationship is clearly difficult to determine. The specialized suids *Phacochoerus* and *Hylochoerus* plot considerably below the line, as was expected. Their postcanine tooth rows consist mainly of enlarged posterior molars.

It will be easy in principle to test whether there is indeed such a bilophodont–bunodont relationship by adding living species not included here because of a lack of data. In lieu of this I have substituted condylobasal skull length for body mass, which allows inclusion of two more species, the suid *Babyrousa babyrussa* and the peccary *Tayassu peccari* (Figures 11.3A,B; Table 11.1). Correlations are even higher than between tooth size and body mass, probably because all the data now come from the same individuals, and the species added fall reasonably close to the line.

If nothing else, this illustrates the crucial importance of choosing the right reference population. If functional groups of this kind are a reality, then estimates based on taxonomic reference populations may be quite inaccurate, or at least distinctly suboptimal.

Conclusions

For body size estimation from teeth, we are firmly stuck with the empirically derived relationships and with the consequent limitations on use. Allometry does not provide the precise anatomical laws that Cuvier dreamed of (Buffetaut 1987; Rudwick 1972). We are still not able to calculate the whole cat from its carnassial, and it seems highly unlikely that we ever will be able to do so. There are, however, reasons for hope as well as for despair. In the final analysis it is a matter of how accurate estimates are needed, that is, for what purpose they are required.

Figure 11.2 (*facing page*). Log-transformed bivariate plot of dental length measurements (mm) against body mass (BM, g) for living, nonselenodont ungulates (for species, see Table 11.2). All lines are major axes (for parameters see Table 11.1). The heavy line is for all the data points shown; the lower thin line is for bovids (not shown). Symbols: E = equid, Hi = hippopotamus, Hy = hyracoid, R = rhinocerotid, S = suid, Tp = tapirid, Ts = tayassuid. (A) TRUL = length of upper postcanine tooth row. (B) TRLL = length of lower postcanine tooth row. (C) M^2L = length of M^2.

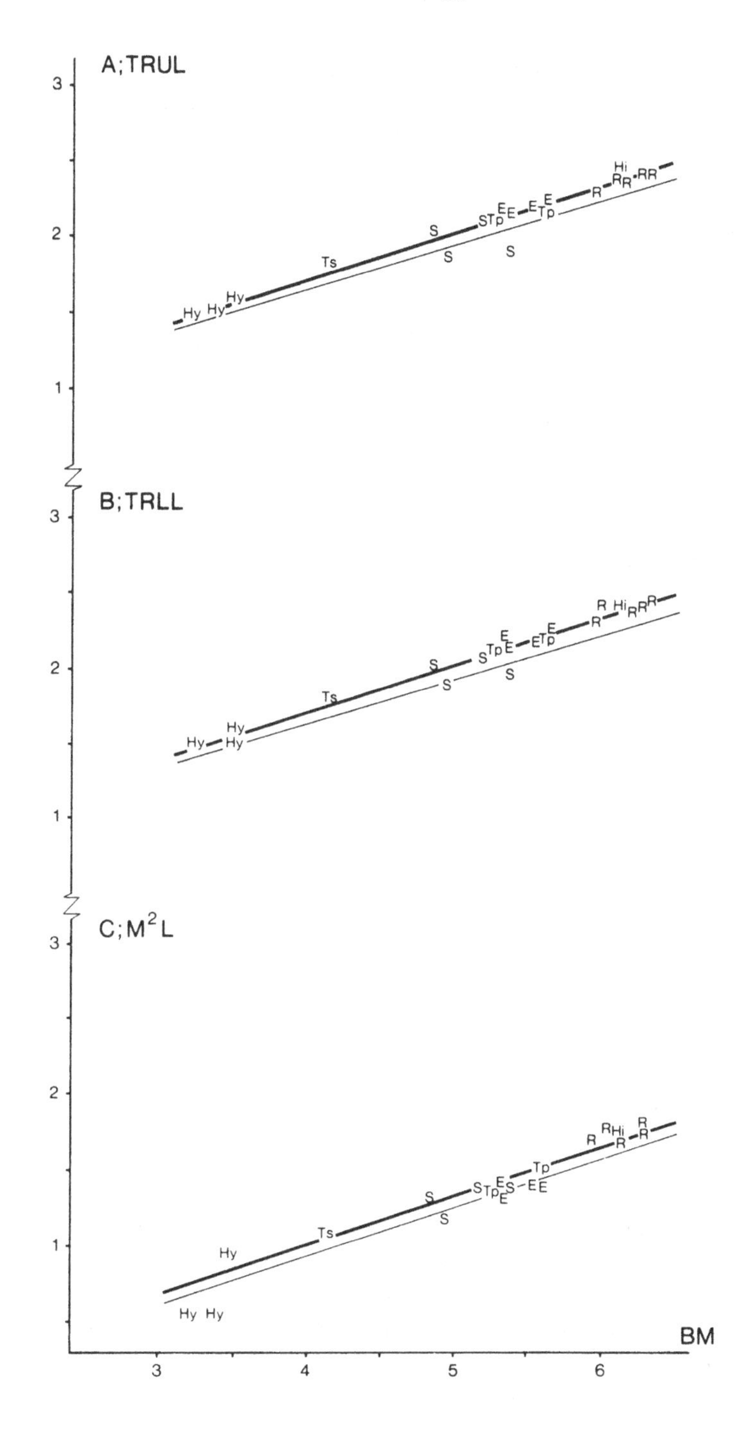
A;TRUL
B;TRLL
C;M²L
BM
Hy
Ts
S
Tp
E
R
Hi

Table 11.2. *Species included in the analyses*

Species	*N*	NOSE	SELE	BOVI	HYRI	BIBU
Diceros bicornis	24	*			*	
Ceratotherium simum	11	*			*	
Dicerorhinus sumatrensis	2	*			*	
Rhinoceros sondaicus	3	*			*	
Rhinoceros unicornis	2	*			*	
Tapirus spp. (S. American)	3	*				*
Tapirus indicus	1	*				*
Sus scrofa	1	*				*
Potamochoerus porcus	4	*				*
Hylochoerus meinertzhageni	2	*				
Phacochoerus aethiopicus	2	*				
Babyrousa babyrussa	4	*				(*)
Tayassu peccari	3	*				(*)
Tayassu tajacu	3	*				*
Hippopotamus amphibius	2	*				
Procavia capensis	3	*			*	
Procavia habessinica	3	*			*	
Heterohyrax spp.	3	*			*	
Dendrohyrax spp. (small)	3	*			*	
Dendrohyrax dorsalis	1	*			*	
Equus grevyi	2	*				
Equus burchelli	2	*				
Equus caballus	2	*				
Equus hemionus	1	*				
Camelus dromedarius	2		*			
Hyemoschus aquaticus	2		*			
Tragulus spp.	6		*			
Giraffa camelopardalis	2		*			
Moschus moschiferus	4		*			
Muntiacus muntjak	3		*			
Hydropotes inermis	3		*			
Dama dama	1		*			
Odocoileus virginianus	1		*			
Rangifer tarandus	4		*			
Alces alces	1		*			
Capreolus capreolus	1		*			

Estimation using allometric equations entails a number of theoretical and practical difficulties. These are worst when there is no simple, or at least known, functional principle involved, as for skulls or teeth. Particularly problematic is the combination of averaging and modernization that will necessarily occur when body masses of fossil species are estimated using an equation derived from a sample of living species.

Table 11.2. (*cont.*)

Species	*N*	NOSE	SELE	BOVI	HYRI	BIBU
Litocranius walleri	1		*	*		
Sylvicapra grimmia	4		*	*		
Madoqua saltiana	3		*	*		
Rhaphicerus campestris	2		*	*		
Tragelaphus scriptus	2		*	*		
Beatragus hunteri	1		*	*		
Alcelaphus buselaphus	2		*	*		
Syncerus caffer caffer	1		*	*		
Syncerus caffer nanus	1		*	*		
Taurotragus oryx	1		*	*		
Oryx beisa	1		*	*		
Oryx gazella	1		*	*		
Hippotragus equinus	2		*	*		
Hippotragus niger	1		*	*		

Note: The data base is that of Fortelius (1985), which the reader should consult for sources. The data are available from the author, but should be treated with caution because of the small sample sizes. The numbering is that of the data base.
Abbreviations: NOSE = nonselenodonts, SELE = selenodonts, BOVI = bovids, HYRI = hyraxes and rhinoceroses, BIBU = bilophodonts and bunodonts.

For weight-bearing bones these problems are less severe, and all would perhaps be well but for the fact that what is available is often nothing but teeth.

There is hope that increased understanding of food comminution and the scaling of other relevant variables (chewing rate, proportion of time spent feeding, energy content of foods, metabolic rate, etc.) will bring increased confidence. Meanwhile both theory and practice suggest that tooth length is a better estimator of body size than is tooth width, and therefore area, which includes width. On the other hand width, and particularly anisodonty, may help to group species into functionally homogeneous categories.

In choosing reference populations among recent species we should pay attention not only to taxonomy, which may sometimes be misleading, but also and perhaps primarily, to dental morphology. In general, variables that are not highly mass dependent, and are thus useless for mass estimation, are probably the best basis for functional groupings.

Both theoretically and practically a major focus seems to be the question of why the commonly observable general relationships gradually change and weaken at successively lower taxonomic levels, until they

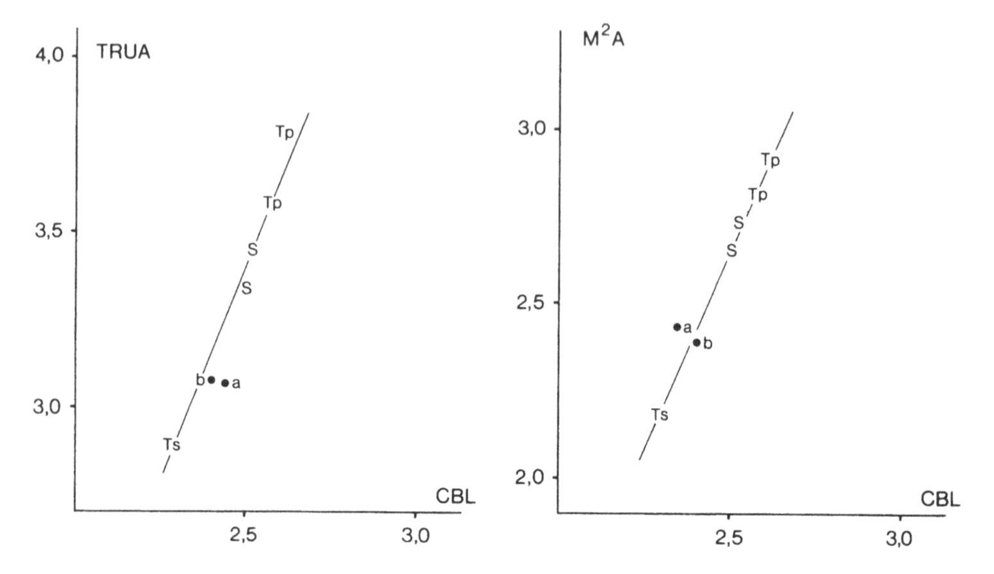

Figure 11.3. Log-transformed bivariate plot of dental area measurements (mm^2) against skull length (CBL, mm) for selected living bilophodont and bunodont species. The lines are major axes for the data points denoted by letter symbols (S = suid, Tp = tapirid, Ts = tayassuid); the black dots are species not included in the computation; *a* is the suid *Babyrousa babyrussa*; *b* is the tayassuid *Tayassu peccari*. (A) TRUA = length of upper postcanine toothrow × width of M^2. (C) M^2A = length × width of M^2.

are practically lost in intraspecific comparisons (e.g., Wolpoff 1985). If we could understand this, we might understand very much more, including how best to estimate body masses.

Acknowledgments

I dedicate this paper to the memory of my teacher and friend Björn Kurtén. I am indebted to Peter Lucas, John Prothero, Mahito Watabe, and the participants in the Gainesville workshop for good discussions. Phil Gibbard checked the English but should not be held responsible for oddities that may remain. This is a contribution from the Valio Armas Korvenkontio Unit of Dental Anatomy in Relation to Evolutionary Theory.

References

Anonymous. 1985. *Manual of Nutrition*. Ministry of Agriculture, Fisheries and Food. London: Her Majesty's Stationery Office.

Buffetaut, E. 1987. *A Short History of Vertebrate Palaeontology*. London: Croom Helm.

Cachel, S. 1984. Growth and allometry in primate masticatory muscles. *Arch. Oral Biol. 29*:287–293.

Creighton, G. K. 1980. Static allometry of mammalian teeth and the correlation of tooth size and body size in contemporary mammals. *J. Zool., Lond. 191*:235–243.

Cuvier, G. 1834. *Recherches sur les ossemens fossiles, où l'on rétablit les caractères de plusieurs animaux dont les révolutions du globe ont détruit les espèces,* Vol. 1, 4th ed. Paris.

Eisenberg, J. F. 1981. *The Mammalian Radiations: An Analysis of Trends in Evolution, Adaptation, and Behavior*. Chicago: University of Chicago Press.

Fortelius, M. 1985. Ungulate cheek teeth: developmental, functional and evolutionary interrelations. *Acta Zool. Fennica 180*:1–76.

Fortelius, M. 1988. Isometric scaling of mammalian cheek teeth is also true metabolic scaling. In *Teeth revisited: Proceedings of The VIIth International Symposium on Dental Morphology, Paris, 1986,* ed. D. E. Russell, J.-P. Santoro, & D. Sigogneau-Russel, pp. 459–562. Paris: Mém. Mus. Natn. Hist. Nat. (sér. C).

Fortelius, M. 1987. A note on the scaling of dental wear. *Evol. Theory 8*: 73–75.

Gingerich, P. D., Smith, B. H., & Rosenberg, K. 1982. Allometric scaling in the dentition of primates and prediction of body weight from tooth size in fossils. *Am. J. Phys. Anthropol. 58*:81–100.

Gould, S. J. 1975. On the scaling of tooth size in mammals. *Am. Zool 15*: 351–362.

Hendrichs, H. 1965. Vergleichende Untersuchung des Wiederkauenverhalten. *Biol. Zbl. 84*:651–751.

Janis, C. M. 1976. The evolutionary strategy of the Equidae and the origins of rumen and cecal digestion. *Evolution 30*:757–774.

Jungers, W. L. (ed.). 1985. *Size and Scaling in Primate Biology*. New York: Plenum Press.

Kay, R. F. 1975. (Comment on Pilbeam & Gould 1974.) *Science 169*:63.

Kay, R. F. 1978. Molar structure and diet in extant Cercopithecidae. *Development, Function and Evolution of Teeth,* ed. R. M. Butler, & K. A. Joysey, pp. 309–339. London: Academic Press.

Kingdon, J. 1979. *East African Mammals. IIIB. Large Mammals*. London: Academic Press.

Lucas, P. W. 1980. Adaptation and form of the mammalian dentition with special reference to primates and the evolution of man. Ph.D. thesis, University of London.

Lucas, P. W., Corlett, R. T., & Luke, D. A. 1986. Postcanine tooth size and diet in anthropoid primates. *Z. Morphol. Anthropol. 76* (3):255–276.

McKay, G. M. 1973. The ecology and behavior of the Asiatic elephant in southeastern Ceylon. *Smithsonian Contrib. Zool. 125*:1–113.

McMahon, T. 1973. Size and shape in biology. *Science 179*:1201–1204.

McNab, B. K. 1980. Food habits, energetics, and the population biology of mammals. *Am. Nat. 116*(1):106–124.

McNab, B. K. 1986. The influence of food habits on the energetics of eutherian mammals. *Ecol. Monogr. 56*(1):1–19.

Peters, R. H. 1983. *The Ecological Implications of Body Size.* Cambridge: Cambridge University Press.

Pilbeam, D., & Gould, S. J. 1974. Size and scaling in human evolution. *Science 186*:892–901.

Prothero, D. R., & Sereno, P. C. 1982. Allometry and paleoecology of medial Miocene dwarf rhinoceroses from the Texas Gulf coastal plain. *Paleobiology 8*(1):16–30.

Prothero, J. 1986. Methodological aspects of scaling in biology. *J. Theor. Biol. 118*:259–286.

Rensch, B. 1952. Neuere Untersuchungen über transspezifische Evolution. *Verh. Deutsch. Zool. Gesell. Freiburg 46*:379–408.

Rudwick, M. J. S. 1972. *The Meaning of Fossils.* London: Macdonald.

Saarinen, E. (ed.). 1982. *Conceptual Issues in Ecology.* Dordrecht: D. Reidel.

Schmidt-Nielsen, K. 1984. *Scaling: Why Is Animal Size So Important?* Cambridge: Cambridge University Press.

Simberloff, D. 1982. A succession of paradigms in ecology: essentialism to materialism and probabilism. In *Conceptual Issues in Ecology,* ed. E. Saarinen, pp. 63–99. Dordrecht: D. Reidel.

Smith, R. J. 1985. The present as a key to the past: body weight of Miocene hominoids as a test of allometric methods for paleontological inference. In *Size and Scaling in Primate Biology,* ed. W. L. Jungers, pp. 437–448. New York: Plenum Press.

Smith, R. J., German, R. Z., & Jungers, W. L. 1986. Variability of biological similarity criteria. *J. Theor. Biol. 118*:287–293.

Van Valen, L. 1960. A functional index of hypsodonty. *Evolution 14*(4): 531–532.

Van Valen, L. 1976. Domains, deduction, the predictive method, and Darwin. *Evol. Theory 1*:231–245.

Van Valen, L. M. 1982. Why misunderstand the evolutionary half of biology? In *Conceptual Issues in Ecology,* ed. E. Saarinen, pp. 323–343. Dordrecht: D. Reidel.

White, T. E. 1959. The endocrine glands and evolution, No. 3: os cementum, hypsodonty, and diet. *Contrib. Mus. Paleont. Univ. Michigan 13*(9): 211–265.

Wolpoff, M. H. 1985. Tooth size–body size scaling in a human population: theory and practice of an allometric analysis. In *Size and Scaling in Primate Biology,* ed. W. L. Jungers, pp. 273–318. New York: Plenum Press.

12

Problems in estimating body masses of archaic ungulates using dental measurements

JOHN DAMUTH

The problem

It might seem that one of the few things we can tell unambiguously about fossil animals is their gross size. The bones and teeth of extant mammals clearly reflect the mass of the whole individual, and multiple regression analyses involving many skeletal parts should allow us to characterize the relationships between morphology and body mass for living animals with some confidence (Jungers, this volume). However, such methods will most often be of limited value in estimating the body masses of fossil mammals. We usually have available only fragmentary remains for fossil species, and must therefore favor techniques of estimation that require as few concurrent measurements as possible. Furthermore, the fact that teeth are so much more well preserved and diagnostic than are other skeletal parts means that much of the time we will need techniques that use dental measurements exclusively. However, teeth are far from ideal as potential body-mass estimators. They have been subject to much adaptive evolution throughout the history of the mammals, and their dimensions may reflect differences in function related to diet or to other biological characteristics, rather than body mass alone. Nevertheless, we are forced to try to get as much information out of teeth as we can.

The present study is an attempt to assess the adequacy of dental measurements for estimating the body masses of fossil ungulates, with particular emphasis on issues involving species of the Oligocene or earlier (the Paleogene). Most of these archaic ungulates have no direct

In *Body Size in Mammalian Paleobiology: Estimation and Biological Implications,* John Damuth and Bruce J. MacFadden, eds.

extant descendents, or their descendents are highly evolutionarily modified and thus are no sure guide to ancestral body-part proportions.

Imprecision and bias

In assessing techniques, it will be useful to distinguish two kinds of unreliability or inaccuracy. *Imprecision* refers to the magnitude of the random error associated with a given prediction method. *Bias* refers to systematic error in the estimation process. Bias arises most insidiously from differences among the mean values of variables in the statistical population that was used to derive the estimation equation (e.g., extant ungulates) and the mean values of those same variables in the population within which we want to predict body mass (fossil ungulates). Bias and imprecision can be independent of each other; an estimate may be subject to very little random error and still be quite wrong. In some applications we need as much precision as possible. (For example, in a study relating morphological measurements on individual fossil specimens, such as measures of brain volume, to estimates of the body masses of those individual specimens.) In other cases we may be able to tolerate considerable imprecision as long as we know that our estimates are unbiased (for example, a study in which the variable of interest is the mean value of body masses for some group of species).

Bias is difficult to anticipate. The usual way that people estimate body masses for fossil mammals is to choose a body part and regress body mass on this part in a sample of extant mammals. If a high correlation results, then this regression is regarded as reliable for estimating body masses of fossil species. Workers have often shown a keen awareness of the problem of variation within samples from a fossil population, which would contribute to imprecision. The arguments for using measurements of M_1 or M_2 as the independent variable in body mass estimation are concerned with this issue (Gingerich 1974a). Following Gingerich (1974b, 1976, 1977), it has become standard to base body-mass estimates of fossil mammals on measurements of individual lower molars, particularly M_1 area (Collinson & Hooker 1987; Conroy 1987; Gingerich, Smith, & Rosenberg 1982; Kay & Simons 1980). However, by far the major source of possible error in estimating body masses of fossil species is that the regressions describing the relationship of some body part to body mass in living species may not be applicable to a sample of fossil species (Conroy, 1987; Gingerich et al. 1982). If the relationship is different among the fossil species, inaccurate estimates may result, regardless of high correlations among extant forms.

Materials and methods

I made dental measurements of specimens representing 95 species of extant ungulates in the collections of the National Museum of Natural History. I measured one male specimen of each species (in the case of incomplete material, data from more than one specimen were used). Measurements were made using Fowler digital calipers, accurate to 0.025 mm. Dental measurements represent maximal values for each dimension, rather than dimensions of the occlusal surface (Janis, this volume). Composite measures (such as [M_{1-3}] length) were calculated from the dimensions for individual teeth. This yielded 54 dental measures for each species (Appendix Tables 16.9 and 16.10). Body mass and head–body length for each species were taken from the literature (Damuth [1987] and references therein). Identical dental measurements and head–body length were made on a sample of mounted skeletons representing 47 species of fossil ungulates, from the collections of the National Museum of Natural History, Smithsonian (USNM), the American Museum of Natural History (AMNH), and Field Museum of Natural History (FMNH). Choice of fossil species was governed by availability. Ages for fossil specimens are approximate and are based on the age ranges of the North American Land Mammal Ages given in Kurtén and Anderson (1980) and Savage and Russell (1983). The taxa range in age from the Pleistocene to the Paleocene (Table 12.1).

Body length for fossil species was measured with tape from the tip of the snout to the base of the tail, following the curve of the vertebral column. For most specimens an appreciable proportion of the skeleton was known to belong to a single individual. Composites were almost always composed of material from the same locality. In some cases, skulls and teeth did not belong to the skeleton, or were incomplete. Skull and dental measurements in these cases were made on conspecific material from the same locality – matched for size whenever possible with whatever material did belong to the skeleton. An unknown amount of inaccuracy is introduced by the different mounting practices used, by incomplete vertebral columns, and by differing ontogenetic ages of the original specimens (obvious juveniles were excluded). These problems may increase random error ("noise" or imprecision) but should not introduce bias. A potential bias does enter because the fossil skeletons

Table 12.1. *Fossil species used and body-mass estimates*

Species	Specimens	Order	Crown morph.	Age (my)	Epoch	LMA	A	B	C
Pantolambda bathmodon	AMNH 2549	Pantodonta	bunodont	62.0	Paleocene	Torrejonian	10.4	14.2[b]	12.2[c]
	AMNH 2550								
	AMNH 16664								
Barylambda faberi	AMNH 32511	Pantodonta	bunodont	54.5	Paleocene	Clarkforkian	183	223	231
	FMNH P14945								
Coryphodon testis	AMNH 2865	Pantodonta	lophodont	52.0	Eocene	Wasatchian	295	468	440
	AMNH 2867								
	AMNH 2872								
	AMNH 2963								
	USNM 21331								
Ectoconus majusculus	AMNH 16500	Condylarthra	bunodont	64.5	Paleocene	Puercan	18.2	19.4	21.9
Phenacodus primaevus	AMNH 4369	Condylarthra	bunodont	52.0	Eocene	Wasatchian	51.2	30.7	41.5
Phenacodus wortmani	AMNH 4378	Condylarthra	bunodont	52.0	Eocene	Wasatchian	10.7	9.4	11.5
Meniscotherium chamense	AMNH 48002	Condylarthra	selenodont	51.0	Eocene	Wasatchian	2.7	3.6	3.4
Uintatherium anceps	AMNH 1692	Dinocerata	bunodont	49.0	Eocene	Bridgerian	867	690	807
	AMNH 12169								
	AMNH 12170								
	AMNH 12174								
	USNM 16662								
Hyracotherium angustidens	AMNH 15428	Perissodactyla	bunodont	52.0	Eocene	Wasatchian	14.5	15.5	17.5
Hyracotherium vasacciense	AMNH 4832	Perissodactyla	bunodont	52.0	Eocene	Wasatchian	6.9	7.4	8.5
Palaeosyops leidyi	AMNH 1544	Perissodactyla	lophodont	49.0	Eocene	Bridgerian	187	304	286
	AMNH 1580								
Palaeosyops paludosus	USNM 16862	Perissodactyla	lophodont	49.0	Eocene	Bridgerian	196	281	276
	USNM 12583								
Helatetes nanus	USNM 12584	Perissodactyla	lophodont	49.0	Eocene	Bridgerian	19.8	17.4	21.0

Hyrachyus eximus	AMNH 5065 USNM 12581 USNM 23641 USNM 23643	Perissodactyla	lophodont	49.0	Eocene	Bridgerian	81.3	88.6	96.9
Orohippus pumilis	AMNH 12648 USNM 26305	Perissodactyla	bunodont	49.0	Eocene	Bridgerian	6.8	6.5	7.7
Dolichorhinus longiceps	FMNH 12200	Perissodactyla	lophodont	44.0	Eocene	Uintan	269	458	421
Menodus giganteus	FMNH P12756	Perissodactyla	lophodont	35.0	Oligocene	Chadronian	472	1600[b]	1339[c]
Brontotherium gigas	USNM 4262	Perissodactyla	lophodont	35.0	Oligocene	Chadronian	1173	2490	2073
Trigonias osborni	USNM 12724	Perissodactyla	lophodont	35.0	Oligocene	Chadronian	238	342	334
Mesohippus bairdi	AMNH 1477 AMNH 1492 USNM 15898	Perissodactyla	lophodont	31.0	Oligocene	Orellan	22.5	21.3	25.0
Metamynodon planifrons	AMNH 546	Perissodactyla	lophodont	31.0	Oligocene	Orellan	544	1002	887
Subhyracodon tridactylus	AMNH 538	Perissodactyla	lophodont	29.0	Oligocene	Whitneyan	383	524	517
Miohippus intermedius	AMNH 1196	Perissodactyla	lophodont	29.0	Oligocene	Whitneyan	48.2	45.1	52.4
Diceratherium cooki	USNM 2800 USNM 10296	Perissodactyla	lophodont	20.5	Miocene	Arikareean	195	245	251
Moropus elatus	AMNH 14375	Perissodactyla	lophodont	20.5	Miocene	Arikareean	1055	1113	1179
Merychippus isonesus	AMNH 14185	Perissodactyla	lophodont	16.5	Miocene	Hemingfordian	114.7	91.5	110.7
Pseudhipparion retrusum	F:AM 70006	Perissodactyla	lophodont	11.0	Miocene	Clarendonian	72.5	68.4	78.7
Protohippus supremus	F:AM 60353	Perissodactyla	lophodont	11.0	Miocene	Clarendonian	171	173	192
Teleocerus fossiger	USNM 2145	Perissodactyla	lophodont	7.5	Miocene	Hemphillian	658	1120	1016
Dinohippus leidyanus	AMNH 17224	Perissodactyla	lophodont	5.5	Miocene	Hemphillian	198	184	210
Equus simplicidens	USNM 12573 USNM 13791	Perissodactyla	lophodont	3.5	Pliocene	Blancan	445	429	476
Merycoidodon culbertsonii	AMNH 594 AMNH 1287 USNM 11909	Artiodactyla	selenodont	31.0	Oligocene	Orellan	20.5	28.1	27.7
Miniochoerus gracilis	USNM 16825	Artiodactyla	selenodont	31.0	Oligocene	Orellan	5.7	8.1	7.9
Agriochoerus antiquus	AMNH 9337	Artiodactyla	selenodont	31.0	Oligocene	Orellan	53.1	53.2	51.3

Table 12.1. (*cont.*)

Species	Specimens	Order	Crown morph.	Age (my)	Epoch	LMA	A	B	C
Poebrotherium sp.	USNM 15917	Artiodactyla	selenodont	31.0	Oligocene	Orellan	35.6	32.4	30.8
Hypertragulus calcaratus	USNM 16744	Artiodactyla	selenodont	31.0	Oligocene	Orellan	1.1	1.5	1.4
Leptomeryx evansi	USNM 16754 USNM 16755	Artiodactyla	selenodont	31.0	Oligocene	Orellan	3.2	3.2	3.0
Protoceras celer	AMNH 1236	Artiodactyla	selenodont	29.0	Oligocene	Whitneyan	19.8	22.4	21.6
Machaeromeryx tragulus	AMNH 20548	Artiodactyla	selenodont	21.5	Miocene	Arikareean	5.1	4.6	4.3
Stenomylus hitchcocki	USNM 16601 USNM 9258	Artiodactyla	selenodont	21.5	Miocene	Arikareean	28.4	37.6	37.0
Oxydactylus longipes	FMNH P12108 FMNH P12113 FMNH P12124	Artiodactyla	selenodont	21.5	Miocene	Arikareean	401	271	257
Promerycochoerus carrikeri[a]	FMNH P12036	Artiodactyla	selenodont	21.5	Miocene	Arikareean	31.9	53.7	54.2
Promerycochoerus carrikeri[a]	FMNH UC1459	Artiodactyla	selenodont	21.5	Miocene	Arikareean	67.8	127.3	131.1
Ramoceros osborni	AMNH 9476	Artiodactyla	selenodont	14.0	Miocene	Barstovian	14.5	16	15.4
Merycodus necatus	USNM 13901	Artiodactyla	selenodont	14.0	Miocene	Barstovian	7.9	9.1	8.7
Procamelus occidentalis	FMNH 15856 FMNH 15858	Artiodactyla	selenodont	9.5	Miocene	Clarendonian	536	581	579
Platygonus vetus	USNM 8200	Artiodactyla	bunodont	1.2	Pleistocene	Irvingtonian	57.8	65.4	71

Note: Tooth crown morphology is broadly classified following Fortelius (1985). LMA = North American Land Mammal Age; A = mass estimate (kg) based upon HBL equation for all ungulates; B = mass estimate based on multiple regression involving HBL and M_{1-3} length, for all ungulates; C = "best" estimate, based upon multiple regressions for nonselenodonts or selenodonts, as appropriate, from Table 12.4.

[a]The two specimens of *Promerycochoerus carrikeri* are treated separately, since this species is geographically (and/or temporally?) highly variable. P12036 (Harrison Beds, Sioux Co., Nebraska, near Agate) is about 20% or more larger in linear dimensions and is much more robust than UC1459 ("Harrison" Beds, Goshen Co., Wyoming, near Van Tassel).

[b]Data for M_{1-3} were not available, but estimates of lower postcanine length (PCRL) could be obtained. Mass is based on the following equation: log Mass (g) = 2.03(log HBL) + 1.26(log PCRL) − 3.91.

[c]Mass based on the following equation: log Mass (g) = 1.75(log HBL) + 1.43(log PCRL) − 3.46.

were articulated without intervertebral discs or other tissue components that might contribute to length measurements on extant specimens. (However, in most mounts the zygapophyses are articulated as they would have been in life, rather than the centra being packed as tightly as possible, partly correcting for the lack of intervertebral discs.) I thus expect that my fossil body-length estimates slightly underestimate the actual values. However, all fossil species were treated the same way, so within that sample all lengths are comparable.

Diets of extant species are based upon Janis (this volume).

Initial considerations

An initial assumption

In order to decide whether estimates based upon dental measurements are reliable, we need to have some estimates for the fossil species that we regard as relatively precise and unbiased. Among extant ungulates, the highest correlation with body mass, and the lowest percent prediction error (%PE) and percent standard error (%SEE) (Van Valkenburgh, this volume) is that for the regression on head–body length, followed by those on a number of dental length measurements (Table 12.2; Figure 12.1; regression equations for all measures in Appendix Tables 16.9 and 16.10). The coefficients for the all-ungulate body-length regression are comparable with the results of previous studies of the scaling of body length (Creighton 1980; Economos 1982; Radinsky 1978; Scott 1985; see also Van Valkenburgh, this volume, for similar observations on carnivores). The relatively high level of precision for body length estimates is not surprising, because body length measures the length of a "cylinder" that encompasses the majority of a mammal's mass – that comprising the head, neck, viscera, and proximal limb musculature. Most differences in body shape among ungulates involve modification of distal elements or of minor components of this cylinder (such as relative neck length, or changes in the distribution of mass; see Scott [1985] and this volume). Among mammals of medium and large size, both theoretical and empirical results suggest that general biomechanical factors (elastic criteria) are the primary determinants of the scaling of body length (Economos 1982, 1983; McMahon 1973; Scott 1985). It is reasonable to assume that fossil species would have been subject to the same biomechanical constraints and that consequently body-length should scale with mass among fossil species similarly to the way it does

Table 12.2. *Correlation coefficients, percent standard errors, and percent prediction errors, of various measurements with body mass*

Measurement		r	%SEE	%PE	N
Head–body	length	.975	45.54	31.42	91
P_4–M_3	length	.967	53.02	34.51	94
M_{1-3}	length	.963	55.82	36.34	94
M^1	length	.961	57.76	39.89	94
P^4–M^3	length	.960	58.41	38.05	93
M_2	length	.960	59.09	38.78	94
P_3–M_3	area	.960	61.68	39.06	89
PCRL	length	.959	59.43	40.39	94
M^1	area	.959	60.07	41.20	95
P_4–M_3	area	.956	63.13	38.39	94
M_{1-3}	area	.955	64.02	39.37	94
M^{1-3}	area	.954	63.28	40.79	93
M^2	length	.954	64.23	42.21	94
P^4–M^3	area	.952	64.25	40.98	93
M_2	area	.952	66.70	41.08	95
P^3–M^3	area	.951	65.07	41.95	93
M^2	area	.948	69.57	46.17	95
M_1	width	.945	73.37	45.65	93
M_1	area	.943	61.94	41.07	94
M^3	area	.942	73.69	44.92	93
M^3	area	.942	75.15	45.55	95
M_3	length	.938	78.57	50.29	94
M^1	width	.936	79.29	52.03	94
M^3	length	.933	81.61	48.60	93
M_2	width	.927	87.79	50.89	94
P_4	area	.925	88.33	57.38	95

Note: Measurements are listed in order of decreasing magnitude of r.
Abbreviations: PCRL, post-canine tooth-row length; N = sample size.

in modern forms. Accordingly, I will regard estimates based upon body length as sufficiently reliable and, in particular, free of bias, that we can use them in an initial evaluation of the usefulness of dental measurements.

The general characteristics of the fossil sample

Body lengths are shown for the fossil sample in Figure 12.2. Body mass estimates based upon head–body length for all ungulates are listed in Table 12.1. There is no overall trend with geologic age, and for the most part the same range of body sizes is represented at each age.

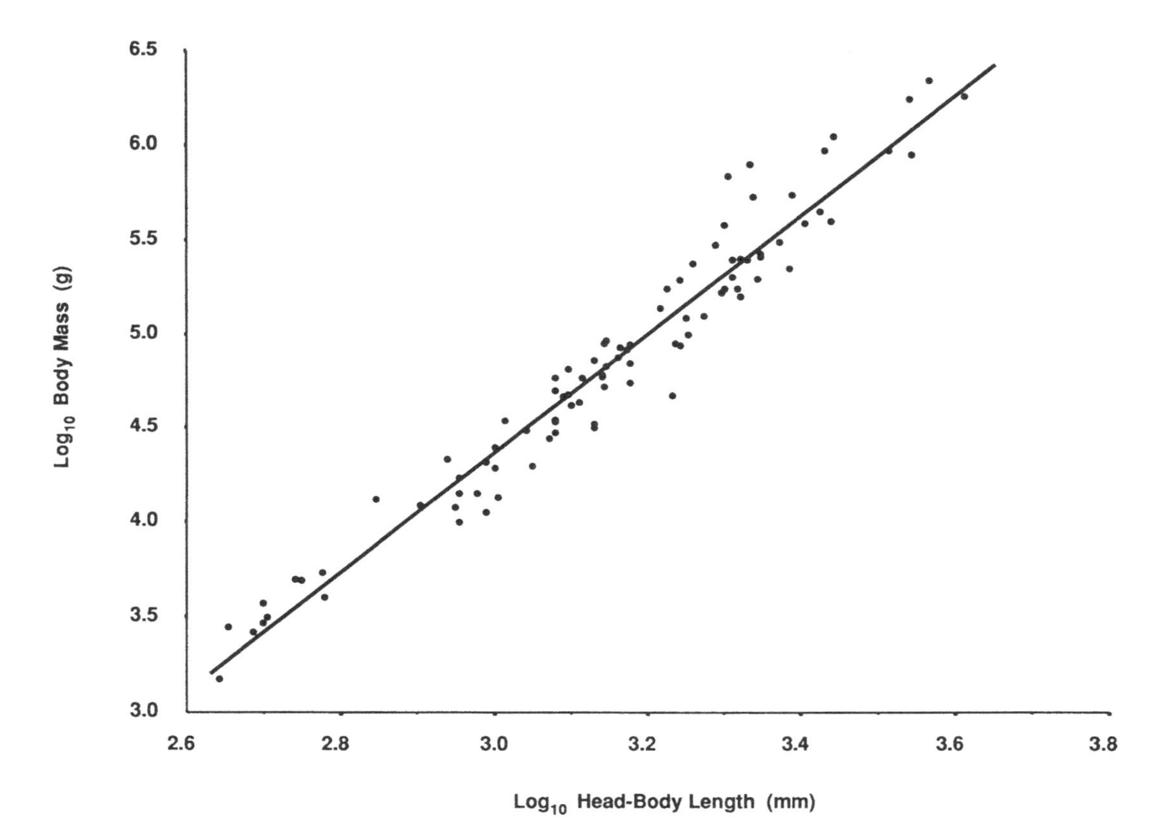

Figure 12.1. Body mass regressed on head–body length for 91 species of extant ungulates.

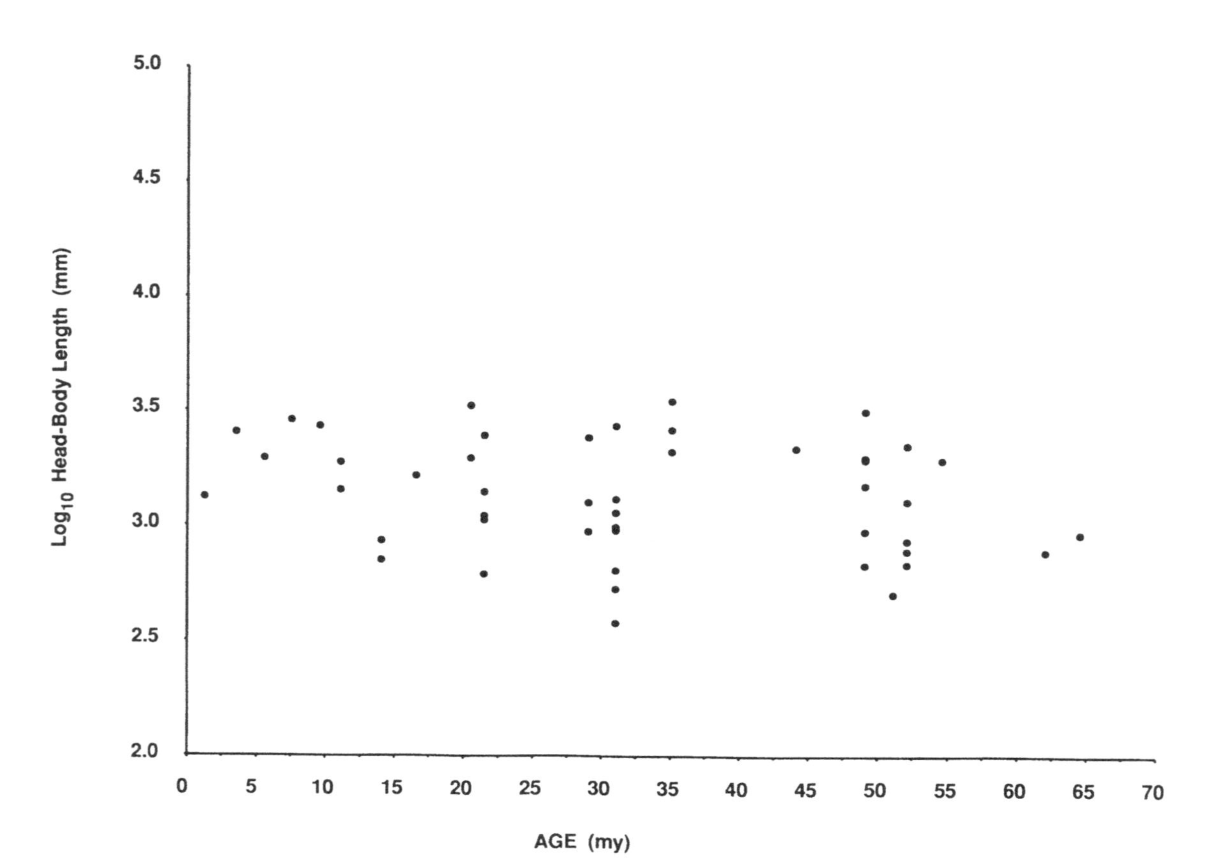

Figure 12.2. Head–body length for 47 species of fossil ungulates, plotted against geologic age (myr = millions of years).

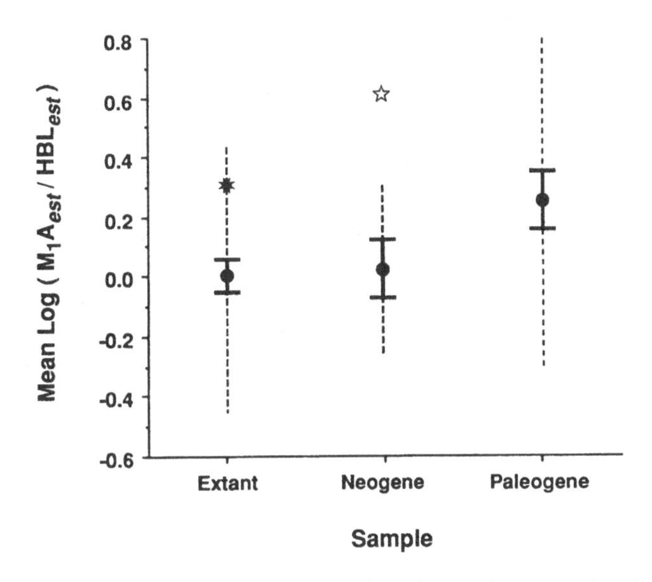

Figure 12.3. Mean values of the logarithm (base 10) of the ratio of body mass estimates based upon M_1 area to estimates based on head–body length, using regressions derived from the total sample of extant ungulates (Table 12.3). Extant = 92 extant species; Neogene = 9 post-Arikareean species; and Paleogene = 36 Puercan-Arikareean species. Filled circles indicate the mean for each sample. Heavy solid vertical lines indicate the 95% confidence intervals for the means, and the dashed lines indicate twice the standard deviation of each sample. Solid star: mean of extant nonselenodont ungulates; open star: *Teleoceras fossiger*.

We can evaluate tooth-based estimates by comparing them with body-length based estimates. Figure 12.3 shows the ratio of estimates based upon M_1 area to estimates based on head–body length, using regressions derived from the total sample of extant ungulates (Table 12.4). A similar plot would result from using any molar-area-based regression. The ratios were transformed to logarithms (base 10) for statistical analysis so that the data would be approximately normally distributed within each sample. The fossil species were divided into Paleogene (Puercan-Arikareean) and Neogene (post-Arikareean) samples. *Teleoceras fossiger,* a Miocene rhinoceros that has been interpreted as having a semi-aquatic habitus (e.g., Romer 1966), is 4.8 standard deviations from the mean of the Neogene sample and is regarded as an outlier. The Neogene sample has been recalculated without *Teleoceras.*

Although there is wide variability in the ratio within all samples, the mean of the sample of Paleogene species is significantly higher than the means of the Neogene or the extant species (one-factor ANOVA, p = .0001). This means that the average Paleogene species has larger molars, relative to body length, than does the average Neogene or Recent species. Assuming that estimates based upon body lengths are relatively unbiased, larger molars result in M_1 area overestimating the masses of Paleogene species by a factor of approximately 1.7, on average, irrespective of any imprecision introduced by random error. Estimates for Neogene species, on the other hand, appear to be relatively unbiased. The mean of extant nonselenodont ungulates, which tend to exhibit molar morphologies and body forms more similar to primitive, Paleogene ungulates, is substantially higher than the mean for all extant ungulates. It resembles closely the mean for Paleogene species. This suggests that dental morphologies and scaling relationships characterizing the modern selenodonts, which dominate the extant sample, may be a major source of the difference between the Paleogene and more recent samples (see below).

Not only do estimates based on M_1 area for all ungulates yield biased values for archaic ungulates, but they are not as precise as we might like for species of any time period. The 95% confidence intervals for predicted values from the M_1-area regression range over a factor of 7 (e.g., for a species estimated as having a mass of 100 kg, the 95% confidence intervals range from 37 to 269 kg). This degree of precision will perhaps be sufficient for some applications, but for others it may be unacceptably large.

Thus we have reason to believe that both bias and imprecision substantially limit our ability to obtain accurate body masses for archaic ungulates, using dental measurements. In order to come up with acceptable body-mass estimates for Paleogene forms, we must eliminate bias and try to increase precision as much as possible.

Eliminating the bias

The average Paleogene ungulate species has larger teeth for its body size than does the average modern one, resulting in an overestimate of its body mass if we use equations based on the ensemble of living species. We can see more specifically what it is about ungulate teeth that has changed over time by considering separately the length and width components of molar area. Figure 12.4 shows lower molar dimensions divided by body length regressed upon geologic age for the fossil species. There is a significant upward trend in relative molar width with age

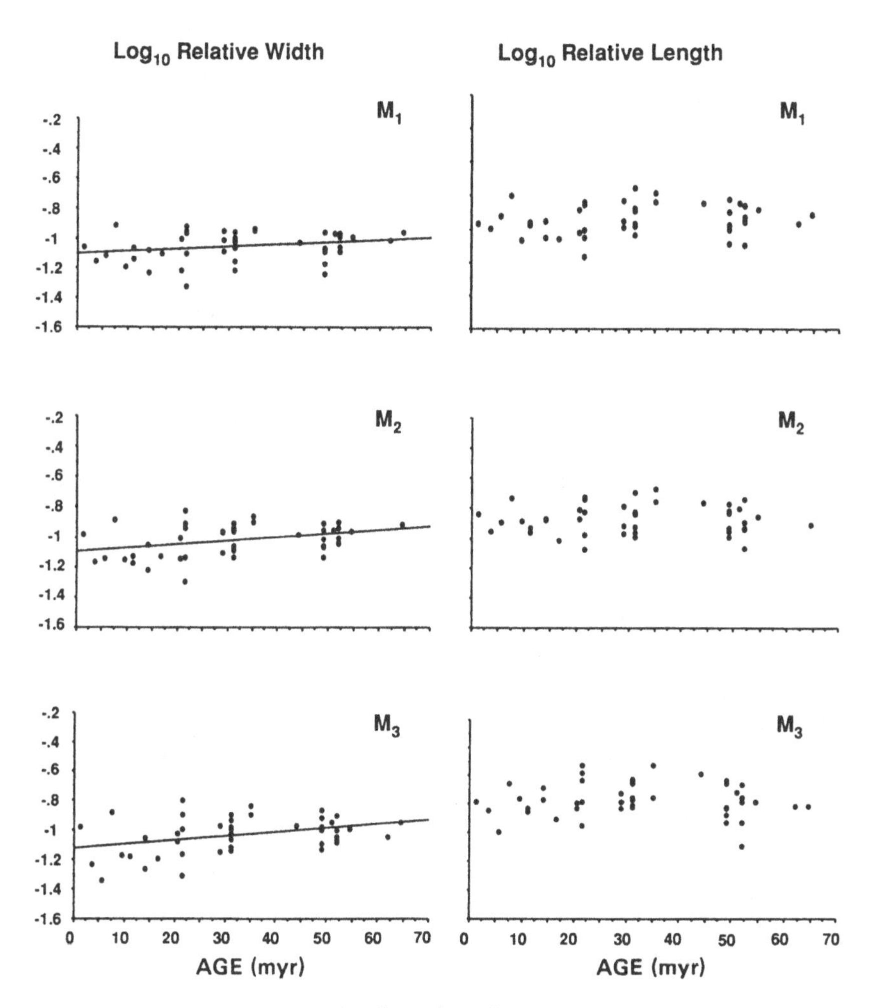

Figure 12.4. Lower molar dimensions divided by body length regressed upon geologic age for the fossil species.

($p < .10$ for M_1, $< .05$ for M_2 and M_3). No such trend is seen for relative molar lengths. (A similar situation is found among the upper molars but the trend in molar width is less pronounced.) Assuming that the fossil sample used here is representative of the fauna at different time periods, we can conclude that the average ungulate molar (especially lower molar) has become narrower throughout the Tertiary.

This suggests that a dietary change may be involved. In extant forms, the ratio of M_1-area-based estimate to body-mass-based estimate (calculated as in the previous section) is strongly related to diet (p = .03, 1-factor ANOVA), with grazing forms having a lower ratio and browsing forms and omnivores a higher one (Figure 12.5A). Among extant species, residuals of the regression of body mass on M_1 area show that grazers tend to be above the regression line, browsers below (Figure 12.5B). For a given body mass, grazers have smaller molar areas than do browsers. Molar width and molar length behave differently when looking across ungulate diets: Molar width is related to diet, but for molar length there is no significant relationship with diet (Table 12.3). Grazers have narrower molars than do nongrazers.

The sample of Recent ungulates upon which our predictive equations are based is dominated by selenodont grazers and mixed-feeding grazer-browsers. The earliest North American evidence of the development of extensive grasslands, and the opportunity to evolve grazing specializations, appears to be from no earlier than the beginning of the Neogene (Daghlian 1981; Thomas & Spicer 1987). Although development of grassland steppe may have occurred only in the Pliocene, there is a distinct shift toward open-country adaptations in the North American fauna in the Miocene (Janis 1984, 1989; Webb 1977). Thus, one would expect extant and Neogene ungulates to possess, on average, narrower molars than those of the Paleogene, and this is what is observed. The evolution of grazing seems to account well for the observed narrowing of ungulate molars over the Tertiary.

The bias due to the narrowing of ungulate molars can be effectively eliminated for Paleogene species by using molar *length* measurements rather than using molar areas or any other measurement incorporating molar widths.

Increasing precision

A 2-factor ANOVA shows that the average of the widths of M_1 and M_2, relative to head–body length, is related not only to diet (grazer vs. browser; p = .0056) but also to order (Artiodactyla vs. Perissodactyla; p = .0093). M_{1-2} length, relative to head–body length, is, as expected, not significantly related to diet (grazer vs. browser; p = .086) but is related to order (Artiodactyla vs. Perissodactyla; p = .022). Relative to browsers, grazers have narrower (but equally long) molars within both taxa, but perissodactyls have larger teeth in both dietary categories than do artiodactyls. Although the relationship to grazing is consistent, teeth are scaling (and presumably functioning) differently in different

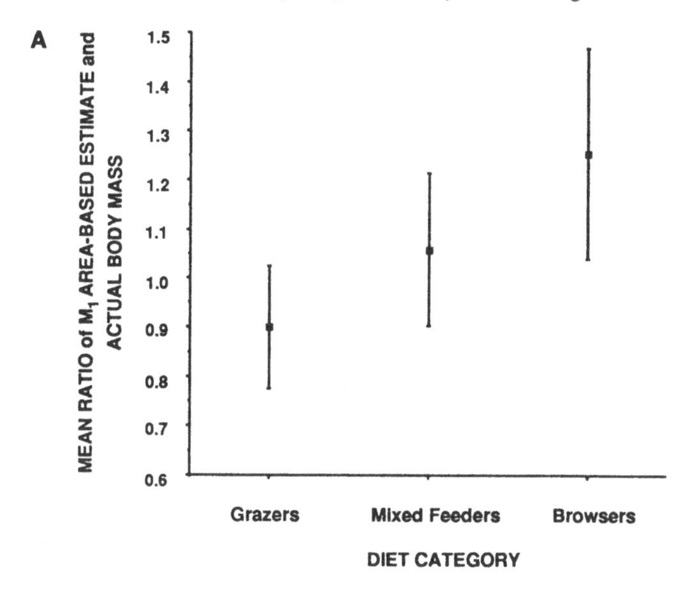

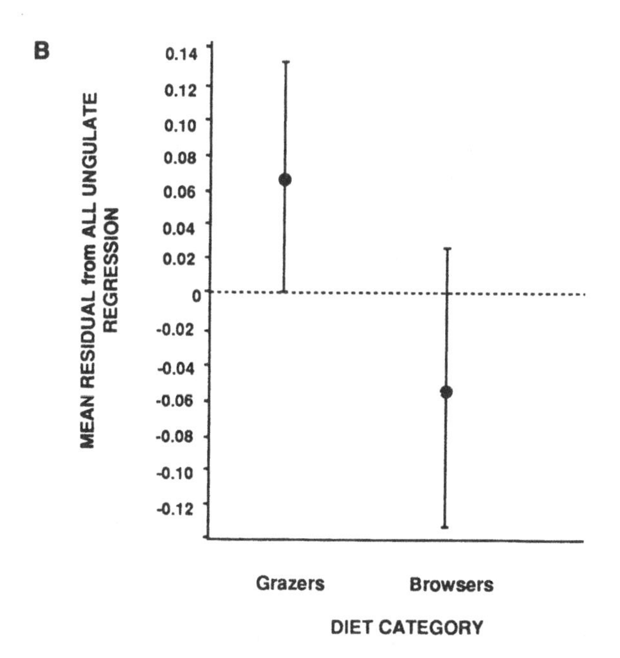

Figure 12.5. Relationship between molar area and diet. (A) Means and 95% confidence intervals for the ratio of body mass estimate based on M_1-area, to actual body mass, for species of three dietary groups. M_1-area estimate is based on the regression for all ungulate species. (B) Means and 95% confidence intervals for residuals of the all-ungulate regression of body mass on M_1-area, plotted for grazers and browsers.

Table 12.3. *Means of the residuals of regressions of body mass on molar dimensions, for extant species classified by diet*

Measurement	Mean of residuals: Grazers	Mixed feeders	Browsers	Probability
M_3 width	0.17	0.028	−0.091	.0008
M_2 width	0.138	0.031	−0.088	.0018
M_1 width	1.03	0.14	−0.056	.030
M_3 length	0.071	−0.056	0.069	.049
M_2 length	0.085	−0.026	0.004	.089
M_1 length	0.061	0.035	−0.068	.33

Note: Probabilities are results of 1-factor ANOVAs.

taxonomic groups. Using a length-based measure thus does not correct for all of the functional variation in ungulate molars. This remaining variation contributes to imprecision when all-ungulate regressions are used to predict body mass.

The problem is now to subdivide the modern ungulates in such a way that we obtain predictive equations, based upon dental-length measures, that exhibit greater precision than overall ungulate regressions are able to provide. The results of the 2-factor ANOVAs suggest that we should divide the extant ungulates into taxonomic groupings or into gross dental morphology categories. We seek functional groups within which relevant aspects of dental scaling are closely similar. Exploration of the data reveals that crown morphology and diet provide groupings that best improve our predictive power (Table 12.4).

Nonselenodonts of all orders exhibit similar scaling of tooth dimensions. Although the different varieties of bunodont and lophodont teeth do not function identically (Fortelius 1985), they nevertheless scale with body size similarly and in this way contrast as a group with selenodont teeth. Fortelius (this volume) and Janis (this volume) independently arrived at essentially the same observation. Nonselenodonts are a taxonomically and morphologically heterogeneous grouping including hyraxes, rhinoceroses, horses, tapirs, hippos, and some pigs (the highly specialized *Phacochoerus* and *Hylochoerus* have been excluded). This is of particular importance for the present study, as the majority of archaic ungulates have nonselenodont teeth. If functional specializations in such teeth tend not to alter significantly the scaling of overall dimensions, we can expect the regressions based on the modern fauna to be

Table 12.4. *Selected regression equations, discussed in text*

Equation	r^2	%SEE	%PE	N
All ungulates				
log Mass = 2.01 (log HBL) + 1.20 (log M_{1-3} Length) − 3.57	.96	38.37	26.19	91
log Mass = 3.16 (log HBL) − 5.12	.95	45.54	31.42	91
log Mass = 4.14 (log M_{1-3} Length) − 0.60	.93	55.82	36.34	94
log Mass = 1.50 (log M_1 Area) + 1.60	.92	61.94	41.07	94
Nonselenodonts				
log Mass = 2.35 (log HBL) + 0.79 (log M_{1-3} Length) − 3.88	.99	25.73	18.18	19
log Mass = 3.17 (log HBL) − 5.06	.99	30.30	22.41	19
log Mass = 3.03 (log M_{1-3} Length) − 0.39	.96	63.78	37.19	19
log Mass = 1.51 (log M_1 Area) + 1.44	.97	51.86	33.36	19
log Mass = 3.17 (log M_1 Length) + 1.04	.98	39.69	26.52	19
log Mass = 3.12 (P_4–M_2 Length) − 0.43	.98	38.71	24.02	19
All selenodonts				
log Mass = 1.93 (log HBL) + 1.32 (log M_{1-3} Length) − 3.54	.94	41.11	27.57	70
log Mass = 3.13 (log HBL) − 5.04	.93	46.79	32.12	70
log Mass = 3.27 (log M_{1-3} Length) − 0.80	.92	53.23	35.24	73
log Mass = 1.58 (log M_1 Area) + 1.47	.90	58.39	38.72	73
Selenodont browsers				
log Mass = 2.39 (log HBL) + 0.69 (log M_{1-3} Length) − 3.90	.98	34.16	22.25	17
log Mass = 2.97 (log HBL) − 4.55	.98	34.80	23.90	17
log Mass = 3.38 (log M_{1-3} Length) − 0.96	.95	57.58	38.66	18
log Mass = 1.53 (log M_1 Area) + 1.51	.93	67.46	46.32	18
Selenodont nonbrowsers				
log Mass = 1.87 (log HBL) + 1.57 (log [Lower postcanine row length]) − 4.08	.93	38.13	27.08	53
log Mass = 3.32 (log HBL) − 5.65	.89	49.65	33.79	53
log Mass = 3.26 (log [Lower postcanine row length]) − 1.42	.88	52.89	34.46	55
log Mass = 1.58 (log M_1 Area) + 1.48	.87	55.5	34.68	55

Abbreviations: r^2 = squared correlation coefficient; %SEE = percent standard error; %PE = percent prediction error; N = sample size; HBL = head–body length. Dental measures are in mm or mm^2; body mass in g.

reliable estimators for a taxonomically diverse array of fossil species. Among nonselenodonts, dental measurements alone can yield estimates that range over a factor of 4 (%SEE < 40; Table 12.4). Of course, for fossil species these values represent the minimum random error associated with the estimates.

Nonselenodont species also exhibit a tight relationship between mass and body length. In cases where we have body length measures for fossil species, we almost always have dental measurements available as well. We should be able to use these measures in combination to obtain more accurate estimates, if dental measures provide any significant additional information. A stepwise multiple regression involving body length and the 54 dental variables reveals that in all groups one dental measurement reflects some aspect of body mass not accounted for by body length itself. All of the dental variables are highly correlated, but in each case it is a length dimension of the lower tooth row that accounts for the most variance over and above that accounted for by body length. A multiple regression for the nonselenodonts involving body length and lower molar-row length accounts for 99% of the variance in body mass, and gives results whose precision approaches that obtained by using limb elements (Jungers; Ruff; Scott, this volume).

For selenodonts we get no substantial improvement from the all-ungulate regressions until we break this group down further into browsers and mixed-feeders-plus-grazers. For browsers, we can get comparable results to those for nonselenodonts, but only for body length and the multiple regression. The nonbrowsing selenodonts seem to remain more of a mixed bag, and even body length proves to be a relatively inaccurate estimator. I did not try to increase resolution further for selenodont ungulates, but further research may be successful in finding subgroupings that provide better estimates.

"Best" estimates of body mass for fossil species

The fossil species of Table 12.1 are ones for which a great deal of morphological information is available. It is thus possible to use the appropriate multiple regressions of Table 12.4 to obtain the most accurate body-mass estimates (Table 12.1, column C). The best estimates reported here are closely comparable to those given by Scott (this volume) for the taxa common to the two studies (*Dinohippus, Stenomylus, Merycodus, Ramoceros,* and *Promerycochoerus*). The body masses reported by Janis (this volume) for *Mesohippus bairdi, Promerycochoerus carrikeri, Miniochoerus gracilis, Protoceras celer,* and *Stenomylus hitchcocki* are larger than those reported here. However, some of these species are highly variable, and none of the actual specimens measured are common to the two studies. Body masses for Paleogene equids are distinctly lower than those reported by MacFadden (1987), though Neogene equid body-masses are closely comparable.

The relatively accurate estimates made possible by such unusually complete specimens can potentially be used to infer correction factors for species of the same functional or taxonomic groups that are not so well represented. For example, if we observe that members of a particular group exhibit unusually large teeth for their body size, based upon complete specimens, we know that estimates for other members of that group that are based on teeth alone are likely to be overestimates.

The effect of sample size on error in the estimator equations

As noted above, the predictor regressions used in this chapter differ from those used by Janis (this volume), who measured the dimensions of the occlusal surfaces of teeth rather than their maximum dimensions (except for molar row lengths, which were measured at the base of the teeth). The measures of error (%PE and %SEE) for the regressions used in this chapter are higher than those for Janis's regressions, for the 20 corresponding measurements reported for the two data sets (sign test, $p = .04$; Table 12.5). The error observed in regressions based on extant species comes from two sources: real biological differences among species, and random errors in estimating the mean values of measurements and body mass for each species. Body masses in both data sets are from the literature and should exhibit comparable errors. However, the magnitude of errors in estimating morphological means depends upon the properties of the sample of extant species that is used to calculate the regressions. A large and representative sample of each species should yield a better estimate of the mean values of morphological characters for each species, and thus result in regressions with lower %PE and %SEE values. Janis's data set is far larger than mine, with at least three and usually more individuals measured for each species, and is one of the best that can be assembled. In contrast, my data come from one, or at most two individuals for each species and represent a "worst-case" situation for sampling the extant fauna. We would expect, on these grounds alone, to see lower error values associated with Janis's regressions. The question is, Are the lower values that are actually observed (Table 12.5) the result of her more adequate sampling or the result of less variation among species in occlusal dimensions, relative to maximum ones?

It is not easy to disentangle these factors. However, it is possible to examine the effect of sample size within Janis's data set. Using her raw data, I created 50 data sets for each of two skull measures (total jaw length, total skull length) and four potentially useful dental measures

Table 12.5. *Comparisons of error in data sets and results of simulations*

Measurement	%PE JD	%PE CMJ	%SEE JD	%SEE CMJ	Mean %PE, fabricated data sets	Mean %PE, all CMJ data[a]	Mean %SEE, fabricated data sets	Mean %SEE, CMJ data
M^2 Length	42.21	34.7	64.23	51.7	41.02	36.86	64.34	57.55
M_2 Length	38.78	31.9	59.09	46.6	45.7	34.02	62.18	49.03
M_1 Length	42.34	34.6	60.73	51.4	41.85	36.36	59.92	50.86
M_{1-3} Length	36.34	32.8	55.82	47.9	39.51	34.83	59.94	52.77
M_3 Length	50.29	41.7	78.57	64.1	—	—	—	—
P_4 Length	59.73	75	89.91	78.6	—	—	—	—
P_3 Length	73.67	84.9	110.69	113	—	—	—	—
P_2 Length	90.62	101	146.95	164	—	—	—	—
M^2 Width	61.02	38.9	94.1	57.8	—	—	—	—
M_3 Width	57.82	42.9	97.52	71	—	—	—	—
M_2 Width	50.89	42.1	87.79	66.3	—	—	—	—
M_1 Width	45.65	38.4	73.37	58.1	—	—	—	—
P_4 Width	61.01	57.5	97.81	79.5	—	—	—	—
P_3 Width	62.21	72.7	95.53	104	—	—	—	—
P_2 Width	75.7	89.8	115.34	145	—	—	—	—
M^2 Area	46.17	32.7	69.57	48.6	—	—	—	—
M_3 Area	45.55	33.1	75.15	54.2	—	—	—	—
M_2 Area	41.08	33.5	66.7	50.7	—	—	—	—
M_1 Area	41.07	33.2	61.94	51	—	—	—	—
P_4 Area	57.38	59.2	88.33	70.2	—	—	—	—

Abbreviations: %PE = percent prediction error; %SEE = percent standard error; JD = data set used in this chapter; CMJ = data set used in Janis (this volume).

[a] Values in this column differ slightly from the values in the corresponding CMJ column reported by Janis, because for computational simplicity I omitted female specimens for sexually nondimorphic species. Thus means for these species are based on a smaller number of specimens than in her study, and should exhibit slightly higher error values. However, these data more closely approximate the composition of the JD data set.

(M_{1-3} length, M_1 length, M_2 length, and M^2 length). Within each fabricated data set each species is represented by one individual, chosen at random from the specimens of that species in the full data set. The 300 fabricated data sets thus resemble the data set used to generate predictor equations in this chapter. Regressions were performed on each data set. The means of the 50 resulting %PE and %SEE values for each measurement can be compared with the results of using the mean value of all individuals of each species (Table 12.5; Figure 12.6). For all four dental measures, the error measures for the 50 fabricated data sets are consistently somewhat higher than in the full data set (paired *t*-tests, $p = .03$ [%PE], $p = .008$ [%SEE]). Some teeth are more variable than others, but in all cases a large sample size for each species is associated with lower levels of random error. In contrast to the dental measures, the error values for skull characters are almost identical in the two data sets, and are little affected by differences in sample size (Figure 12.6).

For the four dental measures used in the simulations, the mean differences in error values between the simulation results and the full data set are 6.5% (for %PE) and 9.0% (for %SEE). The mean differences between my data set and Janis's data set for these same four dental measures are 6.4% and 10.7%, respectively. Thus, these errors differ by an amount that is consistent with the sample-size difference observed in the simulations. The same is true for the other analogous molar dimensions in the two empirical data sets. Assuming that differences in per-species sample size will have similar effects in both data sets, one can conclude that maximum molar dimensions are not appreciably more variable than are molar occlusal dimensions, and the differences in %PE and %SEE observed for molars in the empirical data sets are due primarily to differences in per-species sample size.

For premolar dimensions, however, the situation is different. In spite of the smaller per-species sample size, error values for most premolar measures are lower in my data set than they are in Janis's data set. Thus, the differences in error values for premolars almost certainly reflect a greater variability in premolar occlusal dimensions than in premolar maximal dimensions.

Summary and conclusions

Evolution throughout the Tertiary has changed the average proportions of ungulate teeth, making them relatively narrower. This change can be attributed to the diversification of forms adapted to a grazing diet, which

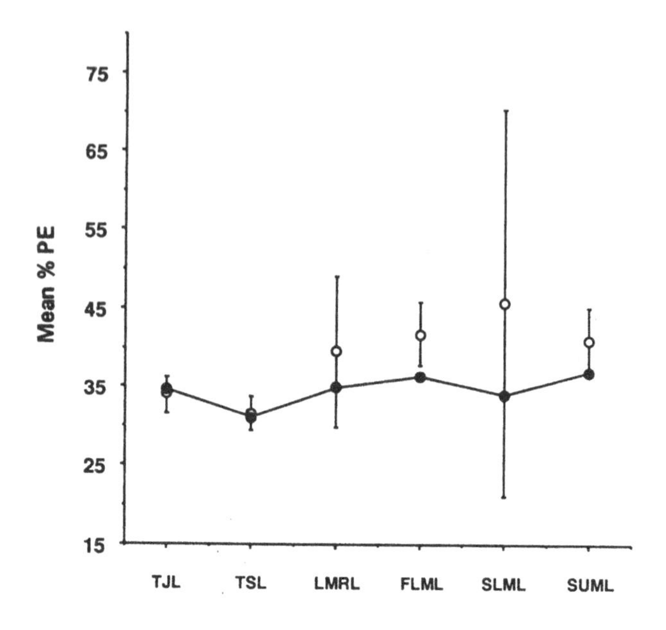

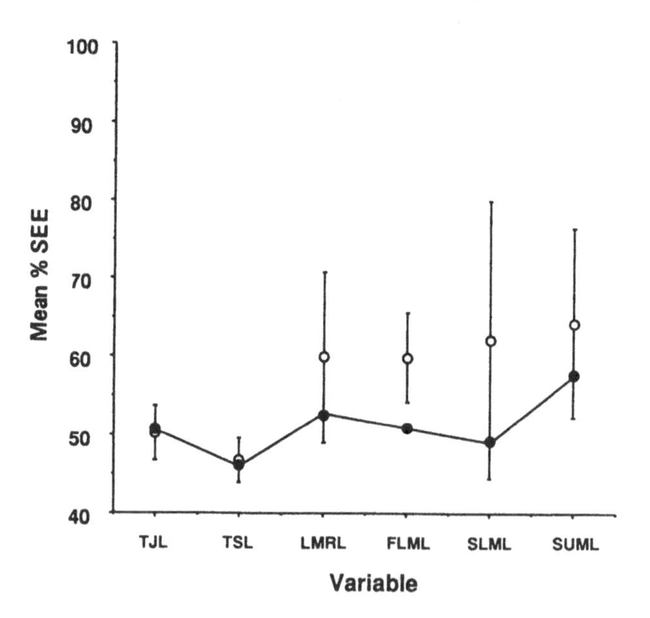

Figure 12.6. % PE (PPE) and % SE (PSE) error values from simulations using fabricated data sets (open circles: means and 95% confidence intervals of 50 runs) compared with result of using the complete data set (closed circles: connected by lines). See text for explanation of simulation. Results are shown for two cranial variables (TJL = total jaw length, and TSL = total skull length) and four dental variables (LMRL = M_{1-3} length; FLML = M_1 length; SLML = M_2 length; and SUML = M^2 length); see Janis (this volume, Table 13.1) for details of measurements.

is characteristic of Neogene and Recent faunas. Molar-area-based regressions (or any regressions involving molar widths) based on living forms thus tend to overestimate the body masses of Paleogene ungulates. This bias can be minimized by using measurements that do not involve molar width (e.g., molar row length).

Extant selenodont artiodactyls tend to have smaller teeth in both length and width dimensions, regardless of diet, relative to the nonselenodont perissodactyls. Although it cannot be demonstrated that this difference contributes to biased estimates for fossil forms, it does contribute to inaccuracy in "all-ungulate" regressions. Substantial improvement in precision can be gained by breaking down the ungulates into gross tooth morphology categories and dietary groupings. However, estimates based upon teeth alone are still fairly inaccurate.

When data on head–body length for fossil species are available, these can be combined with dental measurements to obtain much more accurate estimates of body mass. One purpose for such reconstructions would be to use the well-represented members of a fossil group to improve estimates for members that are more poorly represented and whose estimates must be based on unreliable or evolutionarily variable skeletal parts.

Maximum molar dimensions are not, in general, more variable or less reliable than are occlusal-surface dimensions. For premolars, maximum dimensions are less variable than are occlusal dimensions. Small per-species sample size increases %PE and %SEE values for the resulting regressions, but in the extreme case examined here the increase is on the order of 5% and 10%, respectively.

Finally, it should be clear that it is not possible to separate an understanding of aspects of tooth function from the choice of appropriate scaling relationships for estimating body mass of fossil species. Predictor equations ideally will be based on groups of extant species within which scaling relationships are relatively uniform, and whose grouping criteria are possible to discern in fossil species of interest. In many cases the best groups will be defined on morphological or functional criteria, rather than on the basis of taxonomic affinity.

Acknowledgments

I thank Christine Janis, Bruce MacFadden, Virginia Maiorana, Louise Roth, Kathleen Scott, and Blaire Van Valkenburgh for many fruitful discussions on body size and related topics. I am particularly grateful to Christine Janis for allowing me to use her data for the simulations, and to Kathleen Scott for the loan of measuring equipment. I thank Richard W. Thorington, Robert J. Emry, Michael J. Novacek, and William D. Turnbull for access to specimens in their care. Anna K. Behrensmeyer provided facilities while I was a resident at the Smithsonian. James Griesemer provided access to the computer facilities in the Department of Philosophy, University of California, Davis. Finally, I acknowledge the crucial assistance of Hal Freedman, I. Lorraine Heisler, Carol Hotton, Nicholas Hotton, Ben Jones, and Susan Mazer. This work was supported by National Science Foundation Grant No. BSR–8408030. This paper is contribution number 1 of the Evolution of Terrestrial Ecosystems Consortium.

References

Collinson, M. E., & Hooker, J. J. 1987. Vegetational and mammalian faunal changes in the Early Tertiary of southern England. In *The Origins of Angiosperms and their Biological Consequences,* ed. E. M. Friis, W. G. Chaloner, & P. R. Crane, pp. 259–304. Cambridge: Cambridge University Press.

Conroy, G. C. 1987. Problems of body-weight estimation in fossil primates. *Int. J. Primatol. 8*:115–137.

Creighton, G. K. 1980. Static allometry of mammalian teeth and the correlation of tooth size and body size in contemporary mammals. *J. Zool., Lond. 191*:435–443.

Daghlian, C. P. 1981. A review of the fossil record of monocotyledons. *Bot. Rev. 47*:517–555.

Damuth, J. 1987. Interspecific allometry of population density in mammals and other animals: the independence of body mass and population energy-use. *Biol. J. Linn. Soc. 31*:193–246.

Economos, A. C. 1982. On the origin of biological similarity. *J. Theor. Biol. 94*:25–60.

Economos, A. C. 1983. Elastic and/or geometric similarity in mammalian design? *J. Theor. Biol. 103*:167–172.

Fortelius, M. 1985. Ungulate cheek teeth: developmental, functional, and evolutionary interrelations. *Acta Zool. Fenn. 180*:1–76.

Gingerich, P. D. 1974a. Size variability of the teeth in living mammals and the diagnosis of closely related sympatric fossil species. *J. Paleontol. 48*: 895–903.

Gingerich, P. D. 1974b. Stratigraphic record of Early Eocene *Hyopsodus* and the geometry of mammalian phylogeny. *Nature 248*:107–109.

Gingerich, P. D. 1976. Cranial anatomy and evolution of Early Tertiary Plesiadapidae (Mammalia, Primates). *Univ. Michigan Pap. Paleontol. 15*: 1–140.

Gingerich, P. D. 1977. Correlation of tooth size and body size in living hominoid primates, with a note on relative brain size in *Aegyptopithecus* and *Proconsul*. *Am. J. Phys. Anthropol.* *49*:517–532.

Gingerich, P. D., Smith, B. H., & Rosenberg, K. 1982. Allometric scaling in the dentition of primates and prediction of body weight from tooth size in fossils. *Am. J. Phys. Anthropol.* *58*:81–100.

Janis, C. M. 1984. The significance of fossil ungulate communities as indicators of vegetation structure and climate. In *Fossils and Climate* ed. P. J. Brenchley, pp. 85–104. New York: John Wiley.

Janis, C. M. 1989. A climatic explanation for patterns of evolutionary diversity in ungulate mammals. *Palaeontology* *32*:463–481.

Kay, R. F., & Simons, E. L. 1980. The ecology of Oligocene African Anthropoidea. *Int. J. Primatol.* *1*:21–37.

Kurtén, B., & Anderson, E. 1980. *Pleistocene Mammals of North America.* New York: Columbia University Press.

MacFadden, B. J. 1987. Fossil horses from "Eohippus" (*Hyracotherium*) to *Equus:* scaling, Cope's law, and the evolution of body size. *Paleobiology* *12*:355–369.

McMahon, T. A. 1973. Size and shape in biology. *Science* *179*:1201–1204.

Radinsky, L. 1978. Evolution of brain size in carnivores and ungulates. *Am. Nat.* *112*:815–831.

Romer, A. S. 1966. *Vertebrate Paleontology,* 3rd ed. Chicago: University of Chicago Press.

Savage, D. E., & Russell, D. E. 1983. *Mammalian Paleofaunas of the World.* London: Addison-Wesley.

Scott, K. M. 1985. Allometric trends and locomotor adaptations in the Bovidae. *Bull. Am. Mus. Nat. Hist.* *179*:197–288.

Thomas, B. A., & Spicer, R. A. 1987. *The Evolution and Palaeobiology of Land Plants.* London: Croom Helm.

Webb, S. D. 1977. A history of savanna vertebrates in the New World. Part I: North America. *Annu. Rev. Ecol. Syst.* *8*:355–380.

13

Correlation of cranial and dental variables with body size in ungulates and macropodoids

CHRISTINE M. JANIS

Craniodental elements of ungulates are the most widely found as fossil remains, and are the means by which most species are identified, as is true for most mammals. Ungulate remains are numerous in the fossil record; however, this phenomenon has both good and bad aspects. The relatively large body size of ungulates makes their chances of preservation good, but also reduces the likelihood of discovering complete articulated skeletons. Most living ungulates are herbivorous, and many are gregarious, resulting in large biomasses; this may provide an explanation for the abundance of ungulate fossil remains. But this again limits the possibility of potential assignation of postcranial elements to species defined by craniodental remains. Many closely related species may be represented at the same locality, and it is rare to find an isolated individual where cranial and postcranial remains are definitely associated. Thus craniodental remains are usually the most reliable elements, in a taxonomic and taphonomical sense, for the determination of fossil ungulate body masses (although it is certainly possible to assign postcrania to some extinct ungulate species; see Scott, this volume).

However, there are many problems with using craniodental remains for body mass estimation of ungulates. Primarily, as is true for all animals, craniodental elements are not weight bearing. Thus, although it is reasonable to assume that they scale with body mass in some allometric fashion, they are less constrained by physical laws than elements that must play a role in body support, such as limbs. An additional problem

In *Body Size in Mammalian Paleobiology: Estimation and Biological Implications,* John Damuth and Bruce J. MacFadden, eds.

exists in the case of herbivorous mammals, which subsist on low quality food, and thus must process a greater volume of food per day than most other mammals. In this instance, it might be expected that the skull and teeth would have undergone strong selection for reasons of mechanical design that are not directly related to body size. For example, dietary type (e.g., grazer vs. browser) may play an important role here, as may digestive physiology (e.g., ruminant artiodactyl foregut fermenters have a relatively low daily intake of food in comparison with perissodactyl hindgut fermenters). Thus although craniodental remains are the only elements available for many fossil ungulate species, there are conceptual problems with their potential reliability as body mass estimators. For this reason, in this study artiodactyls (mainly foregut fermenters) were examined separately from perissodactyls and hyracoids (hindgut fermenters), and an additional comparison was made with macropodoid marsupials (kangaroos and rat kangaroos).

The prime aim of comparing macropodoids and ungulates was to see in which respects these marsupials, which evolved herbivory entirely independently from the ungulate condition, were similar to ungulates in the distribution of craniodental variables with body mass. Variables that showed similarly high correlations with body mass in both types of herbivorous mammals might be assumed to be more dependent on physical constraints, and thus could be considered as more "robust" for the determination of body mass in extinct ungulates. The taxa were additionally classified into four different feeding types: grazers, intermediate feeders, browsers, and omnivores, and the distribution of the mean residuals for these different feeding types was examined for all variables. The rationale for this was as follows: If the clustering of taxa of different feeding types around the regression line for a particular variable showed a significant degree of difference (e.g., all grazers above the line, and all browsers below the line), even if this variable had a high correlation with body mass, it might not be a particularly good one to use for the determination of the body mass of an extinct ungulate.

The correlation of craniodental variables with body mass in ungulates and kangaroos was examined by plotting the log of 21 dental and 12 cranial variables against the log of body mass. The data set included all living genera, and most living species from the orders Artiodactyla, Perissodactyla, Hyracoidea (ungulates), Potoroidae, and Macropodidae (kangaroos) (see Table 13.2).

Materials and methods

One hundred thirty-seven species of ungulates (30 grazers, 66 intermediate feeders, 36 browsers, and 4 omnivores) and 53 species of macropodoids (hereafter referred to simply as "kangaroos") (6 grazers, 23 intermediate feeders, 18 browsers, and 6 omnivores) were included in the analyses (see Table 13.1). In order to avoid possible differences due to sexual dimorphism, values for males only were used for dimorphic species. The measurements represent mean values for at least three individuals in the case of almost all ungulate species, and from one to four individuals in the case of the kangaroos. No juveniles (i.e., animals in which the last molar was unerupted) were included in the analyses. In the case of the kangaroos, measurements were taken only on young adults (defined as having M4 erupted, but showing little wear). In the case of the ungulates, adults of all dental wear stages were taken, with the exception of those with highly worn teeth (i.e., little or no enamel left on the occlusal surface).

Regression lines were calculated by least squares method for all ungulates and all kangaroos, and additionally for various subgroups within the ungulates. These were:

1. Ruminant artiodactyls (families Antilocapridae, Bovidae, Cervidae, Giraffidae, Moschidae, and Tragulidae), including 103 species (19 grazers, 63 intermediate feeders, and 21 browsers).
2. Within ruminant artiodactyls, the families Bovidae (72 species: 17 grazers, 42 intermediate feeders, and 13 browsers) and Cervidae (23 species: 2 grazers, 15 intermediate feeders, and 6 browsers) were also considered separately.
3. Perissodactyls plus hyracoids (families Equidae, Tapiriidae, Rhinocerotidae, and Procaviidae), including 19 species (9 grazers, 3 intermediate feeders, and 7 browsers).

Camelid artiodactyls (family Camelidae; five species, all intermediate feeders) and suine artiodactyls (families Suidae, Tayassuidae, and Hippopotamidae; 10 species [2 grazers, 4 browsers, and 4 omnivores]) were not examined separately, owing to the small numbers of taxa in the subsets, and the lack of any representatives of small body size.

Perissodactyls and hyracoids were lumped together because of their convergence of dental morphology (lophodont molars) and digestive physiology (extensive hindgut fermentation). In addition, Fischer's (1986) taxonomic work has shown good evidence for a close taxonomic

Table 13.1. *Complete list of species measured for craniodental dimensions*

Species	No. of obs.[a]	BW (kg) (M/F)	Diet
	UNGULATES		
ORDER ARTIODACTYLA			
Family Antilocapridae			
Antilocapra americana	8	55/45	I
Family Bovidae			
Alcelaphini			
Aepyceros melampus	11	61/45.5	I
Alcelaphus buselaphus	41	136	G
Connochaetes gnou	8	136	G
Connochaetes taurinus	19	239/193	G
Damaliscus dorcas	10	73/66	G
Damaliscus hunteri	3	91/86	I
Damaliscus lunatus	13	155/145	G
Boselaphini			
Boselaphus tragocamelus	6	250/170	I
Tetracerus quadricornis	16	17	I
Bovini			
Anoa depressicornis	4	156/145	I
Bison bison	7	865/450	G
Bison bonasus	4	865/450	G
Bos gaurus	6	1000/510	I
Bos indicus	1	750/450	I
Bos banteng	4	750/450	I
Bubalus bubalis	6	725/400	G
Syncerus caffer	7	400/320	I
Caprini			
Ammotragus lervia	4	113/59	I
Capra ibex	14	87	I
Hemitragus jemlahicus	5	91	I
Ovis canadensis nelsoni	3	73/45	I
Ovis dalli	3	84/59	I
Pseudois nayaur	7	59	I
Cephalophini			
Cephalophus dorsalis	9	20	B
Cephalophus monticola	26	5.5	B
Cephalophus sylvicultor	15	61	B
Cephalophus spadix	5	57	B
Sylvicapra grimmia	11	13	B
Gazellini			
Ammodorcas clarkei	5	31/25	B
Antilope cervicapra	5	45.5/29.5	G

Table 13.1. (*cont.*)

Species	No. of obs.[a]	BW (kg) (M/F)	Diet
Antidorcas marsupialis	9	34/28	I
Gazella dorcas	3	23/18	I
Gazella granti	5	75/50	I
Gazella thomsoni	8	23/18	I
Litocranius walleri	11	45/41	B
Procapra gutturosa	3	35/25	I
Hippotragini			
Addax nasomaculatus	2	118/104	I
Hippotragus equinus	7	280/260	G
Hippotragus niger	8	235/218	G
Oryx gazella	11	177/164	I
Neotragini			
Dorcatragus megalotis	5	9.0	I
Madoqua guentheri	9	3.5	B
Madoqua kirki	11	4.5	B
Neotragus pygmaeus	5	3.5	B
Nesotragus moschatus	27	4.5	B
Ourebia ourebi	10	18	I
Oreotragus oreotragus	19	13.5	I
Raphicerus campestris	28	13.5	I
Raphicerus melanotis	18	10	I
Reduncini			
Kobus ellipsiprymnus	12	227/182	G
Kobus kob	6	70/45.5	G
Kobus leche	5	100/73	G
Kobus vardoni	6	100/73	G
Pelea capreolus	13	41/23	I
Redunca arundium	18	68/57	G
Redunca fulvorufula	20	32/29.5	I
Rupicaprini			
Budorcas taxicolor	10	250	I
Capricornis sumatraensis	8	102	I
Nemorhaedus goral	7	27	I
Oreamos americanus	5	114/80	I
Ovibos moschatus	7	400/364	I
Pantholops hodgsoni[b]	2	45/28	I
Rupicapra rupicapra	4	45/34	I
Saiga tatarica	3	45/40	I
Tragelaphini			
Taurotragus oryx	12	590/432	I
Tragelaphus angasi	4	114/68	I
Tragelaphus buxtoni	4	216/150	I
Tragelaphus eurycerus	6	227/182	B

Table 13.1. (*cont.*)

Species	No. of obs.[a]	BW (kg) (M/F)	Diet
Tragelaphus imberbis	7	91/64	I
Tragelaphus scriptus	15	64/52	I
Tragelaphus spekei	6	91/57	G
Tragelaphus strepsiceros	14	260/170	B
Family Camelidae			
Camelus bactrianus	5	550	I
Camelus dromedarius	8	550	I
Lama guanicoe	5	110/75	I
Lama pacos	2	60	I
Vicugna vicugna	11	50	I
Family Cervidae			
Alces alces	4	450/318	B
Axis porcinus	3	50/35	I
Blastocerus dichotomus	4	140/120	I
Capreolus capreolus	3	35/25	I
Cervus canadensis	3	400/250	I
Cervus duvaucelli	2	212/142	G
Cervus elaphus scoticus[b]	6	200/125	I
Cervus nippon	3	64/41	I
Cervus unicolor equinus	5	215/162	I
Dama dama	6	67/44	I
Elaphodus cephalophus	5	18	I
Elaphurus davidianus	21	190	G
Hippocamelus bisulcus	4	65/55	I
Hydropotes inermis	6	12/9.5	I
Mazama mazama americana	9	20	B
Muntiacus muntjak vaginalis	8	25	I
Muntiacus reevesi	4	14/12	I
Odocoileus hemionus	4	91/57	B
Odocoileus virginianus	6	68/45	B
Ozotoceros bezoarticus	5	40/35	I
Pudu mephistophiles	7	13.5	B
Pudu pudu	4	13.5	B
Rangifer tarandus	7	110/81	I
Family Giraffidae			
Giraffa camelopardalis	12	1150/1000	B
Okapia johnstoni	10	250	B
Family Hippopotamidae			
Choeropsis liberiensis	5	240	B
Hippopotamus amphibius	4	2500	G
Family Moschidae			
Moschus moschiferus	9	11	I

Table 13.1. (*cont.*)

Species	No. of obs.[a]	BW (kg) (M/F)	Diet
Family Suidae			
Babyrousa babyrussa	6	85	B
Hylochoerus meinertzhageni	5	215	B
Phacochoerus aethiopicus	3	80/58	G
Potamochoerus porcus	6	78	O
Sus scrofa cristatus	6	80	O
Family Tayassuidae			
Catagonus wagneri	5	36	B
Tayassu pecari	6	30	O
Tayassu tajacu	6	22	O
Family Tragulidae			
Hyemoschus aquaticus	8	12.5	B
Tragulus javanicus	5	2.0/3.0	B
Tragulus meminna	5	7.0	B
Tragulus napu	6	8.0	B
ORDER PERISSODACTYLA			
Family Equidae			
Equus asinus	4	220	G
Equus burchelli	29	250/220	G
Equus grevyi	10	400	G
Equus hemionus	6	250	G
Equus kiang	6	300	G
Equus przewalskii	4	350	G
Equus zebra	19	260	G
Family Rhinocerotidae			
Ceratotherium simum	13	2800	G
Dicerorhinus sumatrensis	3	1000	B
Diceros bicornis	14	1200	B
Rhinoceros sondaicus	3	1400	B
Rhinoceros unicornis	6	2500	I
Family Tapiriidae			
Tapirus bairdii	6	275	B
Tapirus indicus	5	275	B
Tapirus pinchaque	2	250	I
Tapirus terrestris	7	240	B
ORDER HYRACOIDEA			
Dendrohyrax dorsalis	7	4.5	B
Heterohyrax brucei	8	3.0	I
Procavia capensis	5	4.0	G
	MARSUPIALS		
Family Macropodidae			
Subfamily Potoroinae			
Aepyprymnus rufescens	6	2.1/2.5	B

Table 13.1. (*cont.*)

Species	No. of obs.[a]	BW (kg) (M/F)	Diet
Calopyrmnus campestris	1	0.8	B
Hypsiprymnodon moschatus	6	0.5	O
Bettongia gaimardi	9	1.7	O
Bettongia lesueuri	6	1.7	O
Bettongia penicillata	7	1.3	O
Potorous platyops	1	0.7	O
Potorous tridactylus	7	1.0	O
Subfamily Macropodinae			
Dendrolagus bennetti	1	13/10	B
Dendrolagus dorianus	4	16.5/10.5	B
Dendrolagus goodfellowi	1	7.5	B
Dendrolagus lumholtzi	3	7.4/5.9	B
Dendrolagus matschiei	2	10	B
Dendrolagus ursinus	2	13/10	B
Dorcopsis hageni	4	8/5.5	B
Dorcopsis veterum	3	11/5	B
Dorcopsulus macleayi	2	3.0	B
Dorcopsulus vanheurni	3	2.3/2	B
Lagorchestes conspicillatus	7	3.0	B
Lagorchestes hirsutus	9	2.3	B
Lagorchestes leporoides	1	1.6	B
Lagostrophus fasciatus	5	1.8	I
Macropus agilis	7	19/11	I
Macropus antilopinus	4	37/17.5	I
Macropus bernardus	2	21/13	I
Macropus dorsalis	8	16/6.5	G
Macropus eugenii	3	7.5/5.5	I
Macropus fuliginosus	7	35/23	G
Macropus giganteus	4	43/27	G
Macropus greyi	4	7.0	I
Macropus irma	7	8.0	I
Macropus parma	3	4.9/4	G
Macropus parryi	9	16/11	G
Macropus robustus	6	39/20	I
Macropus rufogriseus	4	19.2/13.8	I
Macropus rufus	6	66/26.5	G
Onychogalea fraenata	5	5.5/4.5	I
Onychogalea lunata	2	4.0/3.0	I
Onychogalea unguifera	4	5.5/4.5	I
Peradorcas concinna	7	1.4	I
Petrogale brachyotis	3	4.2	I
Petrogale godmani	5	5.0	I
Petrogale inornata	7	4.0	I
Petrogale lateralis	6	5.7	I
Petrogale penicillata	11	7.5	I
Petrogale rothschildi	2	5.25	I
Petrogale xanthopus	8	7.0	I

Table 13.1. (*cont.*)

Species	No. of obs.[a]	BW (kg) (M/F)	Diet
Setonix brachyurus	3	3.6/2.9	B
Thylogale brunnei	3	6.0/3.6	B
Thylogale billardierii	3	7.0/3.9	I
Thylogale stigmatica	3	5.1/4.2	B
Thylogale thetis	4	7.0/3.8	I
Wallabia bicolor	4	17/13	I

Note: Not all individuals of each species (especially for those with large sample sizes) comprise a complete set of all measurements.
Abbreviations: B = browser; G = grazer; I = intermediate feeder; O = omnivore.
[a]No. of observations refers to those individuals used in this analysis – males in the case of dimorphic species, both sexes in the case of nondimorphic species – that had the last molars fully erupted, but that did not show heavy dental wear.
[b]Female weights and measurements were used for these taxa (males unavailable).
Sources: Body mass data from Banfield 1974; Haltenorth & Diller 1977; Kingdon 1979, 1982a,b; Lent 1978; Medway 1969; Schaller 1967, 1977; Schaller & Junrang 1988; Scott 1983, 1985; Walker 1983; Whitehead 1972. Information on diets from Banfield 1974; Haltenorth & Diller 1977; Hansen & Clark 1977; Hofmann 1973, 1985; Jarman 1974; Kingdon 1979, 1982a,b; Lamprey 1963; Mackie 1970; Medway 1969; Schaller 1967, 1977; Stewart & Stewart 1970; Walker 1983; Whitehead 1972.

relationship between perissodactyls and hyraxes. In any event, the r^2 values for the regression lines for this lumped grouping are almost uniformly high, confirming my suspicion of similarity of overall craniodental morphology, although these high r^2 values may in part merely reflect the large range in body size distribution in this grouping (see Damuth; Fortelius, this volume).

The distribution of taxa of different feeding types around the regression line was determined by Wilcoxon T-test comparison between the residuals around the line for each pair of feeding types, and are indicated in Appendix Table 16.8. Note was taken primarily of probability values of $p < .01$, although values of $p < .05$ were also noted (marked by an asterisk in the tables). These residuals were calculated from the regression line derived from the calculation with body mass as the independent variable, as were the tests for isometry. These tests are shown in Table 16.8 as isometric, positive, or negative values for isometry, and were calculated by testing if the slope differed significantly from that of 0.33 in the case of linear variables, or from 0.66 in the case of molar areas. Table 16.8 also includes r^2 values, and the slope and intercept for all variables plotted with body weight as the dependent variable, and in addition shows the percentage values of the standard error (%SEE) and the prediction error (%PE) (see Smith 1984; Van Valkenburgh, this volume).

The body masses and dietary types of the different ungulate species were obtained from a variety of literature sources (see Table 13.1). Ungulate body masses are likely to be especially good, owing to the research into this area by K. Scott (1983, 1985). The body masses of kangaroos were obtained primarily from Strahen (1983), with additional field data on certain species provided by Peter Jarman and Tim Flannery (personal communication). The dietary types were defined as follows: *grazer,* taking >90% of the diet as grass on a year-round basis; *browser,* taking <90% of the diet as grass on a year-round basis; *intermediate feeder,* taking between 10% and 90% of the diet as grass; and *omnivore,* taking little in the way of fibrous herbivorous material, relying more on plant storage and reproductive parts, and also including animal material in the diet. Although this definition of "intermediate feeder" might seem rather broad, these ecological categories are clearly reflected in differences in stomach morphology and digestive physiology (see, e.g., Hofmann 1973), and do appear to reflect valid biological categories. It is possible to further divide these three herbivorous categories, as I have done elsewhere (e.g., Janis & Ehrhardt 1988). Grazers can be divided into regular and fresh grass grazers; intermediate feeders, into those in open and those in closed types of habitats; and browsers, into regular and high-level browsers. All these subdivisions can be shown to have validity in aspects of the correlation of craniodental morphology with feeding type. However, for the purposes of this paper, I decided that a four-way comparison of feeding types would be more than adequate for my purposes, and much less confusing than a seven-way comparison!

The morphological variables taken were as follows (see also Figure 13.1 and Table 13.2 key).

Dental variables

1. Lengths, widths, and areas of all lower molars and premolars, and of the upper second molar were measured for all ungulates. Kangaroos are not directly dentally comparable to ungulates. They retain only one permanent lower premolar (P_3), and have four lower molars (although see Archer [1978] for an alternative terminology of macropodid cheek teeth). I have examined only molar variables for kangaroos in this study, as many species show molar progression with early loss of the pre-

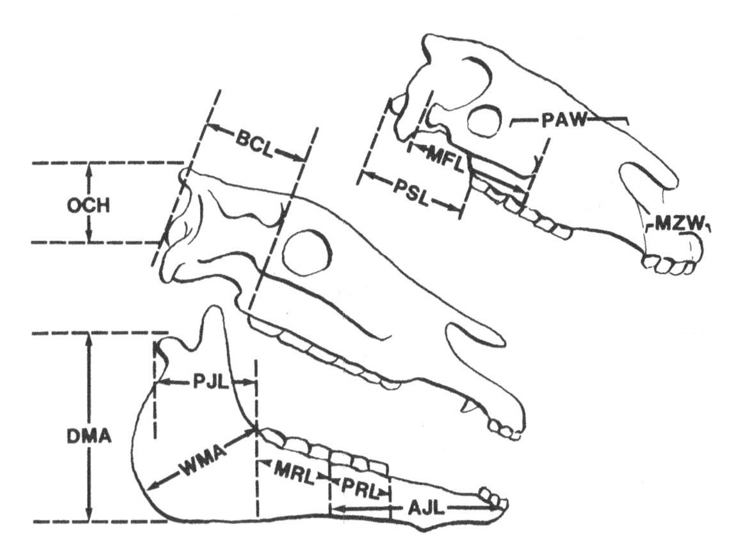

Figure 13.1. Craniodental measurements taken on ungulates and kangaroos. Abbreviations are given in Table 3.2. See text for explanation of measurements.

molar. The lengths and widths were measured as *occlusal* dimensions on the occlusal surface of the tooth. Lower teeth were measured in preference to uppers because of the greater preponderance of lower dentitions in the fossil record.

2. Lower premolar row length (LPRL) and lower molar row length (LMRL) were measured in both cases along the base of the teeth. (The advantage of this measurement, especially for fossil species, is that it can be taken in the case of dentitions with missing teeth.) Premolar row length was not included for kangaroos, for the reason detailed above. In addition, the maximum effective length of the premolar row in kangaroos (calculated from combined data from young individuals) shows almost no correlation with body weight (Janis, in press).

Skull measures

Most cranial measures taken were length variables, rather than width variables, owing to the difficulty of obtaining undistorted measurements on fossil skulls.

Muzzle width (MZW) was measured at the outer junction of the boundary between the maxilla and premaxilla. Palatal width (PAW) was

Table 13.2. *Key to Figure 13.1 and Tables 13.3, 13.4, 16.8, and 16.11 (Appendix)*

INT.	= intercept
%SEE	= percent standard error (see text for calculation)
%PE	= percent prediction error (see text for calculation)
ISO	= isometry of slope of regression line X = isometric; + = positive allometry; − = negative allometry
*	= significant at $p < .05$ (otherwise at $p < .01$)
[a]	= recalculated with the exclusion of suines

Dietary types
G = grazer; I = intermediate feeder; B = browser; O = omnivore

Variables

SLPL	= second lower premolar length
SLPW	= second lower premolar width
TLPL	= third lower premolar length
TLPW	= third lower premolar width
FLPL	= fourth lower premolar length
FLPW	= fourth lower premolar width
FLPA	= fourth lower premolar area
FLML	= first lower molar length
FLMW	= first lower molar width
FLMA	= first lower molar area
SLML	= second lower molar length
SLMW	= second lower molar width
SLMA	= second lower molar area
TLML	= third lower molar length
TLMW	= third lower molar width
TLMA	= third lower molar area

measured between the protocones of M^2. Length of the masseteric fossa (MFL) was measured from the posterior portion of the jaw glenoid to the most anterior extent of the scar for the origin of the masseter muscle. Occipital height (OCH) was measured from the base of the foramen magnum to the top of the occipital region. Length of posterior portion of the skull (PSL) was measured from the occipital condyles to the back of M^3. Basicranial length (BCL) was measured from the base of the foramen magnum to the point in the basicranium where a change in angulation occurred between the basicranium and the palate.

Jaw measures

Anterior jaw length (AJL) was measured from the boundary between P_4 and M_1 to the base of I_1 in ungulates, or to the tip of I_1 in kangaroos,

Table 13.2. (*cont.*)

FOLML	= fourth lower molar length
FOLMLW	= fourth lower molar width
FOLMA	= fourth lower molar area
SUML	= second upper molar length
SUMW	= second upper molar width
SUMA	= second upper molar area
TUML	= third upper molar length
TUMW	= third upper molar width
TUMA	= third upper molar area
LPRL	= lower premolar row length
LMRL	= lower molar row length
MZW	= muzzle width
PAW	= palatal width
MFL	= length of masseteric fossa
OCH	= occipital height
PSL	= posterior length of skull
BCL	= basicranial length
TSL	= total skull length
AJL	= anterior jaw length
PJL	= posterior jaw length
DMA	= depth of mandibular angle
WMA	= maximum width of mandibular angle
TJL	= total jaw length

where the lower incisors form part of the functional length of the lower jaw (see discussion in Janis, in press). Posterior jaw length (PJL) was measured as the horizontal distance from the back of the jaw condyle to the posterior border of M_3. Depth of mandibular angle (DMA) was measured from the top of the jaw condyle to the deepest point of the mandibular angle. Maximum width of the mandibular angle (WMA) was measured from the junction of the posterior part of M_3 with the jaw to the maximally distant point on the angle of the jaw.

Compounded variables

Dental areas were derived from multiplying dental lengths by dental widths. However, total skull length (TSL) and total jaw length (TJL) were derived from compounding other variables, rather than from direct measurements on skulls. The rationale for this was that complete skulls and jaws are rare in the fossil record, but partial skulls and jaws are relatively common, and these values for living ungulates would allow for a more direct comparison with fossil taxa with only partial skull or jaw remains. The variables were derived as follows: Total skull length was derived from adding together posterior skull length, lower molar row length, and anterior jaw length. Total jaw length was derived from adding together posterior jaw length, lower molar row length, and anterior jaw length.

In general, r^2 values were higher for perissodactyls than for other ungulates, and lower for kangaroos than for all ungulates. These differences probably merely reflect the different ranges of body masses in these various groupings. The slopes of the regression lines for the kangaroos are usually greater than for the ungulates.

Results

Extant mammals

Premolar variables. Premolar measurements are relatively poorly correlated with body mass in ungulates, and the more anterior premolars are worse than the more posterior ones in all cases, not only in r^2 values, but also in %SEE and %PE (see Table 16.8). However, this observation holds true principally for artiodactyls: The premolars show much higher values for perissodactyls primarily because perissodactyls retain large, molarized premolars in all dietary types, and premolar measurements are significantly larger in grazers; whereas artiodactyls tend to reduce the size of the premolars in grazing species (the anterior premolars in particular are significantly larger in browsers; see Table 16.8). Premolars might be good estimators of body size in perissodactyls (especially in equids, where isolated premolars are hard to distinguish from molars), but not in other ungulates.

Molar variables.

UNGULATES. Molar lengths (at least for M_1, M_2, and M^2) all show a high correlation with body mass for all ungulates (r^2 = .933–.944), and there is little or no difference in the distribution of feeding types around the regression line in the case of the lower molars (see Figures 13.2 and 13.3). The situation with M_3 is a little different: The correlation is poorer (r^2 = .905), and intermediate feeders (at least for ruminants) and omnivores both have significantly longer M_3's than other feeding types. Molar lengths decrease with wear, whereas molar width increases (see

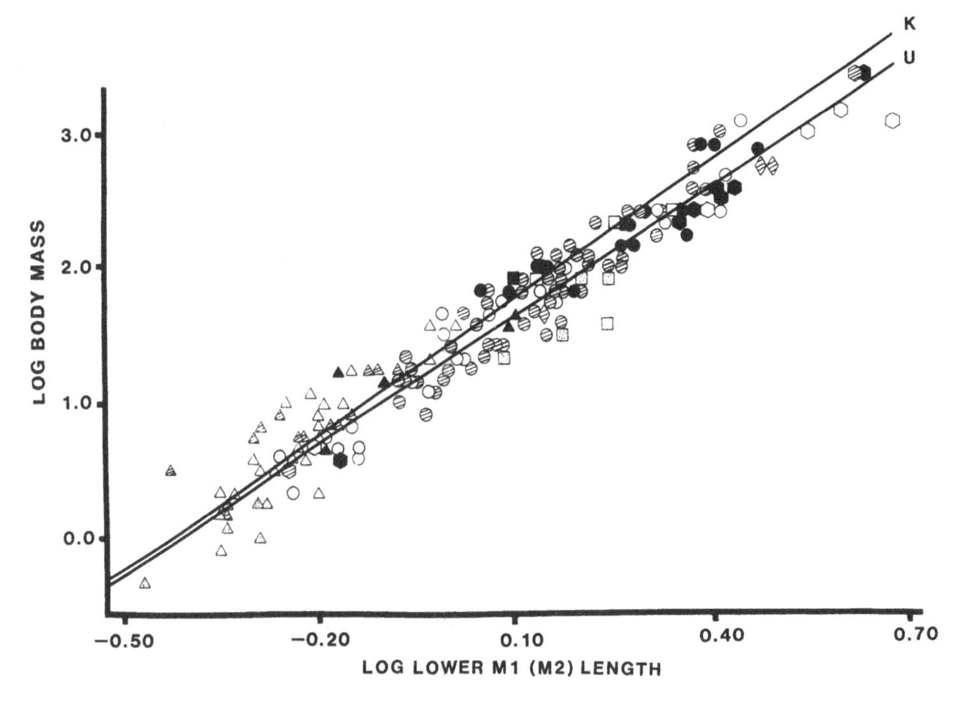

Figure 13.2. Regression of body mass on first lower molar length (second lower molar length in kangaroos). Symbols: circles = ruminant artiodactyls; diamonds = camelid artiodactyls; squares = suoid artiodactyls; pentangles = perissodactyls and hyracoids; triangles = kangaroos; U = ungulate regression line; K = kangaroo regression line. Filled symbols indicate grazers; hatched symbols, intermediate feeders; open symbols, browsers; stippled symbols, omnivores.

Fortelius 1985); therefore, one might intuitively expect molar areas to show better correlation with body mass (if the sampled means were taken from animals at different wear stages) than do the linear dimensions. However, this is not necessarily the case, as can be seen by looking at the correlations with lower molar widths. In all cases, the r^2 values for lower molar widths are lower than for lengths, and the %SEEs and %PEs are greater, owing to the fact that browsers and omnivores have broader lower molar widths than other feeding types (see Table 16.8). Consequently, the lower molar areas, although having better correlations than the widths, also show a similar distribution of feeding types around the regression line. These differences do not apply to the M^2 measurements. Here the widths do not differ between feeding types, although the M^2 length is longer in intermediate feeders (at least in

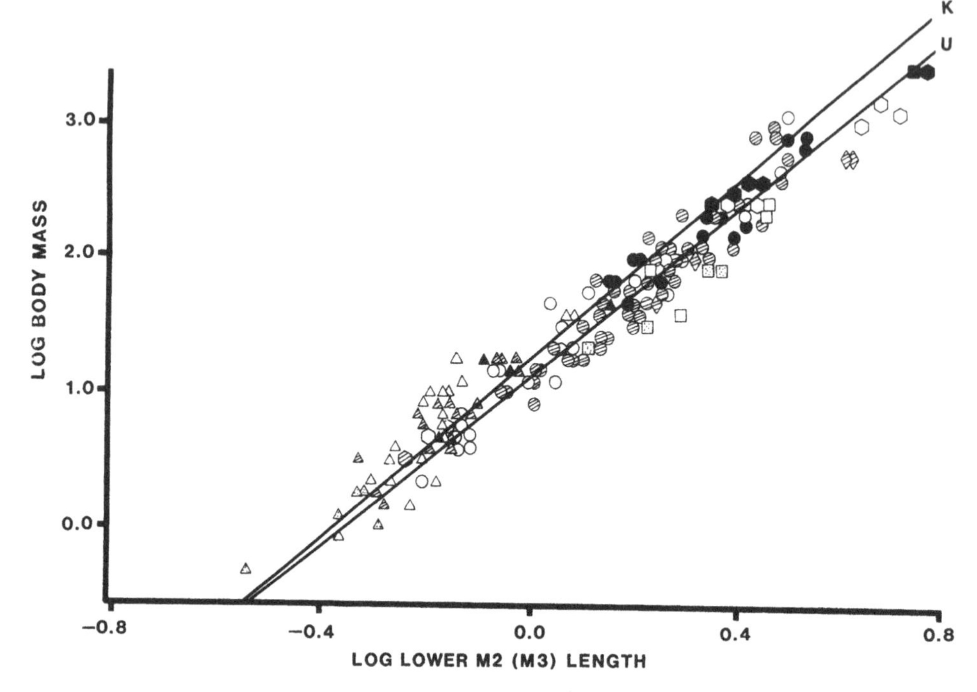

Figure 13.3. Regression of body mass on second lower molar length (third lower molar length in kangaroos). Symbols as in Figure 13.2.

ruminants) and in omnivores than in other feeding types. M^2 width is especially highly correlated with body mass in perissodactyls. M^2 area is well correlated with body mass in all ungulate groups (see Figure 13.4). In all these instances, the r^2 values for perissodactyls are much higher than for all ungulates, or for ruminants considered together, or in separate taxonomic groupings.

KANGAROOS. Kangaroo molar dimensions show a similar correlation with body mass as in ungulates, inasmuch as length measures are usually better correlated than width or areas. However, in comparison with ungulates, the molar measures are relatively more poorly correlated with body mass than are many of the cranial measures, possibly owing to the phenomenon of molar progression in many kangaroos, or to the fact that the mode of occlusion is different from most ungulates (more orthal, with bilophodont rather than trilophodont or selenodont teeth). The best molar correlations with body mass are found with the third molars, supporting the suggestion that I have made elsewhere (Janis,

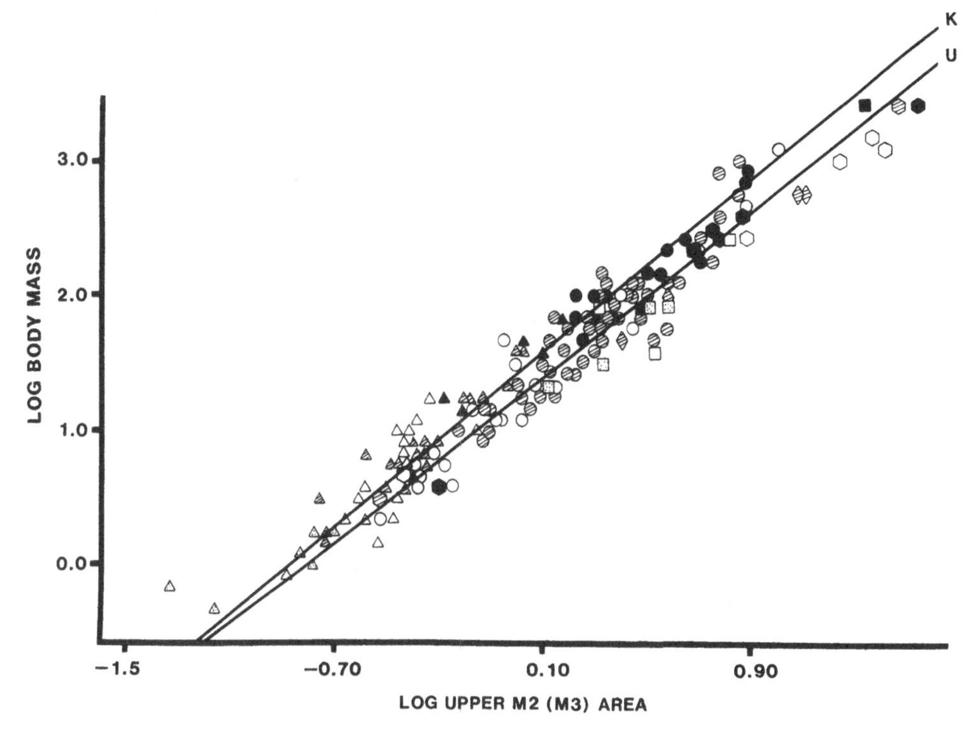

Figure 13.4. Regression of body mass on second upper molar area (third upper molar area in kangaroos). Symbols as in Figure 13.2.

in press) that it is these molars that are analogous to the second molars of ungulates. As in the case of the second molars of ungulates, the third lower molar length and third upper molar area are the variables best correlated with body mass (see Figures 13.2 and 13.3). Among the other molars, the r^2 values for the second and fourth molars are similar, whereas those for the first molar are lower than for the other molars. This suggests a similarity with the values in ungulates to the first molar, the third molar, and the fourth premolar, respectively.

Row length variables.

UNGULATES. Premolar row length correlates poorly with body mass (r^2 = .799 for all ungulates, although it is .932 for perissodactyls). This difference is mainly due to the divergent changes in dental morphology with feeding type between ruminants and perissodactyls: In ruminants, premolar row length is greatest in browsers and reduced in grazers, whereas in perissodactyls it is of a more constant length in all feeding types (although slightly greater in grazers). Premolar row lengths are relatively much shorter in camelids, and the suines are very widely dispersed.

In contrast, molar row length is more highly correlated with body mass in all ungulates (r^2 = 0.940) (see Figure 13.5), and both the %SEE (47.9%) and the %PE (32.8%) are relatively low. The only problem is that intermediate feeders and omnivores have significantly longer lower molar row lengths than other feeding types. (This same trend, although not significant except in the case of the cervids, is seen in ruminants; but in perissodactyls the trend is for the lower molar row length to be longest in grazers.) This is presumably due to the longer M_3 in omnivores and intermediate feeders.

KANGAROOS. Lower molar row length is moderately well correlated with body mass (r^2 = .905), but is relatively less well correlated in comparison with other variables to the condition in ungulates. This difference may again be ascribed to the molar progression seen in many kangaroos. This variable resembles the situation seen with most ungulates in the absence of significant residual differences around the regression line. Interestingly enough, although kangaroos have four lower molars in contrast to the three seen in ungulates, the absolute values for lower molar row length in animals of comparable size are similar (see also Janis, in press) (see Figure 13.5).

Skull variables (widths).

UNGULATES. Muzzle width is relatively poorly correlated with body mass in ungulates (r^2 = .871), and the %SEE (77.4%) and the %PE (50.4%) are relatively high. Suines have relatively broader muzzles than other ungulates. For all ungulates, grazers have significantly broader muzzles than intermediate feeders, and the same trend is also seen in ruminants and perissodactyls (although it is not significant in the latter case). Palatal width is also fairly poorly correlated with body mass (r^2 = .858), although this is apparently due to the fact that all suines have relatively narrow palates for their body size (indeed all suine points fall above those for all other ungulates, and omnivores have significantly narrower muzzles than all other feeding types) (see Figure 13.6). However, if the regression line for ungulates is recalculated with the exclusion of the suines, the correlation coefficient is much better (r^2 = .917), the %SEE (56.3%) and the %PE (38.2%) are considerably lower, and there

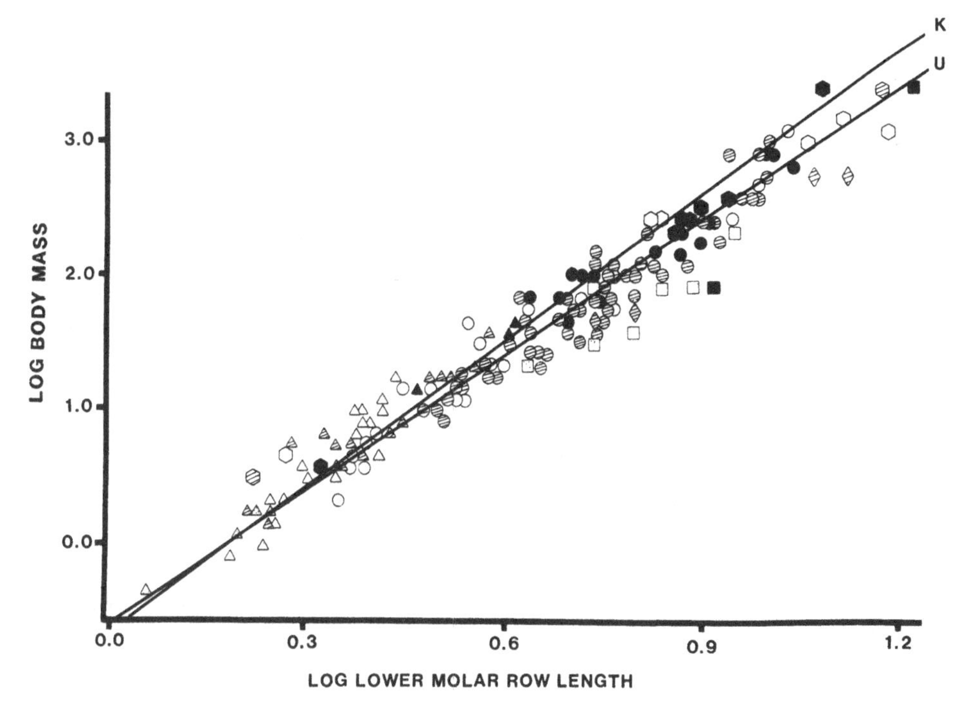

Figure 13.5. Regression of body mass on lower molar row length. Symbols as in Figure 13.2.

is no difference in the distribution of feeding types around the regression line.

KANGAROOS. Muzzle width is relatively more poorly correlated with body mass in kangaroos than it is in ungulates (r^2 = .566; and the values for the %SEE and %PE are also greater). In contrast with the condition with ungulates, the largest muzzle widths are found in browsers, and grazers have relatively narrow muzzles (see discussion in Janis, in press). Palatal width is relatively better correlated with body weight than in ungulates (even omitting suines) (r^2 = .919; and the %SEE and %PE are also lower). As with ungulates, there is no significant difference in the distribution of residuals around the regression line (see Figure 13.6).

Skull variables (lengths).

UNGULATES. Several skull measures show a fairly high correlation with body mass (r^2 = .901–0.938), but are problematical with regard to

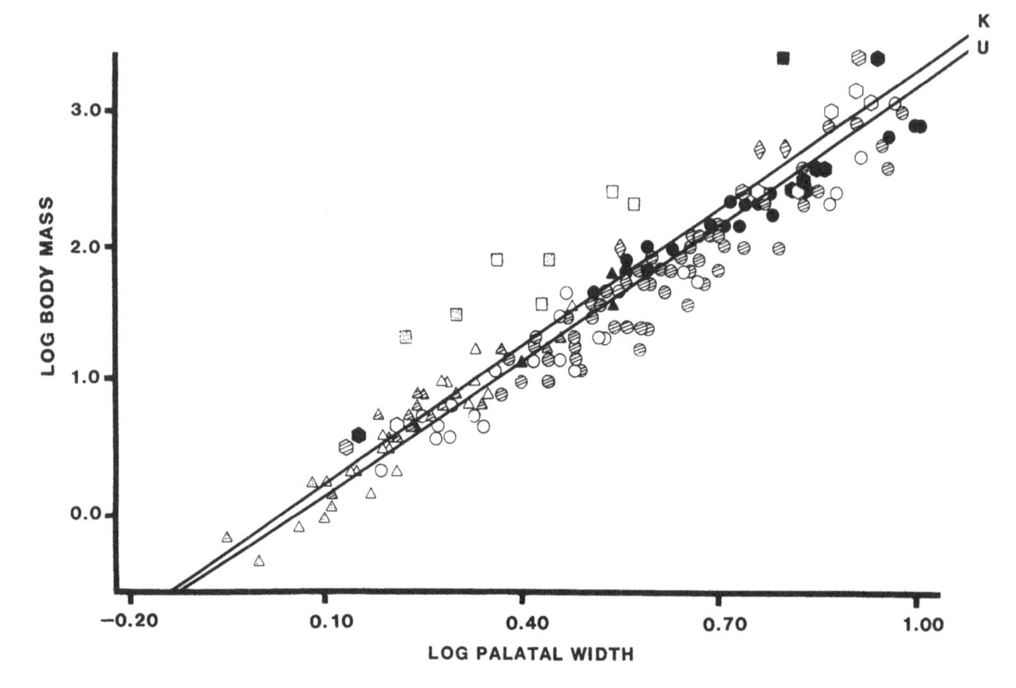

Figure 13.6. Regression of body mass on palatal width. Symbols as in Figure 13.2.

the distribution of feeding types around the line. In ungulates of all types, grazers have longer masseteric fossae than other feeding types, and omnivores have shorter fossae. Most suines have a significantly higher occiput than other ungulates, and within other ungulates browsers have higher occiputs than other feeding types. However, if occipital height is recalculated for all ungulates with suines excluded, this variable shows one of the best correlations with body mass ($r^2 = .948$), and the lowest %SEE (42.5%) and %PE (28.1%) values, although browsers still have higher occiputs than other feeding types (see Table 16.8). Grazers again have significantly longer posterior skull lengths than other feeding types, although here suines cluster well with other ungulates. Despite these differences with the dietary type, all these variables (with the exclusion of suines) have similarly relatively low %PEs (around 35%). The percent errors are lower for these variables in the more restricted taxonomic ungulate groups, with the exception of the length of the masseteric fossa in perissodactyls.

Basicranial length has been used to estimate body size in mammals

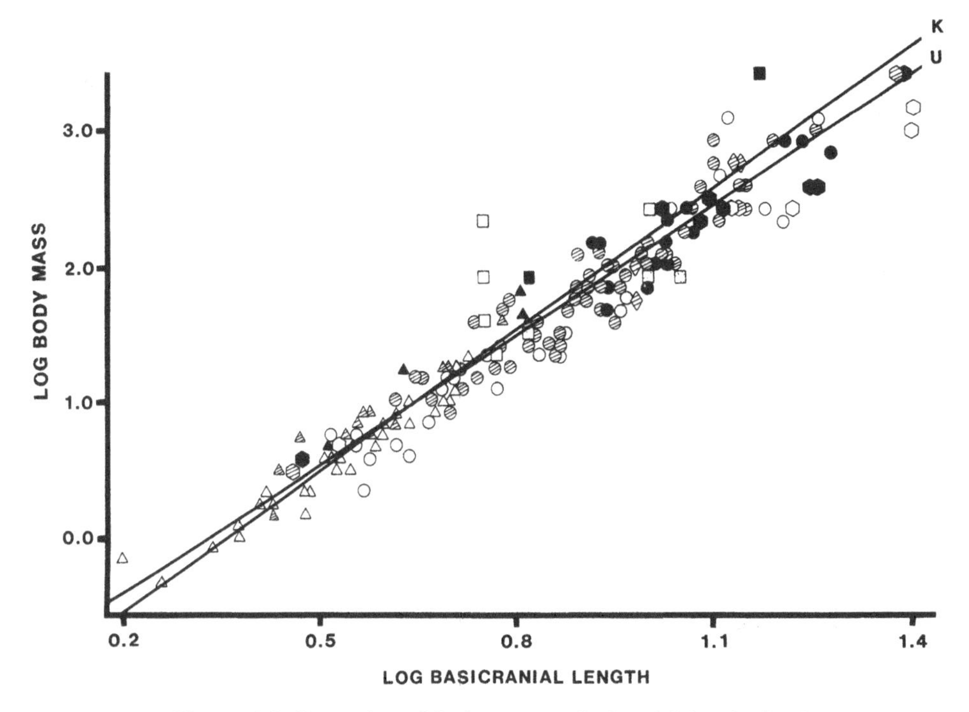

Figure 13.7. Regression of body mass on basicranial length. Symbols as in Figure 13.2.

(e.g., Gould 1974; Radinsky 1984). Here, basicranial length is not especially well correlated with body mass in the case of all ungulates (r^2 = .901), with a high %SEE (65.6%) and %PE (51.8%), although these low values are due at least in part to the divergence of the suines from the main axis (shorter basicranial length than in other ungulates) (see Figure 13.7). The correlation coefficients are slightly better for the case of ruminants only (r^2 = .913) or perissodactyls (r^2 = .928), and cervids in particular show a strong correlation of basicranial length with body mass (r^2 = .967). Thus basicranial length may be a good variable to use in the case of more restricted taxonomic groups, with a greater uniformity of cranial morphology. Although there is no significant difference in the distribution of the residuals for different feeding types around the regression line, there is a nonsignificant trend in all groups for browsers to have longer basicrania than other feeding types. In another test that I performed, using pairs of Wilcoxon *T*-tests between transformed variables (variables divided by lower molar row length to make

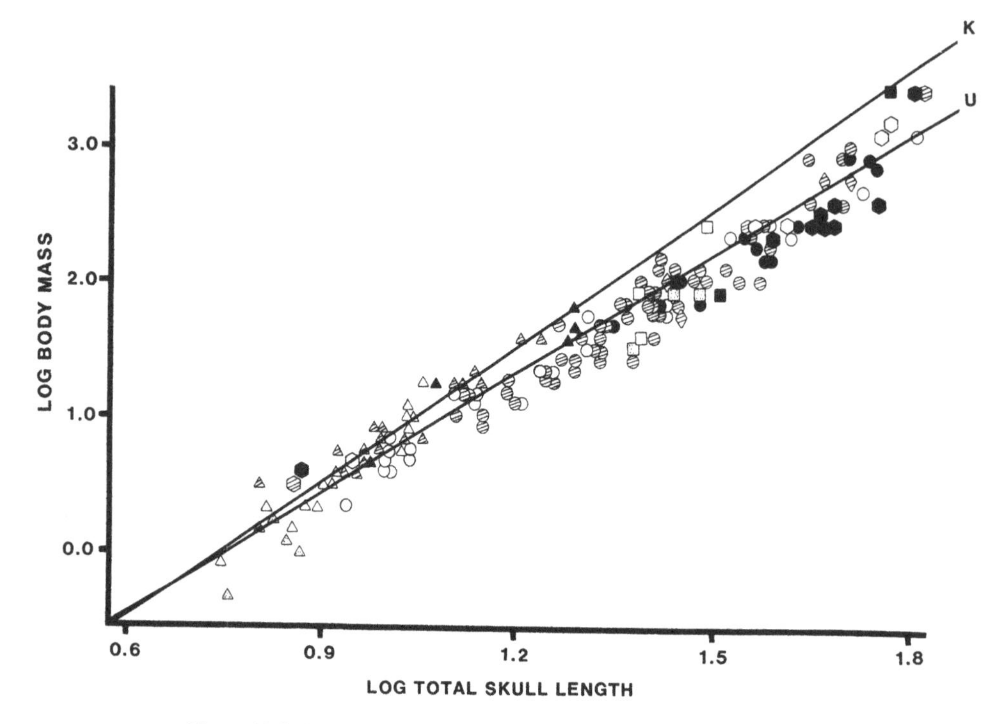

Figure 13.8. Regression of body mass on total skull length. Symbols as in Figure 13.2.

them largely independent of body mass), browsers could be seen to have significantly longer basicrania in the case of all ungulates and ruminants alone ($p < .01$).

Total skull length shows the highest correlation with body mass of any variable in all ungulates ($r^2 = .950$), and the lowest %SEE (43.7%) and %PE (30.5%) of any variable with the exception of occipital height calculated without the suines. In the case of this variable, suines cluster well with other ungulates (see Figure 13.8). However, omnivores have significantly longer skulls than other feeding groups, and there is a trend for grazers to have longer skulls than other feeding types, although this is significant only in the case of the bovids. Hyracoids have a short anterior jaw length in comparison with other ungulates, which additionally results in a short total skull length (see Figure 13.8).

KANGAROOS. The length of the masseteric fossa shows the highest correlation with body mass of all craniodental variables in kangaroos

(r^2 = .944), and the %SEE and %PE are also low. The masseteric fossa length is not especially well correlated with body mass in any ungulate group, although the r^2 value is usually fairly high. This difference between kangaroos and ungulates may be related to the morphology of the masseter muscle in kangaroos: Macropodids have a masseteric fossa in the angle of the dentary, and the overall form, size, and orientation of the masseter muscle may thus be more constrained, resulting in greater uniformity in the area of its origin from the masseteric fossa (see also discussion in Janis, in press). However, kangaroos resemble ungulates in the distribution of the residuals around the regression line for the masseteric fossa length: In both, the trend is for grazers to have a longer masseteric fossa than other feeding types (although this is not significant in the case of the kangaroos).

Occipital height is moderately well correlated with body mass (r^2 = .912), and the %SEE and %PE are better than for ungulates. There is a significant difference in the residuals around the regression line in kangaroos, but this takes the opposite form to the condition in ungulates; that is, occipital height is greater in grazers rather than in browsers. Posterior skull length is also well correlated with body mass (r^2 = .938), and again the %SEE and %PE are lower than they are in ungulates. There is a similar trend to that seen in ungulates, although not significant, for grazers to have a longer posterior portion of the skull than do other feeding types.

Basicranial length is one of the best correlates of body mass (r^2 = .926), and the %SEE and %PE are much lower than in ungulates. However, a similarly high correlation of basicranial length is seen in cervids, as previously mentioned. Despite this high correlation, there is a significant difference in the distribution of the residuals, with browsers having longer basicrania than other feeding types (similar to the nonsignificant trend seen in ungulates) (see Figure 13.7). Total skull length has a high correlation with body mass (r^2 = .935), and the %PE is lower than in ungulates. Kangaroos again resemble ungulates in that there is a similar trend (although nonsignificant) for grazers to have a longer skull length than other feeding types.

Lower jaw variables.

UNGULATES. Anterior jaw length is fairly poorly correlated with body mass (r^2 = .877), and the %SEE (75.4%) and the %PE (51.9%) are high, and there is a nonsignificant trend for grazers to have a longer

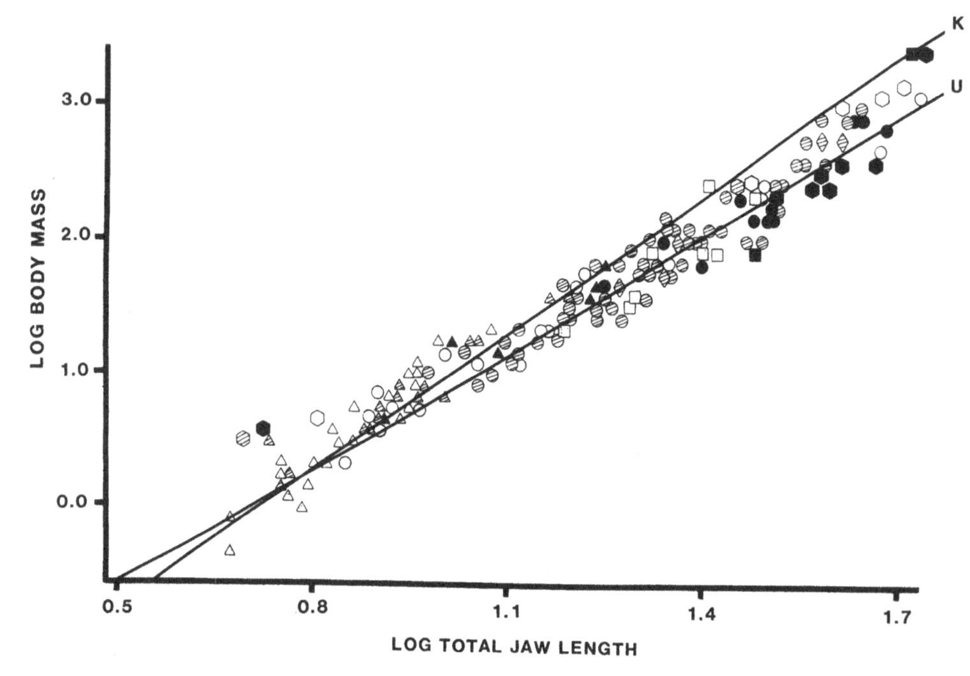

Figure 13.9. Regression of body mass on total jaw length. Symbols as in Figure 13.2.

anterior part of the jaw than other feeding types. However, a stronger correlation of this variable with body mass is seen in ruminants (r^2 = .919 in all ruminants, .932 in bovids, and .907 in cervids), and in bovids this difference between grazers and other feeding types is significant. Posterior jaw length has a somewhat better correlation for all ungulates (r^2 = .931), and both %SEE (52.8%) and %PE (36.5%) are lower than for anterior jaw length. However, grazers and omnivores have significantly longer posterior jaw lengths than other feeding types in all ungulate groups. Total jaw length shows a similarly high correlation with body mass to total skull length (r^2 = .942), and a similarly low %SEE (47.2%) and %PE (33.4%), although again omnivores have significantly longer jaws than other feeding types for all ungulates, and there is a nonsignificant trend for grazers to have longer total jaw lengths than other feeding types. Suines cluster with other ungulates, although hyracoids have relatively short jaws (see Figure 13.9).

The depth and width of the mandibular angle are both moderately well correlated with body mass (r^2 = .879 and r^2 = .916, respectively).

However, both show a significant difference in the distribution of the residuals; grazers and omnivores have significantly deeper and wider mandibular angles than other feeding types in most ungulate groups. The width of the mandibular angle shows higher correlation coefficients, and has lower %SEEs and %PEs, in the various ungulate groups considered separately. This is mainly because the absolute values are different in the different groups (greater in perissodactyls than in ruminants, and in ruminants greater in cervids than in bovids), but within each group the correlation with body mass is fairly high. However residual differences (wider mandibular angle in grazers) are seen in the case of all ruminants and in bovids alone. The depth of the mandibular angle shows relatively low correlation coefficients, high %SEEs and %PEs, and significant differences in residual values (higher in grazers), for all ungulate groups except for the case of cervids alone.

KANGAROOS. Anterior jaw length is relatively better correlated with body mass in comparison with other variables than in ungulates (r^2 = .860), and the %SEE and %PE are much lower. There is a difference in the residuals similar to the condition seen in ungulates (longer anterior jaw length in grazers). Posterior jaw length is relatively well correlated with body mass (r^2 = .927), and again the %SEE and %PE values are lower than in ungulates. There are no significant residual differences, but the trend in their distribution is similar to the ungulate condition: that is, longer posterior jaw length in grazers. Total jaw length also shows a relatively high correlation with body mass (r^2 = .933). As in ungulates, there is a nonsignificant trend for grazers to have longer jaws than other feeding types (see Figure 13.9).

The correlations of mandibular angle depth and width show a similar relation to the condition in ungulates. The width of the mandibular angle shows a better correlation with body mass (r^2 = .941 as opposed to .868), and the %SEE and %PE for both variables are somewhat lower than in ungulates. Although there is no significant difference in the distribution of the residuals, the trend is the same as for the significant difference seen in ungulates: that is, deeper and wider mandibles in grazers.

In summary, ungulates and kangaroos show a similar correlation of craniodental variables with body mass. The lower r^2 values in kangaroos are probably due to the smaller range in the distribution of body masses. The difference in the number of significant residual differences observed among the variables for kangaroos is probably due to the smaller sample size, and the trend in the distribution of the residuals is similar in all cases except for muzzle width and occipital height. The variables in kangaroos generally have a lower %SEE and %PE than the condition seen in all ungulates or all ruminants. However, these values are usually only slightly lower than those for the bovids considered alone, and are somewhat higher than those for cervids alone or perissodactyls alone, suggesting that this is related to both the relatively small sample size and the more restricted taxonomic grouping.

The correlations of palatal width and basicranial length with body mass are relatively better in kangaroos than in most ungulates. However, basicranial length is also highly correlated with body mass in cervids, suggesting that this measurement is less variable in a group with a more restricted type of cranial morphology. Occipital height is less well correlated with body mass in kangaroos than in ungulates, whereas the length of the masseteric fossa is considerably better correlated with body mass in kangaroos than in any ungulate grouping (as previously suggested, this may be related to the morphology of the masseter muscle in kangaroos). Finally, dental measures (including lower molar row length) appear to be less well correlated with body mass in kangaroos than in ungulates. As previously suggested, this may be related to the molar progression seen in many kangaroos, or to the type of occlusion.

Fossil ungulates

Body masses were predicted for a variety of North American fossil ungulates, mostly Miocene species. A number of species were examined from the families or subfamilies Dromomerycinae (Palaeomerycidae, ruminant artiodactyls), Merycodontinae (Antilocapridae, ruminant artiodactyls), Camelidae (tylopod artiodactyls), Merycoidodontidae (tylopod artiodactyls), and Equidae (perissodactyls), and a single species was examined from the Blastomerycinae (Moschidae, ruminant artiodactyls) and Protoceratidae (tylopod artiodactyls). The masses were predicted from either a single individual, or the mean values from two to four individuals from the same locality. A limited range of craniodental variables was chosen for comparison of predicted body masses, based on those variables that had showed the best correlation with body mass in extant ungulates (see Table 13.3). (Basicranial length, although not especially well correlated with body mass in extant forms, was also included, as it has been used to estimate body mass in the literature, as

Table 13.3. *Summary of "best" variables in ungulates*

Variable	r^2	%PE	Dietary differences
FLML	.930(R)–.985(P)	34.6%(R)–16.5%(P)	Never
FLMA	.941(R)–.983(P)	33.2%(U)–20.4%(P)	Rarely (U)
SLML	.928(B)–.986(P)	33.6%(U)–16.9%(P)	Sometimes (U,C)
SLMA	.935(U)–.983(P)	33.6%(U)–21.1%(P)	Sometimes (U,P)
SUML	.912(B)–.964(P)	36.5%(B)–18.3%(C)	Sometimes (U,C)
SUMA	.931(R)–.985(P)	32.7%(U)–18.6%(P)	Sometimes (U,C)
TLMA	.927(U)–.980(P)	33.1%(U)–19.9%(C)	Often (U,R,B)
LMRL	.934(B)–.978(P)	32.8%(B)–19.9%(C)	Sometimes (U,C)
MFL	.918(P)–.953(R)	51.4%(P)–21.8%(C)	Usually (U,R,B,P)
OCH[a]	.944(R)–.971(P)	30.1%(P)–20.7%(C)	Usually (U,R,B,P)
PSL	.933(B)–.959(P)	33.6%(P)–16.7%(C)	Often (U,B,P)
TSL	.945(P)–.960(R)	42.1%(P)–19.4%(C)	Sometimes (U,B)
PJL	.922(B)–.945(P)	41.9%(P)–20.3%(C)	Often (U,R,B)
WMA[b]	.916(U)–.984(P)	40.5%(U)–19.4%(P)	Often (U,R,B)
TJL	.940(P)–.956(B)	44.1%(P)–23.1%(C)	Rarely (U)

Key: U = all ungulates; R = all ruminants; B = bovids; C = cervids; P = perissodactyls and hyracoids.

[a]Variable considered without suines in data set.

[b]WMA appears to be an important variable within the different taxonomic groups, but is absolutely larger in perissodactyls than in ruminants, and within ruminants is larger in cervids than in bovids, so does not show up with a strong correlation in all ungulates or all ruminants considered together.

previously discussed.) Two masses were calculated for each variable for each species: one from the regression line for all ungulates, and one from the more restricted regression line: the ruminant-only line in the case of the artiodactyls, and the perissodactyl + hyracoid-only line in the case of the equids. The species examined, and the craniodental variables used are shown in Table 13.4.

Table 13.4 also shows the mean predicted body mass for each species, which may or may not have any real "validity," as it is obviously dependent on the diversity of the variables examined. However, it does seem to be useful in examining the standard deviation obtained for the various mass estimates, and I have also expressed the standard deviation as a percentage of the mean value in Table 13.4 (%SD). Surprisingly, the %SD is generally worse for the more restricted regression line than it is for the line for all ungulates (although this is less true in the case of the equids). Although I would have expected this to be the case for those fossil ungulates with few or no living representatives (i.e., camelids

Table 13.4. *Body mass predictions for fossil ungulates*

Species	FLML	SLML	TLML	SUML	SUMA	LMRL	AJL	PJL
Dromomerycids (ruminant artiodactyls)								
Aletomeryx marshlandensis	23.9[a]	38.7	28.8	38.5	42.4	35.4	26.9	32.1
	24.5[b]	39.4	26.4	38.9	44.4	34.6	24.1	31.1
Aletomeryx scotti	27	29.2	36.2	23.8	42.2	34.1	26.7	49.1
	29.1	27.9	33.6	23.5	44.2	33.3	23.9	47.7
Barbouromeryx trigonocorneus	39.4	24.8	32.4	20.9	26.5	33	34.4	24.1
	41.6	24.8	29.6	20.4	26.6	32.3	30.9	23.3
Bouromeryx trinitiensis	74.7	59.5	39.6	55.6	50.2	63.8	60.1	36.2
	82.1	61.7	36.8	57.3	53.3	63.2	54.7	35.1
Cranioceras sp.	146	105.3	111			113.1	84.6	56.5
	160	111.7	105.9			113.4	76.7	55
Dromomeryx whitfordi	198.5	194	206.6	108.5	163.8	182.1	196.3	151.4
	232	211.4	200.3	116.1	193.3	184.4	183.6	141.8
Drepanomeryx sp. (Observ. Q)	92.1	130.8	184.7	126	142.3	143.3	186.5	243.8
	102.5	140.1	178.5	135.9	165.9	144.4	174.2	240.7
Drepanomeryx sp. (Trinity R)	115.8	172	183.3	264.8	222.9	172.7	316.3	246.9
	130.8	186.4	177.2	297.5	270.4	174.7	299.2	243.8
Pediomeryx hamiltoni	243.3	214.5	224.7	209.3	189.6	222.4		192.8
	288	234.7	218.3	232.1	226.7	226.1		189.9
Pediomeryx hemphillensis	87	107.1	114.6	92	108.5	129	216.6	184.7
	96.5	113.7	109.5	97.4	123.5	129.7	203	181.9
Procranioceras skinneri	107	91.7	86	82.3	116.4	103.8	88.3	45.8
	120.2	96.8	81.5	86.7	133.3	103.9	81.1	44.4
Rakomeryx sinclairi	152	72.4	144.1	170.1	168.7	146	110.2	91.6
	174.6	75.6	138.4	186.5	199.7	147.2	101.7	89.6
Sinclairomeryx sinclairi	65.1	73.7	84.8	94.4	88.4	73.3	45.9	117.3
	70.9	77.1	80.3	100.2	93.4	72.9	41.6	115
Subdromomeryx scotti	65.9	65.7	76.7	52.4	67.6	64.3	42.6	38.6
	71.8	68.4	72.5	53.8	73.8	63.7	38.4	37.5
Merycodontine antilocaprids (ruminant artiodactyls)								
Cosoryx furcatus	20.3	28.3	23	22.1	17.9	16.9	13.5	20.6
	20.6	28.4	21.1	21.6	17	16.3	11.9	19.8
Merriamoceras coronatus	149	15.9	11.9	13.1	10.2	12.8	10.7	
	14.7	15.4	10.7	12.5	9.4	12.3	9.4	
Meryceros nenzelensis	14.5	23.4	30	24	18.8	20.6	29.1	38.3
	14.3	23.3	27.6	23.7	18.3	20	26.1	37.1
Merycodus sabulonis	10.2	11.2	11.7	13.6	14.7	9	14.3	18.1
	9.9	10.8	10.6	12.9	14	8.6	12.6	17.4
Paracosoryx wilsoni	18.1	22	22.9	20.1	18.3	16.8	16	12.7
	18.2	21.8	21	19.6	17.7	16.2	14.1	12.2

WMA	TJL	PAW	BCL	MFL	OCH	PSL	TSL	Mean	SD	%SD
25.7	28.8	19.4	19.8	25	22.8	23.1	25.3	28.5	7	24.5
28.2	27.5	15.2	18.1	23.5	23.9	23.2	24.6	28	7.9	28.2
33.6	32.9	34.2	13.8	14	28.3	26.8	26	29.7	9	30.4
37.5	31.4	27.3	12.4	13.1	26.8	26.7	25.9	29	9.4	32.2
21.7	29.2	44.6	29.3	2.66	50.7	32.8	32	31.4	8.1	25.8
23.3	27.9	36	27.3	25	55.3	32.8	31.1	30.5	8.5	27.7
42.8	50.7	43.8	46.7	49.4	56.6	44.8	52.2	51.7	9.8	19.1
48.4	48.8	35.3	44.4	46.7	62.2	44.6	50.7	51.6	12.6	23.4
59	78.1							93.3	29	31.1
68.1	75.6							95.8	33.8	35.3
94.7	179.2	168.7	155.7	249.4	158.6	213.7	204.6	176.6	38.8	22
112	175.5	143.3	156.6	238.6	182.7	206.2	198.4	180.2	36.8	20.4
121	196.8	228.9	36.3		79.2			147.1	59.7	40.6
146	192.3	196.8	34.1		88.3			149.3	58	35.5
163	265.3		61.3		108.3			191	74.9	39.2
100	261.2		59		122.5			201.8	75.1	37.2
129		84.6						190	51.1	26.9
156		69.9						204.6	61.7	30.2
	187		19.7	56.2	97.9			116.7	56.3	48.2
	184		18	53.2	110.3			118.4	53.1	44.9
59	74.1	27.3	76.8	78	96.1	65	78.8	79.8	22.8	28.6
68.1	71.7	21.6	74.8	73.9	108.1	64.2	76.6	81.7	27.3	33.6
67.4	109.8	145.3	32.3	78.1	41.5	89.6	105.9	107.8	43.4	40.2
78.3	106.8	122.7	30.2	74	44.9	87.9	102.8	110.1	48.7	44.2
82.6	69.2	42	33.8	50.7	33.8	57.6	53.9	66.4	23.1	34.8
97.2	66.9	33.7	31.7	47.9	36.2	57	52.4	67.2	25.9	38.4
33	43.8	62	84.7	62.8	81.1	63.9	53.5	59.9	15	25
36.8	42.1	50.6	82.9	59.5	90.6	63.1	52	59.8	16.5	27.5
16	15.1	15.5	12.3	13	12	11.3	12.2	16.9	4.8	28.7
17	14.3	12	11	12.2	12.3	11.6	11.9	16.2	5	30.9
								12.8	2.1	16.1
								12.1	2.4	20
25.4	27.9	17.9	16.3	25.8	20.6	25.8	24.6	23.9	5.9	24.5
27.9	26.7	13.9	14.7	24.2	21.6	26	23.9	23.1	6	26
12.5	12.8		9.7		11.8	12.3	11.3	12.3	2.3	18.7
13.1	12.1		8.6		12.1	12.6	11.1	11.9	2.3	19.3
18.8	14.1	17.4	7.2	12.2	14.2	9.6	12.2	15.8	4.3	18.7
20.2	13.3	13.5	6.3	11.4	14.6	9.8	11.9	15.1	4.4	29

Table 13.4. (*cont.*)

Species	FLML	SLML	TLML	SUML	SUMA	LMRL	AJL	PJL
Blastomerycine moschids (ruminant artiodactyls)								
Parablastomeryx gregori	24.7	14.8	20.9	14.4	21.7	17.4	19.4	14.1
	25.2	14.3	18.5	13.8	21.5	16.7	17.2	13.6
Camelids (tylopod artiodactyls)								
Oxydactylus campestris	90.5	103.5	127.6	85.9	112.1	97.7	80.6	99.6
	101.8	109.8	122.2	90.7	128	97.7	73.9	97.5
Poebrotherium sp. (Douglas)	18.4	23	36.2	29.1	28.5	25.4	11.5	10.2
	18.4	28.4	33.6	29.5	28.8	24.7	10	9.8
"*Pseudalticamelus*" sp. (Love)	415.3	473.9	341	741.2	406.8	375.6	213.6	162.9
	508.5	536.2	334.8	881	523.1	385.9	200.2	160.2
Stenomylus hitchcocki	59.3	162.3	200	177.3	107.4	102.6	14.5	24.5
	64.2	175.4	193.8	194.9	122.1	102.7	12.7	23.7
Protoceratids (tylopod artiodactyls)								
Protoceras celer	45	45.2	46.4	38.9	37.1	42.8	50.1	31.9
	47.5	46.4	43.3	39.4	38.4	42.1	45.4	30.9
Oreodonts (tylopod artiodactyls)								
Merychyus arenarum	34.3	48.4	89.6	43.9	57.9	56.6	8.5	42.8
	35.9	49.7	85.1	44.7	62.4	55.9	7.4	41.6
Merycochoerus proprius	215.5	294.4	586.5	760.9	798.4	412.7	32.3	143.1
	253.1	326.5	583.9	906.4	1085	424.8	29	140.5
Minicochoerus gracilis	21.7	18.8	20.6	59.6	87.6	21.2	3.5	13.5
	22.1	18.6	18.9	61.6	97.8	20.5	3	13
Promerycochoerus carrikeri	240.8	228.3	481.9	179.8	299.8	347.4	56.9	104.4
	284.8	250.4	476.7	197.7	373.4	356.4	51.7	102.3
Ustatochoerus sp. (Valentine)	100.7	183.4	224.7	219.5	194.6	192.3	16.6	89.3
	112.7	199.4	218.3	224	233.3	194.9	14.6	87.3
Equids (perissodactyls)								
Archaeohippus blackbergi	21.7	16.5	8.7	20.3	37.9	11.4	32.3	17.5
	18.4	19.9	19.6	25.5	35.5	17.9	61.4	24.5
Cormohipparion plicatile	280.5	178.3	72.3	109.4	177.5	147.2	216.6	330.9
	223.4	186.3	144.2	118.2	148.8	194.3	304.9	359
Dinohippus leidyanus	248.4	188.7	165.7	174.9	325.8	209.3	280.2	259
	198.4	196.5	315	181.1	261.6	269.6	378.7	287
Mesohippus bairdi				23.2	35.3	26.2	38.9	
				28.8	33.2	38.9	71.9	
Neohipparion trampasense	122.1	76.9	72	156.2	189.8	104.2	233.2	236.[illegible]
	99.2	84.5	143.7	163.5	158.4	140.8	324.4	264.[illegible]
Parahippus cognatus				108.5	213.7	84.8	186.1	
				117.3	176.8	116.3	268.4	

WMA	TJL	PAW	BCL	MFL	OCH	PSL	TSL	Mean	SD	%SD
25.3	16.1	10.8	17.2	8.1	24.8	19.6	17.9	17.9	4.9	27.4
27.7	15.3	8.2	15.6	7.5	26.2	19.8	17.4	17.4	5.8	33
133	88.7	45	125.3	71.9	56.4	92.9	87.7	93.7	24.1	25.8
162	86	36.3	124.7	68.1	61	91.1	85.2	96	29.8	31
	12.7	4.2				14.3	13.9	19.4	9.8	50.5
	12	3.1				14.5	13.5	18.8	9.7	51.8
179	226.5	225.4	186.8	94	101.7	125.9	195.4	297.2	170.9	61.2
221	222.5	193.7	189.4	89.3	114.7	122.8	189.5	304.5	213.4	70.1
22.1	28.5	31.2	20.9	17	18.2	13.3	21.5	63.8	64.8	102
24	26.9	24.8	19.2	16	18.9	13.6	21	65.9	68.7	104
40.3	40.6	15.3	39.2	70.9	50.7	29.3	37.5	41.3	11.3	28.3
45.3	39	11.9	37	67.2	55.3	29.4	36.5	42	12	29.2
70.4	23.2	35	42.2	18.1	54.4	25.1	19.9	41.9	21.1	50.3
82	22.1	27.9	40.2	17	59.4	25.3	19.4	42.3	22.6	53.5
188	107.3	268.2	102.2	112.5	241.8	62.3	77.8	275.2	242.7	88.2
232	104.3	231.9	100.8	107	284	61.6	75.6	309.2	306.8	99.3
38.6	8.3	13.5	20.7	16.5	18	10.1	7.8	23.5	21.4	91.4
37.5	7.8	10.4	19	15.5	18.7	10.4	7.6	23.9	24.1	101
206	110.2	77.1	160.3	83.2	253.9	100.8	107.8	189.8	115.8	61
256	107.2	63.5	161.4	78.8	298.9	98.7	104.7	203.9	128.5	63
53	56.5	55.5	129.2	59.1	162.1	72.9	55.5	122.8	67.7	55.2
87	54.5	45.1	128.8	55.9	186.9	71.8	54	130.5	77.5	59.3
32	20.3	34.6	9	18.1		27.6	23.8	22.1	9.2	41.5
12.3	32.6	43.4	12.5	21.1		26.9	32.4	26.9	12.8	47.8
60								201.1	79.9	39.6
54								212.6	80.6	37.9
31	263.6	187.4	381.3	319.2	159.6	232.2	251.7	251.5	71.1	39.6
43	318.9	258.2	365.3	326.2	150.5	227.4	286.2	266.5	65.5	37.9
		20.8	33.8	20.1	32.4	25.7	29	28.5	6.4	22.5
		25.3	41.1	23.3	28.4	25.1	38.7	35.5	14.3	40.4
14	204.3	104.4	111.1	190.5	111.9	170.8	181	155	54.8	35.4
22	254.3	139.2	120.6	199.3	103.7	167.2	210.9	168.5	66.6	39.5
		87	130.4	78.5	105.5	101.5	126.5	122.2	44.7	36.6
		115.7		138.8	85.5	97.6	99.3	151.4	136.6	53.8

Table 13.4. (*cont.*)

Species	FLML	SLML	TLML	SUML	SUMA	LMRL	AJL	PJL
Parahippus leonensis	75.5	54.6	45.6	68	107.1	49.1	81.8	92.9
	62	61.2	93.4	76.7	93.1	70	134.3	112.6
Pliohippus sp. (Valentine)	179.5	163.5	132.4	113.9	218.8	153.1		
	144.4	171.7	245.5	122.5	180.5	201.5		
Pseudhipparion sp. (Love)	59.3	45.3	51.3			41.6	74.5	91.1
	48.9	51.4	104.3			59.8	124.2	118.3

Note: Abbreviations as in Table 13.1.
[a] *First line*: predictions from ungulates-only regression line.
[b] *Second line*: predictions from ruminants-only regression line, except for Equids section, in which second line = perissodactyls + hyracoids-only regression line.

and oreodonts), I am surprised to find that this is also the case for the dromomerycids and merycodontines, which appear to bear a close resemblance to extant ruminants.

In the case of the dromomerycids, the predicted masses that are closest to the mean estimates are usually derived from total jaw length and total skull length. High predicted values are usually obtained from first lower molar length, second upper molar length, and second upper molar area. Palatal width and basicranial length show the greatest inconsistency in predicted values, yielding either very high or very low estimates, depending on the species. Additionally, the estimates from palatal width show the greatest discrepancy in the values obtained from the two different regression lines (giving higher estimates in the case of the line for all ungulates). The lowest variation in estimated masses between the two regression lines is seen for lower molar row length and posterior skull length.

In the case of the merycodontine antilocaprids, the predicted masses that are closest to the mean value are usually derived from the lower molar row length, the second upper molar area, and the width of the mandibular angle. High estimates are usually obtained from the second and third lower molar lengths and the posterior jaw length, low estimates from the basicranial length and first lower molar lengths, and variable estimates from the palatal width. As with the dromomerycids, palatal width yields the greatest discrepancy in the masses predicted from the

WMA	TJL	PAW	BCL	MFL	OCH	PSL	TSL	Mean	SD	%SD
106	75.1	43.1	27.6	63.5	49	70.7	68.8	67.4	22.4	33.4
52.2	104.5	54.6	34.2	69.9	43.7	69	86.2	71.6	26.6	34.9
		199.9		138.5	109.1	106.3	158.2		152.1	36.9
		276.4		146.5	117.1	98.3	154.2		169.9	55.9
85.2	70.9							65.8	19.9	30.3
40.2	99.2							80.8	34.1	42.2

two regression lines, with higher values predicted from the all-ungulates line. The greatest similarity in mass predictions between the two regression lines is seen with the second upper molar area, occipital height, posterior skull length, and total skull length. The one blastomerycine moschid examined, *Parablastomeryx gregori,* appears to be similar to the other ruminants in the distribution of the predicted masses with the mean value.

As might be expected, camelids have generally higher %SD values than the ruminant artiodactyls or the equids. The variables that yield mass estimates that are closest to the mean values are usually the lower molar row length (a notable exception being the genus *Stenomylus,* with its bizarrely elongated third molars) and the posterior skull length. Consistently low estimates are obtained with the variables palatal width, anterior jaw length, length of masseteric fossa, and occipital height, suggesting that camelids as a group have relatively lower values for these variables than do most extant ungulates. Consistently high estimates are obtained for third lower molar length, second upper molar length, and second upper molar area. This is not surprising, as camels have a reduced premolar row, and show consequent elongation of the posterior molars (although all molar lengths yield high body mass estimates). Given the high estimations obtained from the molar lengths, it is surprising that the estimation from lower molar row length is usually the one closest to the mean value. However, first lower molar lengths are usually also closer to the mean than other length estimates, and lower molar length is measured along the base of the teeth, rather than along the occlusal surface. The greatest similarity between mass estimates from the two regression lines is seen from posterior skull length, and the greatest discrepancy is seen with palatal width (higher estimates produced from the ungulates-only line).

In contrast with the situation in the camelids, the single protoceratid, *Protoceras celer,* has %SDs that are more similar to the ones obtained for the fossil ruminants. Lower molar row length, total skull length, and total jaw length yield estimates that are close to the mean value; high estimates are obtained from the length of the masseteric fossa, the occipital height, and the anterior jaw length (which also shows the greatest discrepancy in the estimates derived from the two regression lines), in contrast to the low estimates for these variables in camelids, and low estimates are obtained from the palatal width. All molar variables yield estimates reasonably close to the mean value. This range of estimates seems to confirm the subjective observation that protoceratids (at least the early ones) have a cranial design that is closer to that seen in ruminants than to that of camelids, despite their closer taxonomic affiliation with camelids.

The oreodonts are the group with the greatest %SDs in their predicted masses, as might be expected since they have no living relatives. Although they show at least a superficial resemblance to hyraxes in cranial design, their skull is obviously built on a different plan from that seen in the majority of extant ungulates. The variables that produce the body mass estimates closest to the mean value are basicranial length and occipital height, although with the great range observed in the predicted masses it is difficult to assign any real meaning to this observed "mean value." High estimates are obtained from third lower molar length, second upper molar length, and second upper molar area, as also seen in the camelids, and the estimates from lower molar row length are also usually fairly high. Very low estimates are obtained from anterior jaw length, which is not surprising, as oreodonts have a short jaw and lack a diastema. However, this means that the estimates obtained from total skull length and total jaw length are also low. Low estimates are also obtained from the length of the masseteric fossa, and the posterior length of the skull. The greatest similarity in estimates between the two regression lines is seen in the case of the lower molar row length and the posterior skull length, whereas the greatest discrepancy is seen in the second upper molar area and the width of the mandibular angle (in both cases, the estimates obtained are higher for the ruminants-only line).

Finally, in the case of the equids, the %SDs are comparable to those obtained for the fossil ruminants. The variables that yield estimates closest to the mean value are second upper molar length, posterior skull length, and total skull length. High estimates are obtained for anterior jaw length, especially from the perissodactyls-only regression line (as might be expected, as equids are particularly long-jawed in comparison to tapirs, rhinos, and hyracoids). Low estimates are obtained from basicranial length, occipital height, and third lower molar length (especially in the case of the ungulates-only line, as perissodactyls lack the posterior extension of the last molar typical of ruminants). The greatest similarity in mass estimates between the two regression lines is seen with the variables posterior skull length and second upper molar length. The greatest discrepancy is seen with the estimates from the third lower molar length, the lower molar row length, and the palatal width (in all cases, the predicted values are higher for the perissodactyls-only line).

A quantitative comparison was made of the deviation of the masses estimated from the two different regression lines for each variable, and of the deviation of these estimated masses from the mean value mass estimate from all the variables. The aim of this comparison was to determine which variables were the most consistent in their estimations of body mass in individual species across the range of taxa. (Application of the same methodology to living ungulates of known body mass will hopefully yield a percentage reliability estimate.) The most constant variables were found to be lower molar row length, posterior skull length, and total skull length. Table 13.5 compares the body mass estimates from these three variables for the range of fossil ungulates examined. (The percent deviation values in parentheses show the extent to which the mass estimate for a given variable deviated from the mean of all the estimates calculated for that species in Table 13.4.)

Discussion

Results from extant mammals

Molar lengths seem to be the best dental predictors of body mass in ungulates; in general, they are slightly better than molar areas, and considerably better than molar widths (this conclusion is also supported

Table 13.5. *Mean body mass predictions for fossil ungulates from more "constant" craniodental variables*

Species	LMRL	PSL	TSL	Other
Dromomerycids				
Aletomeryx marslandensis	35.0 (23.9)	23.1 (18.1)	25.0 (12.2)	22.8
Aletomeryx scotti	33.7 (14.9)	26.8 (8.7)	26.0 (10.6)	
Barbouromeryx trigonocorneus	32.7 (5.5)	32.8 (6.1)	31.0 (1.8)	
Bouromeryx trinitiensis	63.6 (23.0)	44.7 (13.5)	51.5 (1.3)	50.8
Cranioceras sp.	113.2 (10.6)			
Dromomeryx whitfordi	183.3 (2.7)	210.0 (17.7)	178.4 (25.9)	
Drepanomeryx sp. (Observ. Q)	143.8 (2.9)			
Drepanomeryx sp. (Trinity R)	173.7 (11.3)			
Pediomeryx hamiltoni	224.3 (13.8)			
Pediomeryx hemphillensis	129.3 (13.3)			
Procranioceras skinneri	103.9 (28.7)	64.6 (23.4)	77.7 (6.1)	
Rakomeryx sinclairi	146.6 (34.7)	88.8 (18.5)	104.4 (4.2)	
Sinclairomeryx sinclairi	73.1 (9.5)	57.3 (14.3)	53.2 (20.4)	
Subdromomeryx scotti	64.0 (6.9)	63.5 (6.0)	52.8 (11.9)	
Merycodontines and blastomerycids				
Cosoryx furcatus	16.6 (0.3)	11.4 (30.7)	12.1 (27.2)	
Merriamoceras coronatus	12.6 (1.1)			
Meryceros nenzelensis	20.3 (13.6)	25.9 (7.0)	24.3 (4.8)	
Merycodus sabulonis	8.8 (27.3)	12.5 (3.1)	11.2 (7.6)	18.8
Paracosoryx wilsoni	16.4 (6.7)	9.7 (37.1)	12.1 (22.2)	
Parablastomeryx gregori	17.1 (3.5)	19.7 (11.7)	17.7 (0.2)	

Taxon	LMRL	PSL	TSL	Other
Camelids and protoceratids				
Oxydactylus campestris	97.7 (3.1)	92.0 (3.0)	86.5 (8.8)	
Poebrotherium sp. (Douglas)	25.1 (31.0)	14.4 (24.5)	13.7 (28.3)	
"Pseudalticamelus" sp. (Love)	380.8 (30.7)	124.4 (57.3)	192.5 (33.9)	
Stenomylus hitchcocki	102.7 (58.5)	13.4 (79.3)	21.3 (67.1)	53.0
Protoceras celer	42.5 (3.1)	29.4 (28.6)	37.0 (10.0)	
Oreodonts				
Merychyus arenarum	56.3 (33.7)	25.2 (40.2)	19.7 (53.2)	
Merychochoerus proprius	418.8 (43.7)	62.0 (78.8)	76.7 (73.6)	61.0
Minichochoerus gracilis	20.9 (11.9)	10.3 (51.8)	7.7 (67.5)	
Promerychochoerus carrikeri	351.9 (78.9)	98.8 (49.3)	106.3 (46.0)	100.3
Ustatochoerus sp. (Valentine)	193.6 (53.0)	72.4 (42.9)	54.8 (56.8)	
Equids				
Archaeohippus blackbergi	14.7 (41.2)	27.3 (12.4)	28.1 (14.0)	
Cormohipparion plicatile	170.1 (17.7)			
Dinohippus leidyanus	239.5 (8.5)	229.8 (10.9)	270.0 (5.2)	242 (230)
Mesohippus bairdi	32.6 (8.9)	25.4 (19.7)	33.9 (5.3)	(42)
Neohipparion trampasense	112.5 (24.7)	169.0 (5.5)	161.8 (21.0)	
Parahippus cognatus	100.6 (17.6)	100.4 (22.2)	140.0 (7.1)	113.6 (111)
Parahippus leonensis	60.0 (17.6)	69.8 (7.0)	77.5 (7.7)	(76.6)
Pliohippus sp. (Valentine)	177.3 (9.6)	156.3 (6.5)		175.6 (155.3)
Pseudhipparion sp. Love	50.7 (31.4)			

Note: Mass estimates represent average values from the two predictions from the ungulates-only regression line and the more restricted taxonomic regression line (see Table 13.4).

Key: LMRL = lower molar row length; PSL = posterior skull length; TSL = total skull length; Other = body weight estimates from Scott (this volume) or MacFadden (1986). MacFadden estimates are in parentheses. Figures in parentheses after body mass estimates are percent deviation of that estimate from the mean value (see text for explanation).

by both Damuth and Fortelius, this volume). However, molar dimensions are relatively poorer estimators of body mass in kangaroos. Second molar measures are best for the case of all ungulates considered together, but this is not necessarily true for the more restricted taxonomic groupings. The third molar appears to be the tooth in kangaroos that is analogous to the second molar in ungulates, and these teeth (upper and lower) in general yield the best predictions of body mass in macropodoids. Premolar variables, and molar widths and areas, have higher correlation coefficients and lower percent errors in perissodactyls (and hyracoids) than in ruminants. I speculate that the reason is that hindgut fermenters are more constrained than ruminants in their dental dimensions, because they process a greater amount of food per day (Janis 1976), and chew the food more thoroughly on initial ingestion (see data in Pollock 1980).

First and second lower molar lengths have the advantage over other dental measurements with r^2 values in that they show little or no residual differences by dietary type around the regression line. (This is also the case for the second and third lower molar lengths, respectively, in kangaroos.) Whereas other dental variables, such as second upper molar width, may show higher correlation coefficients and lower percent errors in perissodactyls, first and second lower molar lengths show a consistency in correlation and percent errors across all ungulate groups, and I suggest that these should be the preferred variables chosen for body mass prediction if only single teeth are available. However, better r^2 and percent error values are usually obtained by total lower molar row length. This may be superior to any single dental length because, as the occlusal lengths of M_1 and M_2 decrease with the degree of wear, both the occlusal and the basal length of M_3 increase with age (see Fortelius [1985]; this conclusion is also supported by my work in preparation). Thus lower molar row length may be a superior variable to use across a range of ages of mature individuals over any single dental dimension. However, it should be noted that lower molar row length is a relatively poorer estimator of body mass in kangaroos than in ungulates.

All skull and jaw variables presented here, with the exception of muzzle width and anterior jaw length, have a fairly high correlation with body mass, and fairly low percent prediction errors. (In general, these are better for the skull variables, excluding palatal width and basicranial length, than the jaw variables.) However, all of these variables (with the exception of palatal width if suines are excluded, and basicranial length) have the same problem in that there are significant residual differences of the different feeding types around the regression line. Basicranial length shows a better correlation and prediction error in certain groups (cervids, kangaroos, and perissodactyls), but despite the popularity of this variable as an estimator of body mass (e.g., Radinsky 1984), I would be reluctant to use this variable for mass prediction except maybe in the case of restricted taxonomic groups with a relatively constrained cranial morphology. (For example, fossil cervids might be compared with living forms; e.g., Gould 1974.)

Anterior jaw length is better correlated with body mass in ruminants than in perissodactyls, whereas muzzle width shows a better correlation in perissodactyls than in ruminants. Both these observations may be related to the fact that ruminants are more selective feeders than perissodactyls (Gwynne & Bell 1968); thus ruminants may be more constrained in the length of the jaw (i.e., they are all fairly "long-faced"). In contrast, hindgut fermenting perissodactyls rely on a greater daily intake of food than ruminants (Demment & Van Soest 1985), and so tend to have a greater constancy of muzzle width regardless of dietary type (see also data in Janis & Ehrhardt 1988). Anterior jaw length is also moderately well correlated with body mass in kangaroos. In comparison with ungulates, kangaroos show a relatively poorer correlation of occipital height with body mass, and a far superior correlation of the length of the masseteric fossa. Posterior skull length and posterior jaw length appear to be fairly good predictors of body mass in all herbivorous mammals, although all of these variables have values that are significantly greater in grazers. Total skull length and total jaw length have a high correlation with body mass in all ungulates (although they are less well correlated, in comparison with other variables, in perissodactyls than in ruminants). Although grazers tend to have relatively higher values for these variables than other feeding types, these differences are not significant.

Fossil ungulates

Although it may seem circular to derive conclusions about the value of craniodental variables for body mass estimation from a consideration of the values predicted for fossil species, some patterns clearly emerge from the range of predicted masses presented here, that have some bearing on choosing the "best" craniodental variables for the estimation of body mass. For a start, it is apparent that the range of predicted body masses is much greater for those fossil mammals that have few or no close living relatives, the camelids and the oreodonts. In both groups, anterior jaw length yields masses that are consistently below the range of the other predicted masses (although this is far more apparent for oreodonts than for camelids). The variability of predictions based on anterior jaw length throws into considerable doubt the value of total skull length or total jaw length for body mass estimation in fossil ungulates that lack a diastema. (This is true, for example, for most of the endemic South American ungulates.) In contrast with the situation in these tylopod artiodactyls, equids appear to have especially long lower jaw lengths, and masses estimated from this variable are high in comparison to estimates based on other variables. The length of the masseteric fossa and occipital height also yield consistently low estimates for camelids. Interestingly enough, these two skull variables also show a difference in their correlation with body mass in kangaroos and ungulates, suggesting that they may be unreliable estimators across the range of living and extinct herbivorous mammals.

Molar lengths also give relatively high estimates of body mass in camelids and oreodonts. This seems to be due to an elongation of the posterior cheek teeth in camelids, and relatively large overall cheek tooth size in oreodonts. Other mammals can be shown to have relatively large cheek teeth for their body size. For example, this is the case for the dromomerycid *Aletomeryx,* as seen by a comparison in fossil ruminants with body mass predictions determined from craniodental dimensions and postcranial dimensions (work in preparation by K. Scott and myself). Interestingly enough, lower molar row length gives lower body mass estimates than dental measures in these animals with enlarged cheek teeth, which may be the result of taking this measurement along the base of the teeth. Alternatively, it seems that in many cases an enlargement of the posterior cheek teeth is accompanied by a reduction in size of the first molar, thus resulting in an averaged estimate in the case of the total molar row length. Again, it is interesting to note in this context that molar dimensions are relatively more poorly correlated with body mass in kangaroos than in ungulates.

Considering other variables in extinct ungulates, lower molar row length and posterior skull length appear to yield mass estimates that are close to the "mean value" a good proportion of the time (although oreodonts are an aberrant example here in having a relatively short posterior skull length). In contrast, palatal width and basicranial length appear to be the most "erratic" predictors: They yield estimates that are exceptionally high or exceptionally low, and also frequently show the greatest discrepancy of estimates between different regression lines (although in the case of palatal width, this may be an artifact created by the extremely low values of palatal width for suids, or by the fact that fossil palatal widths are often distorted and thus difficult to estimate accurately).

Comparisons of mass estimates derived in this paper with those derived by other researchers have only a limited application at present. Only a few of the species that I have considered here have been examined by other workers, and ideally the same individual fossil specimens should be compared for mass estimates from different elements. However, I have included in Table 13.5 the average body masses given by Scott (this volume) from postcranial dimensions for a variety of ungulate species, and by MacFadden (1986) from a combination of cranial and dental dimensions for some equids.

Scott's average mass estimation for fossil ruminants (as derived from her all-ruminants regression line) suggests that skull measurements bear a closer relationship to postcrania in mass estimation than does lower molar row length. (The estimates from lower molar row length are high in the case of the dromomerycids, and low in the case of the merycodontines.) Her average mass estimate for the camelid *Stenomylus* (derived from the all-artiodactyl regression line) is very different from any of my mass estimates in Table 13.5, but accords fairly closely with the "mean value" shown in Table 13.4, and with the estimate from the first lower molar length. In the case of the oreodonts (average mass estimates from the suid regression line), the estimates from the skull measurements accord fairly well with those from the postcrania, but the ones derived from the lower molar row length are considerably higher. In the case of the equids, Scott's postcrania values (derived from the equid regression line) accord better with my estimates from lower molar row length than with those from skull measurements, but MacFadden's estimates show a closer resemblance to those derived from the skull.

Summary

Lower molar row length appears to be fairly well correlated with body mass in all extant herbivores, and to yield "reasonable" and consistent body mass estimates in fossil ungulates. Certainly lower molar row length is superior to any single dental variable across the range of living and extinct herbivores, and appears to remain a consistently good predictor in herbivores with enlarged individual cheek teeth. Although total molar rows are less common in the fossil record than single teeth, this variable may also be estimated from edentulous jaws, or from those with missing teeth. If single cheek teeth are all that is available for body mass estimation, it is apparent that molar lengths are more reliable than are other dimensions (with the possible exception of second upper molar area). In most cases, the upper or lower molar that shows the best correlation with body mass (and which yields "middle value" estimates for fossil ungulates) is the one that is the second to last: that is, the second molar in ungulates, or the third molar in kangaroos.

Although many length measurements of the more posterior portions of the skull and jaw appear to be good estimators of body mass in living herbivores (although problems exist with the length of the masseteric fossa and the occipital height, as previously discussed), few of them are readily available in the fossil record. Even if partial skulls and jaws are preserved, the critical areas for measuring these variables (e.g., the top of the occiput, the jaw condyle, the outer edge of the mandibular angle) tend to be the parts that have been broken off. Palatal width and basicranial length are more often found preserved in fossils, and palatal width is superior to basicranial length in terms of the r^2 value and %PE if suines are excluded from the regression line. However, the mass predictions obtained for the range of fossil ungulates suggest that both these variables should be used with extreme caution.

Although total skull length and total jaw length are exceptionally well correlated with body mass, with correspondingly low prediction errors, in all extant herbivores (with skull length being the slightly better predictor), these may not be the best variables to use for the estimation of body mass in fossil ungulates. Complete skulls and jaws are not common in the fossil record, although as these variables are derived from compounding smaller variables (see Methods) it may be possible to obtain them by combining results from a variety of partial specimens. However, an obvious problem remains in the fact that anterior jaw length is extremely variable among herbivores. Certain living herbivores have jaw lengths that are either slightly (kangaroos) or considerably (hyraxes) shorter than is the "norm" for extant ungulates, whereas others have relatively long jaws (equids). It is clear from the diversity of body mass estimates for oreodonts, that total skull and jaw length measures from herbivores that lack a diastema will result in underestimates of body mass. However, one skull variable that does show consistently high correlation with body mass in extant herbivores, and a reasonable degree of constancy in the estimates determined for fossil ungulates, is the posterior length of the skull. This may prove to be the most reliable cranial variable for body mass estimation in species that lack a diastema. However posterior skull length, like many other cranial measures, also suffers from the problem that in living ungulates grazers have longer values than other feeding types.

In conclusion, dental variables from single teeth appear to be unreliable estimators of body mass across the range of living and fossil ungulate species, although dental lengths are preferable to other dimensions. Total skull length and total jaw length appear to be good variables for estimating the mass of fossil ruminants, but should be used with caution for other ungulates. Lower molar row length and posterior skull length appear to be the most consistent variables for weight estimation across a wide range of skull morphologies. Of these two, lower molar row length is more readily available in the fossil record, and has the additional advantage that the variation in this measurement in living herbivores is only slightly correlated with dietary type in certain instances.

Acknowledgments

I am indebted to the following persons and institutions for the opportunity to measure specimens in their collections: Dr. M. Rutzmoser, Museum of Comparative Zoology, Harvard University; Dr. R. Thorington, Smithsonian Institution, National Museum of Natural History, Washington, D.C.; Dr. G. Musser, American Museum of Natural History, New York; Drs. J. Clutton-Brock and K. Bryan, British Museum of National History, London; Dr. A. Friday, University of Cambridge, Cambridge; Dr. L. Jacobs, National Museum of Kenya, Nairobi; Dr. E. Vrba, Transvaal Museum, Pretoria, South Africa; Dr. Q. Hendy, South African Museum, Cape Town; Dr. R. Molnar, Queensland Museum, Brisbane, Australia; Dr. T. Flannery, Australian Museum, Sydney; Dr. C. Kemper, South Australian Museum, Adelaide.

I am indebted to the following people for their help in input of data: Marna Dolinger, David Ehrhardt, Lyuba Konopasek, Viyada Sarabanchong, and Kathleen Yang. Loren Mitchell aided with both data input and data analysis. Useful information on diets and body weights was provided by Dr. T. Flannery, Dr. P. Jarman, and Dr. K. Scott, and this paper has benefited from discussion with Dr. J. Damuth, Dr. M. Fortelius, Dr. W. Jungers, Dr. K. Scott, and Dr. B. Van Valkenburgh. This paper was supported by National Science Foundation Grant No.

BR 84–18148 and by Brown University Biomedical Research Grant No. RR0708–22.

References

Archer, M. 1978. The nature of the molar–premolar boundary in marsupials and a reinterpretation of the homology of marsupial cheek teeth. *Mem. Queensland Mus. 18*:157–164.

Banfield, A. W. F. 1974. *The Mammals of Canada*. Toronto: University of Toronto Press.

Demment, M. W., & Van Soest, P. J. 1985. A nutritional hypothesis for body size patterns of ruminant and non ruminant herbivores. *Am. Nat. 125*: 641–672.

Fischer, M. S. 1986. Die Stellung der Schliefer (Hyracoidea) im phylogenetischen System der Eutheria. *Courier Forshungsinst. Senckenberg 84*:1–132.

Fortelius, M. 1985. The functional significance of wear induced change in the occlusal morphology of herbivore cheek teeth, exemplified by *Dicerorhinus etruscus* upper molars. *Acta Zool. Fenn. 170*:157–158.

Gould, S. J. 1974. The origin and function of "bizarre" structures: antler size and skull size in the "Irish elk," *Megaloceros giganteus*. *Evolution 28*: 191–220.

Gwynne, M. D., & Bell, R. H. V. 1968. Selection of vegetation components by grazing ungulates in the Serengeti National Park. *Nature 220*:390–393.

Haltenorth, T., & Diller, H. 1977. *A Field Guide to African Mammals Including Madagascar*. London: William Collins.

Hansen, R. C., & Clark, R. C. 1977. Foods of elk and other ungulates at low elevation in Northwestern Colorado. *J. Wildlife Mgmt. 41*:76–80.

Hofmann, R. R. 1973. *The Ruminant Stomach*. East African Monographs in Biology, ed. T. R. Odhiambo. East African Literature Bureau.

Hofmann, R. R. 1985. Digestive physiology of deer: their morphophysiological specialization and adaptation. *Bull. R. Soc. N.Z. 22*:393–407.

Janis, C. 1976. The evolutionary strategy of the Equidae, and the origin of rumen and cecal digestion. *Evolution 30*:757–774.

Janis, C. M. in press. Correlation of cranial and dental variables with dietary preferences in mammals: a comparison of macropodids and ungulates. In *Problems in Vertebrate Biology and Phylogeny: An Australian Perspective,* ed. S. Turner, R. A. Thulborn, & R. Molnar. Memoirs of the Queensland Museum, special volume.

Janis, C. M., & Ehrhardt, D. 1988. Correlation of relative muzzle width and relative incisor width with dietary preferences in ungulates. *Zool. J. Linn. Soc. 92*:267–284.

Jarman, P. J. 1974. The social organization of ungulates in relation to their ecology. *Behaviour 48*:213–267.

Kingdon, J. 1979. *East African Mammals,* Vol. IIIB. London: Academic Press.

Kingdon, J. 1982a. *East African Mammals,* Vol. IIIC. London: Academic Press.

Kingdon, J. 1982b. *East African Mammals,* Vol. IIID. London: Academic Press.

Lamprey, H. F. 1963. Ecological separation of the large mammal species in the Tarangire Game Reserve. *East Afr. Wildlife J. 1*:63–92.

Lent, P. C. 1978. Musk ox. In *Big Game of North America,* ed. J. L. Schmidt & D. L. Gilbert. Harrisburg: Stackpole Books.

MacFadden, B. J. 1986. Fossil horses from "Eohippus" (*Hyracotherium*) to *Equus*: scaling, Cope's law, and the evolution of body size. *Paleobiology 12*:355–382.

Mackie, R. J. 1970. Range, ecology and relation of mule deer, elk and cattle in the Missouri River Breaks, Montana. *Wildlife Monogr. 20*:1–79.

Medway, G. G-H. 1969. *The Wild Mammals of Malaya.* Oxford: Oxford University Press.

Pollock, J. 1980. Behavioural ecology and body condition in New Forest ponies. *R. Soc. Protection Anim. (U.K.) Sci. Publ. 6*:1–118.

Radinsky, L. B. 1984. Ontogeny and phylogeny in horse skull evolution. *Evolution 38*:1–15.

Schaller, G. B. 1967. *The Deer and the Tiger.* Chicago: University of Chicago Press.

Schaller, G. B. 1977. *Mountain Monarchs.* Chicago: University of Chicago Press.

Schaller, G. B., & Junrang, R. 1988. Effects of a snowstorm on Tibetan antelope. *J. Mammal. 69*:631–634.

Scott, K. M. 1983. Body weight prediction in fossil Artiodactyla. *Zool. J. Linn. Soc. 77*:199–215.

Scott, K. M. 1985. Allometric trends and locomotor adaptations of the Bovidae. *Bull. Am. Mus. Nat. His. 179*:197–228.

Smith, R. J. 1984. Allometric scaling in comparative biology: problems of concept and method. *Am. J. Physiol. 246*:R152–R160.

Stewart, D. R. M., & Stewart, J. 1970. Food preference data by faecal analysis from African plains ungulates. *Zool. Afr. 15*:115–129.

Strahen, R. 1983. *Complete Book of Australian Mammals.* Sydney: Angus & Robertson.

Walker, E. P. 1983. *Mammals of the World,* 3rd ed. Baltimore: Johns Hopkins University Press.

Whitehead, G. K. 1972. *Deer of the World.* London: Constable.

14

Postcranial dimensions of ungulates as predictors of body mass

KATHLEEN M. SCOTT

Although many body dimensions of animals show a general relationship with body mass, there are good biomechanical reasons why postcranial dimensions and mass should be highly correlated. Bones must be of sufficient thickness to withstand the compressive and bending forces generated in weight-bearing elements. In a series of animals of increasing size and similar shape, skeletal dimensions would be expected to increase in a regular way to accommodate increasing body mass. Size-imposed limitations on the skeleton probably operate most strongly on terrestrial cursorial mammals of large body size, where the skeleton would be continuously exposed to stresses during locomotion. The elucidation of the relationship between skeletal dimensions and body mass in a variety of ungulate taxa should provide a basis for estimation of mass in fossil taxa. This study will examine the scaling relationships between various postcranial dimensions and body mass in a variety of living artiodactyls and perissodactyls, calculate predictive equations based on these relationships, and test them on living and fossil ungulate taxa.

Many dimensions of the postcranial skeleton of ungulates scale closely with body mass, and thus provide a potential means of evaluating body mass in fossil ungulates. Studies by Alexander (1977) and Scott (1985) on African bovids, Scott (1987) on cervids, Karp (1987) on suids and tayassuids, and Scott (unpublished data) on equids have all shown close relationships between various skeletal dimensions and body mass. In general, nonlength dimensions and proximal limb elements show higher correlations with body mass than do lengths of distal bones, although

In *Body Size in Mammalian Paleobiology: Estimation and Biological Implications,* John Damuth and Bruce J. MacFadden, eds.

distal bones may also scale closely with body mass in less diverse taxa (i.e., cervids). Although these studies indicate that a variety of skeletal measurements do scale closely with body mass in several ungulate taxa, exponents and intercepts vary among the groups. Selection of an appropriate living taxon from which to calculate body mass of fossil taxa thus becomes an important issue. Although the choice may be obvious for fossil taxa that belong to families with many extant species (e.g., bovids), the choice is less apparent for fossil groups with no (e.g., dromomerycids) or few (e.g., camels) close living relatives. Selection of appropriate living groups on which to base predictive equations for such fossil taxa will depend on determining why and to what degree skeletal dimensions scale differently in different groups.

Living ungulate families do differ in morphology in certain important respects; for example, suoids do not have a cannon bone, and perissodactyls and artiodactyls differ in number of functional digits and distribution of body mass. These differences are presumably responsible at least in part for observed scaling differences. Similar biomechanical constraints such as compressive and bending forces have been shown to play a role in defining skeletal dimensions in a variety of ungulates (Alexander 1977; McMahon 1975a,b; Scott 1985); however, these forces probably act differently on bones of different shape. For example, McMahon (1975a) showed that elastic properties of bone were critical in determining how bones change with increasing body mass, but stated that these constraints would result in regular elastic scaling only if all bones were the same shape. Scott (1985) showed that even for a relatively uniform group such as Bovidae, bones did change in shape under the influence of both habitat and size. Even these relatively small shape changes can distort regular scaling relationships; the effect of shape changes would be even more marked between dissimilar families. Nevertheless, changes may still be regular enough to predict body mass. And although bones of different shapes probably change differently with increasing size, there may be skeletal dimensions that are relatively uniform for a wide variety of ungulates.

Scott (1983) demonstrated that scaling relationships based on bovids can be used successfully to predict body mass in a number of living and fossil ruminant taxa. It is not surprising that body mass prediction would be successful for cervids and antilocaprids based on bovid regressions, since these taxa are at least superficially similar in skeletal morphology. However, camels differ morphologically in a number of ways related to their secondarily digitigrade posture, and it might be anticipated that this would affect skeletal morphology sufficiently for ruminant predictive equations to give erroneous results. Although this was true for those dimensions that are obviously modified (i.e., distal width of metapodials), it was not true for all dimensions. This suggests that ungulate scaling relationships can be generalized to some degree for many dimensions, and can thus be used to construct more widely applicable ungulate predictive equations. However, in order for these to be useful, those dimensions that are likely to scale similarly among taxa must be identified.

The applicability of predictive equations based on various combinations of living ungulates can be tested by using them to predict body mass of living taxa and comparing the results from differently derived predictive equations to actual body mass. These comparisons will provide a basis for deciding how taxa should be grouped in creating and using predictive equations. When predictive equations are used to predict masses of the taxa on which they have been based, the pattern of biases can be instructive in evaluating the effects of the scatter that occurs in all scaling relationships. It is more difficult to "test" the effectiveness of the predictive equations in fossil taxa. However, if body mass predictions for a fossil taxon are consistent for a variety of measurements from different limb elements, and if the magnitude of the standard deviations is similar to those of living test species, it would suggest that the skeleton of that taxon is constructed in a similar way to that of the groups used to construct the predictive equations. Scott (1983) predicted body masses based on bovid regressions for a variety of fossil ruminants in the families Antilocapridae, Dromomerycidae, and Cervidae and found that predictions were consistent from bone to bone. However, certain dimensions gave aberrant results (i.e., proximal articular surface of the humerus in cervids), and would have resulted in aberrant results if they were used as the sole basis of predictions. Identification of such characters in fossil taxa is thus critical to accurate mass prediction.

Materials and methods

Body mass determinations

The construction of accurate body mass prediction equations depends on having good body mass data for living taxa. Body mass data are widely available for small mammals, but are rare for large species, which are seldom weighed in the field. Because field masses are so seldom available, mass prediction regressions cannot be based on individual masses and measurements. However, since there are many problems associated with the use of one-time masses of wild ungulates (see Scott 1983), the use of average masses of a species is probably both more realistic and more accurate. This is especially true since the object of the predictive regressions is to predict average body masses of fossil species, not to compare masses of individuals.

Average body masses are readily available for many taxa, especially African species. However, many of the widely used body masses for ungulates are based on those in Walker (1968), and these data are problematic in two ways. First, mass ranges given are for a genus, and few genera of ungulates are monotypic. For example, species in the genus *Cephalophus* range in body mass from about 5 to 140 kg, and other relatively speciose genera such as *Tragelaphus* and *Gazella* also exhibit wide ranges in body mass. Second, many of the masses given are taken from *Rowland Ward's Records of Big Game* (Best, Edmond-Blanc, & Whitting 1962) and are actually trophy records for the largest recorded male. Although these may be useful when the trophy specimens are in fact known to have been placed in museum collections (for example, the record male *Ovis ammon* described by Clark [1963] are in the collections of the American Museum of Natural History), in general they overestimate the mass of an average male by approximately 10–15%. Such records can be used if they are adjusted accordingly, but their influence on the masses given in standard references should be recognized.

The best body mass data to use as a basis for predictive equations are means of large samples of wild collected individuals. Using this kind of data, a reasonable estimate of the average adult mass of each sex can be made. These data are widely available for African ungulates (e.g., Meinertzhagen 1938; Sachs 1967; and other studies summarized in Von la Chevallerie 1970) and for North American and European species. However, Asian ungulates have been poorly studied in this regard, and in many cases trophy, zoo, or anecdotal data are all that is available (see references and discussion in Scott 1983, 1985, 1987). Where these are the only kinds of data available, I have used them, supplemented if possible by field data. The masses of zoo and trophy animals have been adjusted downward by approximately 10%. Most of the masses available are for males; where species are sexually dimorphic, I have deleted female specimens if body masses were not available. Although female body masses could be estimated from differences in skull length, body length, or shoulder height, I have preferred not to use estimated masses as a basis for the regressions.

The lack of good body mass data on Asian taxa is more problematical for some families than others. Body mass data are available for most bovid species, since the majority are African. However, since most cervids, suids, and half the living species of equids are found outside well-studied areas, the mass estimates used are probably not as accurate as those for bovids. This problem is compounded by the smaller size range and lower species numbers of these taxa; errors in mass estimation may produce a large effect in regressions based on only a few points of similar sizes. For equids, I have increased the size range of the family by including domestic breeds. However, this may have compounded the problem since body mass data for domestic horses may also be inaccurate. Many breeds vary in size (e.g., the three grades of Welsh ponies), and breed standards may change over time. I suspect that some of the masses I have used for domestic breeds are not very accurate.

Measurements and analyses

Measurements were chosen to be nearly equivalent among taxa, but this was not possible in all cases. The original measurements taken on bovids are described and illustrated in Scott (1983); these measurements were also used for cervids, antilocaprids, giraffids, and moschids since these taxa are very similar in gross morphology of the postcranial skeleton. They were also used for camelids, but some measurements, such as width across the distal ends of the metapodials, are clearly not equivalent even though they can be taken at the same points. The measurements on proximal bones were also used for suoids since they are similar in overall morphology, even though they clearly differ in proportions. Since the metapodials are unfused, a different set of measurements was used for these elements; these are shown in Figure 14.1. Equids differ from artiodactyls in a number of ways, and many measurements, although equivalent, differ slightly in the way they were taken. Some artiodactyl measurements, such as F2 and F3, could not be taken on equids. These were replaced with other measurements that seemed more appropriate to the taxon. All equid measurements are illustrated in Figure 14.2.

Measurements were taken on approximately 1000 individuals belonging to 160 species. All measurements were recorded in centimeters. The specimens used were wild-caught individuals in the collections of the

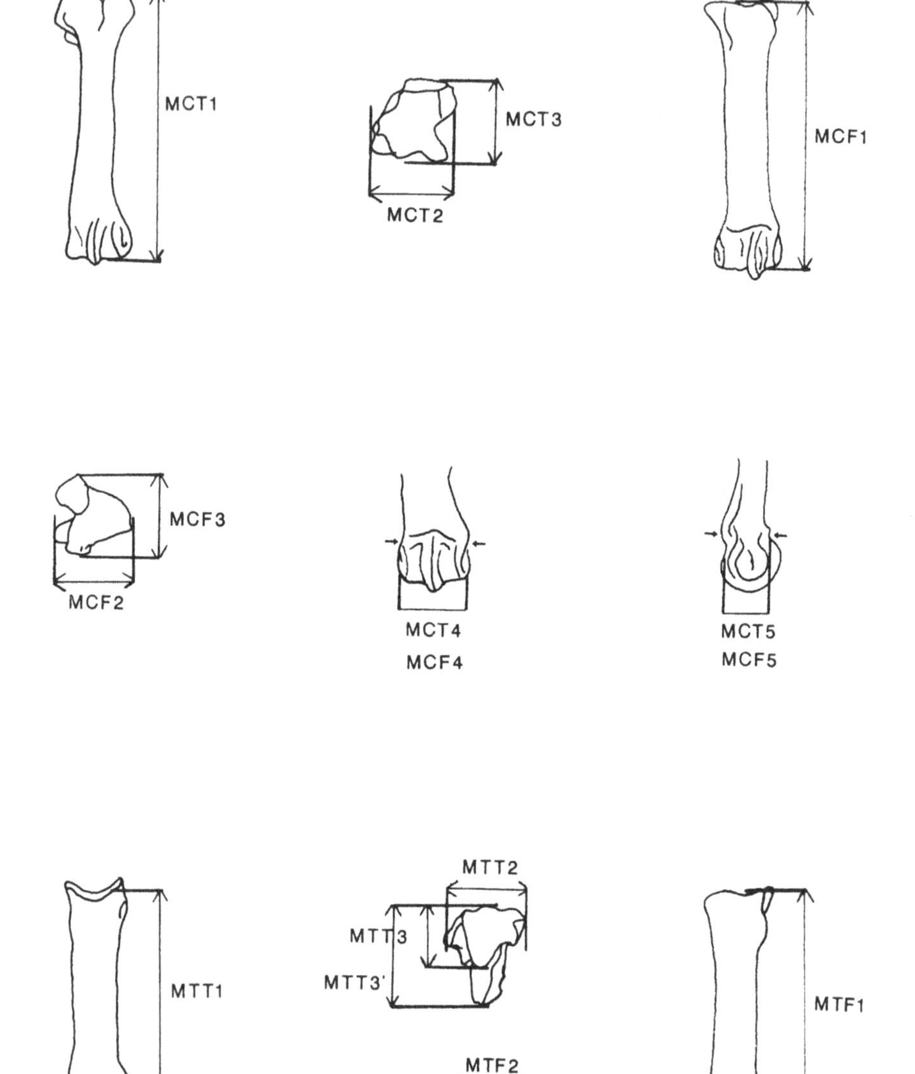

Figure 14.1. Definitions of the metapodial measurements of suoids. MCT, metacarpal III; MCF, metacarpal IV; MTT, metatarsal III, MTF, metatarsal IV.

American Museum of Natural History; British Museum of Natural History; Museum of Comparative Zoology; the Field Museum of Natural History; the United States National Museum; Cambridge University; the University of Connecticut; the Museum National d'Histoire Naturelle, Paris; the Senkenberg Museum, Frankfort; and the Rijksmuseum van Natururlijke Historie, Leiden. Zoo or captive animals were included only if no wild specimens were available (e.g., *Equus przewalskii* and *Elaphurus davidianus*). Each species was separated by sex and subspecies if there was more than a 10–15% dimorphism in mass between the sexes or subspecies. This resulted in the creation of approximately 314 size-specific units (169 bovid, 94 other ruminants, 20 suids and tayassuids, and 31 equids).

For each of these units the mean of each measurement was calculated for all included specimens. For each unit the mass of an average individual (in kilograms) was estimated from the literature. Because of missing mass data or differences in measurements taken among taxa, fewer units were actually used in the calculations. For each measurement I performed a least squares regression on the relationship between that measurement and body mass using log-transformed data. Since body mass is the quantity to be estimated, it was designated the dependent variable in each case. The results of these regressions were used to construct a series of individual predictive equations for body mass based on each measurement. Each equation was of the form: log body mass $= b \log X + m$, where b is the slope and m is the intercept of the regression. The predictive equations were calculated for several different subsets of the data: bovids only, cervids only, suoids (suids plus tayassuids) only, and equids only, ruminants (bovids, cervids, giraffids, *Antilocapra, Moschus,* and camelids), artiodactyls (ruminants plus suoids), and ungulates (all taxa). For the combined artiodactyl regressions, length of the third metapodial of suids was used as the equivalent of the ruminant cannon bone. However, for transverse measurements of the metapodials (at the proximal and distal ends and at midshaft) the dimensions of the third and fourth metapodials were added to give a roughly equivalent measurement. For the ungulate regressions equid measurements on the single metapodial were combined with their artiodactyl equivalents. However, since F2 and F3 are completely different measurements in the two groups, equids were dropped from the calculations for the ungulate regressions for these two measurements.

For each predictive equation, two measures of the predictive ability were calculated in addition to r^2; these are percent prediction error

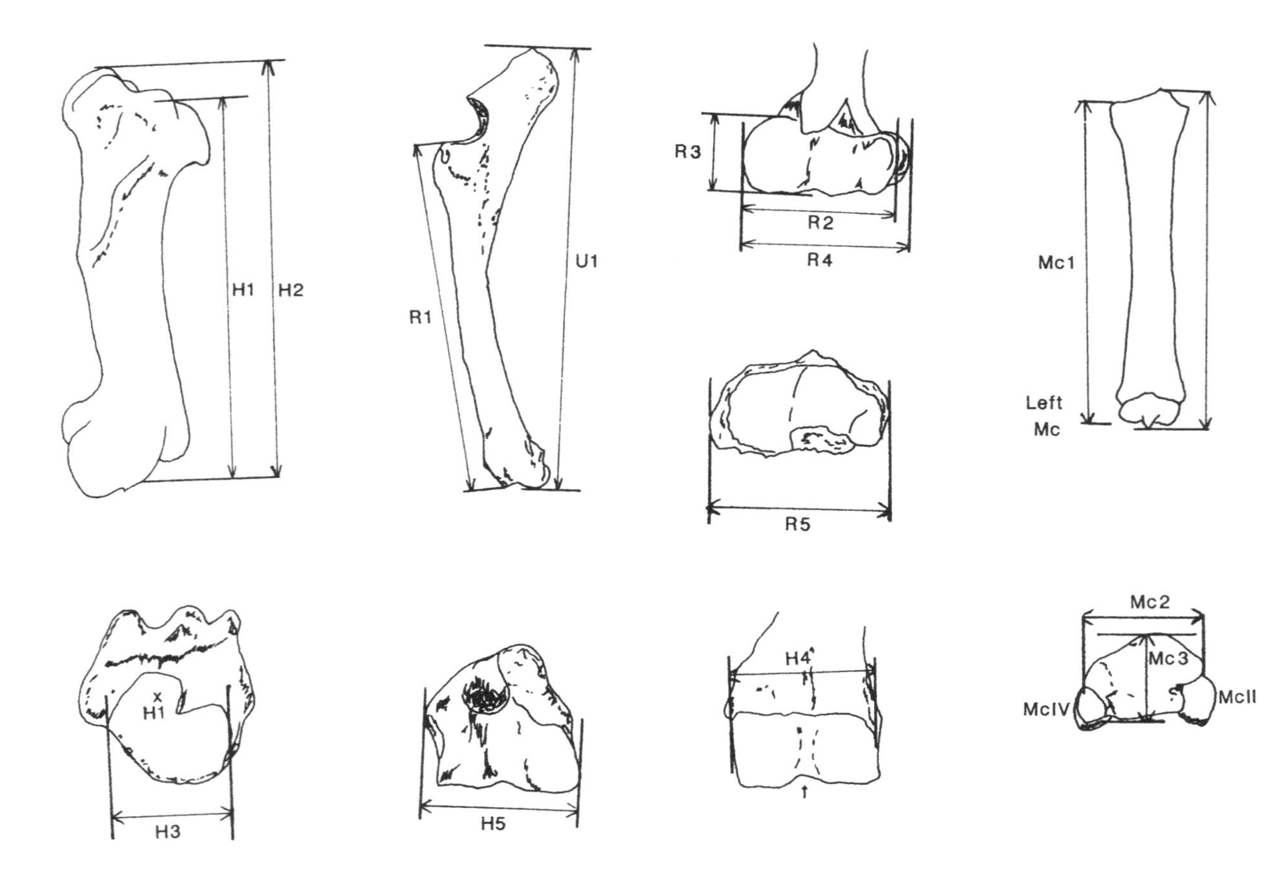
H1
H2
R1
U1
R3
R2
R4
R5
Mc1
Left
Mc
x
H1
H3
H5
H4'
Mc2
Mc3
McIV
McII

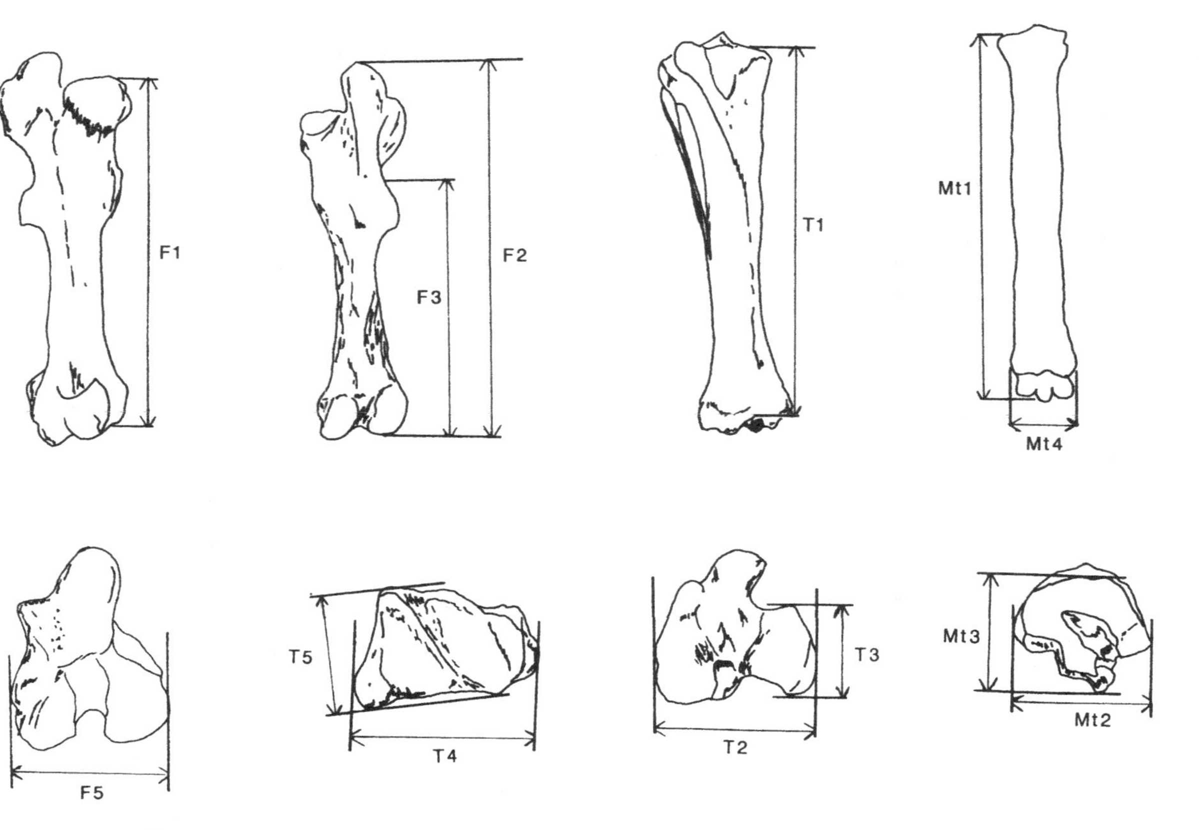

Figure 14.2. Definitions of the measurements used to predict body mass in equids. H, humerus; U, ulna; R, radius; Mc, metacarpal; F, femur; T, tibia; Mt, metatarsal. Not illustrated are transverse diameters at midshaft (H7, R6, Mc6, F6, T6, Mt6) and anteroposterior diameters at midshaft (H8, R7, Mc7, F7, T7, Mt7).

(%PE) and percent standard error of the estimate (%SEE). The method of calculation and rationale for each is described by Van Valkenburgh (this volume).

The accuracy of these equations for predicting body mass was tested by predicting body mass for a test data set of living ungulates of various sizes. Because previous work (Scott 1983) had shown that the mean of all the nonlength predictions was a better predictor than any individual measurement, I calculated a mean predicted mass for each long bone. The mean predicted mass was the arithmetic mean of all of the nonlength measurements taken on that bone. I also calculated the body mass predicted by the length of each long bone. For each species in the test data set, body masses were calculated for all long bones based on all of the different sets of predictive equations. The percent prediction error was calculated for each of these predictions. The applicability of the predictive equations was tested by predicting body masses for a variety of fossil ungulates in the families Dromomerycidae, Antilocapridae, Moschidae, Equidae, and Merycoidodontidae. Again, for each species masses were predicted by use of several of the predictive equations.

Results

Predictive equations

The various sets of equations developed to predict body mass for different combinations of taxa are given in Appendix Table 16.7. For each equation, the *R*-square value, %PE, and %SEE are also given. Plots of some of the relationships are given in Figures 14.3–14.5. The exponents and intercepts are similar for the bovid-only, cervid-only, and ruminant regressions. Within taxa, most nonlength dimensions scale closely with body mass, with r^2 values above .9 for most dimensions of bovids, cervids, and suoids. For the bovid-only regressions, %SEE ranges from about 28% to 45% for most nonlength measures, with corresponding %PE of about 18–30%. For cervid-only nonlength regressions, %SEE and %PE are similar, with %SEE ranging from 25% to 40% for most dimensions, and %PE ranging from 17% to 35%. For the all-ruminants regressions %SEE and %PE are generally lower than for either the bovid- or cervid-only regressions.

Suoid exponents are also similar to those of bovids, cervids, and ruminants, although in general they are more similar to cervids. However, a number of suoid intercepts differ markedly from those of the

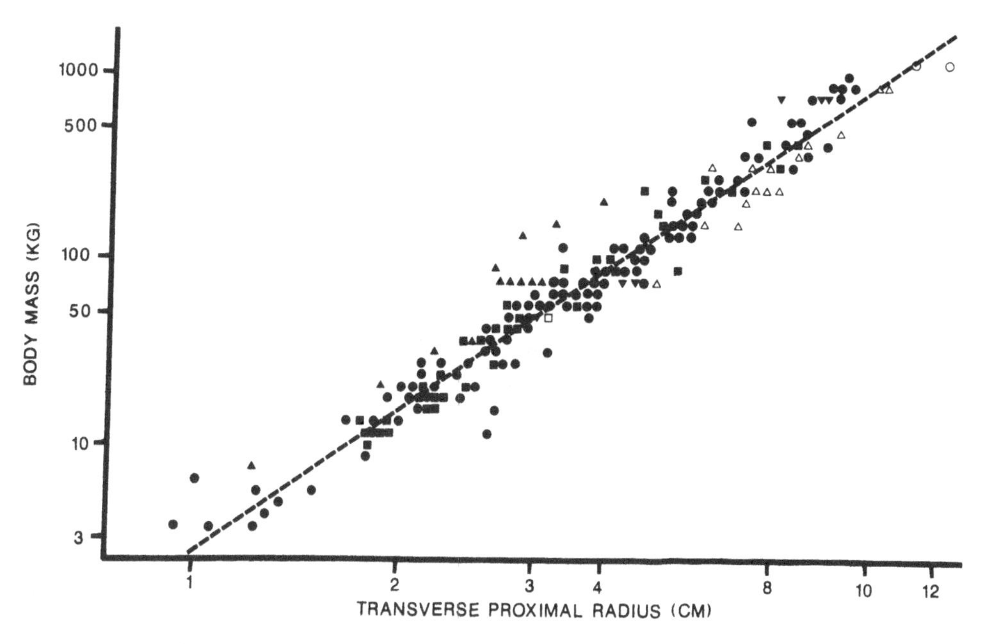

Figure 14.3. Log–log plot of width of proximal articular surface of the radius against body mass. (●) Bovidae; (■) Cervidae; (○) Giraffidae; (□) *Antilocapra*; (▲)Suidae and Tayassuidae; (△) Equidae; (▼) Camelidae.

ruminants. The differences between suoids and ruminants can be seen in plots of the various dimensions against body mass (see Figure 14.3). Although the slopes of the lines are similar, suoids tend to be consistently larger or smaller in a given dimension (depending on the dimension) than a ruminant of the same body size. The r^2 values for suoids are generally high and values of %SEE are slightly lower than for other artiodactyl-based regressions. Predictive errors are generally in the range of 20% except for a few isolated measurements.

The artiodactyl equations are similar to those for bovids, cervids, and ruminants, which is to be expected since these taxa dominate the data. %SEE and %PE are similar to those of the ruminant regressions, except that distal long-bone lengths have much higher %PE with the addition of suoids, since their distal elements are much shorter than other artiodactyls of the same size.

Equid nonlength dimensions also scale closely with body mass, although r^2 values are in general lower, ranging from .79 to .89 for nonlength dimensions. Equid nonlength scaling relationships differ from artiodactyls, as might be expected, but the differences between equids

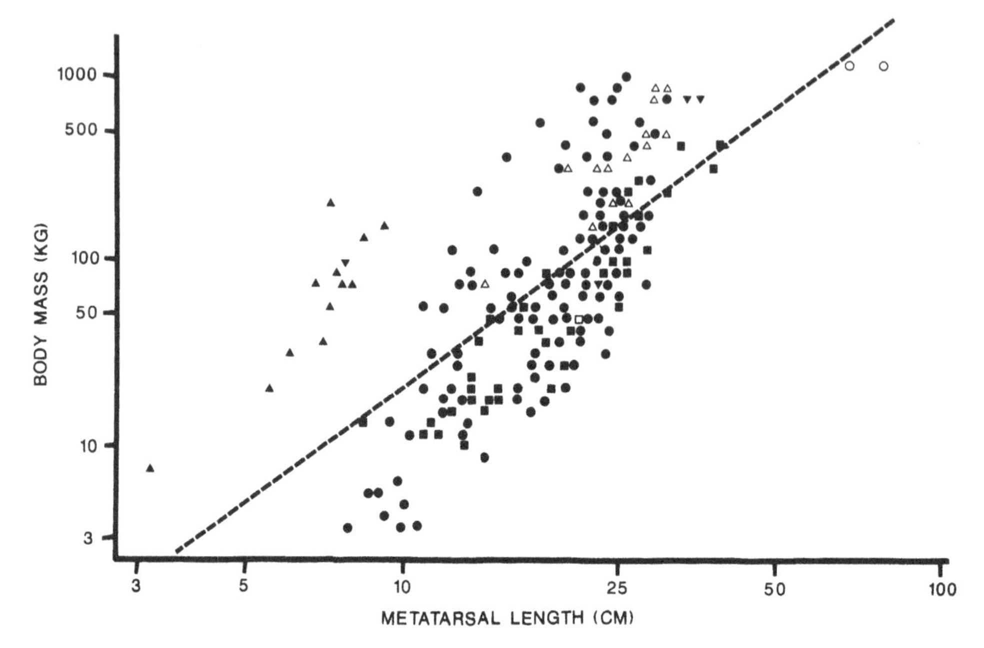

Figure 14.4. Log–log plot of metatarsal length against body mass. Symbols as in Figure 14.3.

and ruminants are less than between suoids and ruminants. This can be seen clearly in plots of nonlength dimensions (see Figure 14.3); suoids are much more likely to be outliers than are equids, and in many cases all suoids lie along the edge of the scatter of the other taxa. %SEE values are somewhat higher than for any of the other sets of regressions, ranging from 30% to about 100%. Despite the higher values of %SEE, %PE values are mostly in the range of 15–20%.

Lengths of long bones, other than humerus and femur, are in general less highly correlated with body mass than nonlength dimensions and show high degrees of scatter (compare Figures 14.4 and 14.5). For length measurements, in all equations, both %SEE and %PE may be much higher; for example, bovid metatarsal length has %SEE equal to 134% and %PE equal to 96%. Humerus and femur lengths for bovids and all ruminants have %SEE and %PE in the same range as nonlength measurements and would be acceptable mass predictors. These results were to be expected since lengths of distal long bones are known to vary with habitat (Scott 1983, 1985). In cervids and suoids, lengths of distal long bones show higher corre-

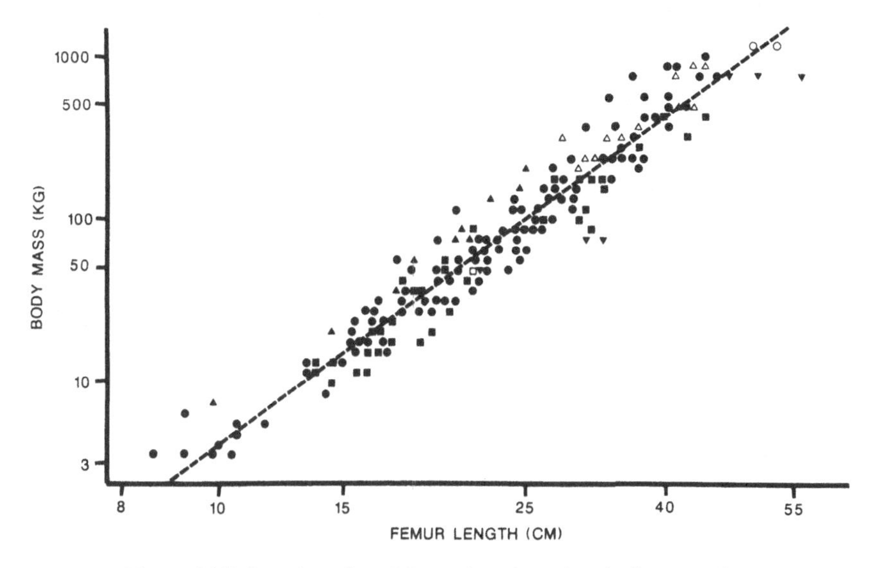

Figure 14.5. Log–log plot of femur length against body mass. Symbols as in Figure 14.3.

lations with body mass and lower values of %SEE and %PE. These taxa inhabit a much narrower range of habitats than do bovids, or ruminants generally; additionally, the range of variation in suoids may be mechanically limited by the unfused metapodials (Karp 1987). For lengths of proximal bones, r^2 values are high, even for the combined regressions (artiodactyls and ungulates), and %PE values range from about 15% to 25%. Although this is undoubtedly due in part to the fact that these regressions are dominated by bovids and cervids, the plot of femur length versus body mass (Figure 14.5) clearly shows that femur length of most ungulate taxa scales similarly with body mass. The only taxon that differs markedly from the other ungulates is the Camelidae. The trends for the humerus are similar.

Test data

Estimated body masses for a number of living ungulate species are given in Tables 14.1–14.4. For each predicted body mass the predictive error is also given. Table 14.1 shows predicted masses for cervids, suoids, and equids based on the regressions for their respective families. Table 14.2 gives predicted masses for ruminant taxa based on ruminant regressions.

Table 14.3 gives predicted masses for artiodactyls based on artiodactyl regressions, and Table 14.4 gives predicted masses for artiodactyls and equids based on ungulate regressions.

"Own taxon" predictions. The mean predicted body masses based on nonlength dimensions in general gave accurate mass predictions for the test taxa, as was found by Scott (1983) for bovids (see Table 14.1). For most bovids and cervids the over- or underestimates were within 5–15% of the actual average body mass. Large species are the most problematical, with masses of large Bovini tending to be consistently underestimated, but most of these estimates are within 20%. Lengths of long bones other than humerus and femur are generally poor predictors of body mass in ruminants. Distal long-bone lengths may over- or underestimate body mass by as much as 700%. Even within relatively uniform taxa, such as the cervids, distal long bones generally give unreliable mass estimates, although they give closer estimates than do distal long bones for many bovid species. For most taxa, masses predicted by humerus and femur lengths are within 15% of actual masses, and thus give estimates comparable to nonlength measures. However, for a few taxa, %PE from femur length may be as much as 75% over- or underestimated. For smaller suids mass estimates are generally within 10–15% of actual masses, and humerus and femur lengths are both good estimators of body masses. For large suids, like large bovids, predicted masses tend to be underestimates. Equid mass estimates are more variable than for the other families, with higher standard deviations. However, most of the mass estimates are within 20% of actual masses.

Ruminant predictions. The accuracy of mass predictions for bovids and cervids is similar when ruminant regressions are used, although the standard deviation within a bone tends to be higher, and the estimates for any one species may be slightly better or worse in one method versus the other (compare Tables 14.1 and 14.2). Mass estimates for *Antilocapra* and *Moschus* are close to actual masses, and the standard deviations consistent with those of bovids and cervids. However, mass estimates for the giraffe, even for the nonlength measurements, consistently overestimate body mass by as much as 75%. However, for each long bone there is a single measurement (either transverse or anteroposterior diameter) that gives the largest overestimate. Elimination of these obviously aberrant dimensions would improve mass prediction for

Table 14.1. *Body mass predictions for living ungulate genera, based on regressions derived from their own family*

	Mass	HM ± SD	%PE	RD ± SD	%PE	MC ± SD	%PE	FM ± SD	%PE	TB ± SD	%PE	MT ± SD	%PE
Cephalophus	5.5	4.9 ± .5	12.2	4.9 ± .2	12.2	5 ± .4	10.0	5.2 ± .5	5.8	4.9 ± .7	12.2	4 ± .8	37.5
monticola		3.8	44.7	7.7	−28.6	5.1	7.8	3	83.3	6.5	−15.4	6.5	−15.4
Madoqua kirki	5	5.1 ± .7	−2.0	5.5 ± .5	−9.1	5.2 ± .5	−3.8	5.4 ± .5	−7.4	6 ± .8	−16.7	5.3 ± .3	−5.7
		5.6	−10.7	5.9	−15.3	14.1	−64.5	4.9	2.0	5.9	−15.3	10.4	−51.9
Redunca fulvorufula	32	35 ± 4	−8.6	34 ± 3	−5.9	35.7 ± 5	−10.4	32 ± 2	0.0	37 ± 3	−13.5	34.7 ± 4	−7.8
		33	−3.0	40	−20.0	68	−52.9	42	−23.8	52	−38.5	60	−46.7
Ovis dalli	84	91 ± 10	−7.7	80 ± 11	5.0	81 ± 9	3.7	82 ± 8	2.4	76 ± 10	10.5	77 ± 11	9.1
		103	−18.4	88	−4.5	71	18.3	98	−14.3	118	−28.8	80	5.0
Gazella thomsoni	23	28 ± 3	−17.9	27 ± 3	−14.8	28 ± 2	−17.9	26 ± 3	−11.5	28 ± 3	−17.9	31 ± 5	−25.8
		19	21.1	31	−25.8	63	−63.5	24	−4.2	30	−23.3	52	−55.8
Damaliscus korrigum	123	131 ± 9	−6.1	138 ± 4	−10.9	140 ± 14	−12.1	128 ± 14	−3.9	136 ± 10	−9.6	136 ± 13	−9.6
		132	−6.8	235	−47.7	173	−28.9	118	4.2	150	−18.0	167	−26.3
Syncerus caffer	820	721 ± 89	13.7	770 ± 71	6.5	697 ± 154	17.6	764 ± 100	7.3	699 ± 133	17.3	735 ± 140	11.6
		486	68.7	278	195.0	81	912.3	470	74.5	329	149.2	103	696.1
Mazama americana	20	19 ± 2	5.3	19 ± 2	5.3	19 ± 2	5.3	21 ± 2	−4.8	23 ± 3	−13.0	22 ± 2	−9.1
		24	−16.7	18	11.1	26	−23.1	22	−9.1	21	−4.8	33	−39.4
Axis porcinus	47	44 ± 5	6.8	45 ± 9	4.4	40 ± 4	17.5	49 ± 10	−4.1	50 ± 5	−6.0	43 ± 3	9.3
		44	6.8	25	88.0	28	67.9	42	11.9	35	34.3	30	56.7
Alces alces	450	504 ± 82	−10.7	518 ± 79	−13.1	498 ± 55	−9.6	582 ± 160	−22.7	595 ± 73	−24.4	588 ± 140	−23.5
		874	−48.5	635	−29.1	346	30.1	624	−27.9	696	−35.3	596	−24.5
Odocoileus virginianus	45	43 ± 3	4.7	43 ± 3	4.7	48 ± 8	−6.2	43 ± 6	4.7	48 ± 4	−6.2	49 ± 7	−8.2
		53	−15.1	51	−11.8	66	−31.8	53	−15.1	53	−15.1	85	−47.1
Phacochoerus	80	89 ± 3	−10.1	73 ± 3	9.6	90 ± 18	−11.1	79 ± 3	1.3	89 ± 14	−10.1	61 ± 29	31.1
aethiopicus		93	−14.0	80	0.0	75	6.7	75	6.7	83	−3.6	89	−10.1
Hylochoerus	215	186 ± 24	15.6	170 ± 14	26.5	152 ± 22	41.4	169 ± 21	27.2	166 ± 14	29.5	142 ± 29	51.4
meinertzhageni		155	38.7	115	87.0	88	144.3	154	39.6	91	136.3	68	216.2

Table 14.1. (*cont.*)

	Mass	HM ± SD	%PE	RD ± SD	%PE	MC ± SD	%PE	FM ± SD	%PE	TB ± SD	%PE	MT ± SD	%PE
Tayassu pecari	30	37 ± 5	−18.9	35 ± 4	−14.3	38 ± 9.5	−21.1	31 ± 2	−3.2	35 ± 4	−14.3	33 ± 5	−9.1
		34	−11.8	36	−16.7	42	−28.6	35	−14.3	35	−14.3	41	−26.8
Equus asinus	245	231 ± 14	6.1	248 ± 7	−1.2	263 ± 35	−6.8	255 ± 43	−3.9	253 ± 32	−3.2	261 ± 19	−6.1
		272	−9.9	301	−18.6	285	−14.0	265	−7.5	304	−19.4	275	−10.9
Equus kiang	230	252 ± 21	−8.7	237 ± 13	−3.0	241 ± 33	−4.6	221 ± 22	4.1	240 ± 21	−4.2	245 ± 36	−6.1
		242	−5.0	275	−16.4	342	−32.7	235	−2.1	272	−15.4	349	−34.1
Equus burchelli	230	267 ± 8	−13.9	276 ± 34	−16.7	275 ± 14	−16.4	273 ± 34	−15.8	286 ± 29	−19.6	276 ± 23	−16.7
		236	−2.5	219	5.0	256	−10.2	258	−10.9	260	−11.5	232	−.9
Equus grevyi	390	362 ± 35	7.7	375 ± 21	4.0	361 ± 26	8.0	352 ± 25	10.8	382 ± 36	2.1	340 ± 30	14.7
		362	7.7	373	4.6	385	1.3	394	−1.0	324	20.4	378	3.2
Equus caballus (thoroughbred)	490	571 ± 36	−14.2	565 ± 73	−13.3	545 ± 48	−10.1	567 ± 83	−13.6	552 ± 47	−11.2	552 ± 59	−11.2
		685	−28.5	664	−26.2	570	−14.0	641	−23.6	524	−6.5	581	−15.7
Equus zebra hartmanae	315	338 ± 15	−6.8	353 ± 43	−10.8	312 ± 23	1.0	327 ± 30	−3.7	337 ± 43	−6.5	302 ± 29	4.3
		303	4.0	290	8.6	285	10.5	334	−5.7	589	−46.5	262	20.2
Equus caballus (arabian)	420	427 ± 14	−1.6	391 ± 56	7.4	393 ± 35	6.9	397 ± 65	5.8	380 ± 24	10.5	404 ± 33	4.0
		496	−15.3	519	−19.1	457	−8.1	487	−13.8	432	−2.8	488	−13.9

Note: The top line for each species gives the mean (± SD) of all measurements for that bone, and the percent predictive error (%PE) for that estimate. The lower value for each bone is the mass predicted by length of that bone, followed by %PE of the prediction.
Abbreviations: HM, humerus; RD, radius; MC, metacarpal; FM, femur; TB, tibia; MT, metatarsal.

Table 14.2. *Body mass predictions for living ruminant genera, based on ruminant regressions*

	Mass	HM ± SD	%PE	RD ± SD	%PE	MC ± SD	%PE	FM ± SD	%PE	TB ± SD	%PE	MT ± SD	%PE
Cephalophus	5.5	4.9 ± .5	12.2	5 ± .2	10.0	5 ± .3	10.0	5 ± .5	10.0	4.9 ± .7	12.2	4 ± .7	37.5
monticola		6.3	−12.7	4.8	14.6	7.8	−29.5	5.1	7.8	3	83.3	6.8	−19.1
Madoqua kirki	5	5.5 ± .7	−9.1	5.5 ± .5	−9.1	5.3 ± .4	−5.7	5.4 ± .4	−7.4	6 ± .9	−16.7	5.5 ± .5	−9.1
		5.7	−12.3	7.2	−30.6	14	−64.3	4.9	2.0	6	−16.7	10.6	−52.8
Redunca	32	35 ± 5	−8.6	35 ± 3	−8.6	35 ± 5	−8.6	31 ± 2	3.2	37 ± 3	−13.5	34 ± 4	−5.9
fulvorufula		31	3.2	41	−22.0	66	−51.5	40	−20.0	51	−37.3	56	−42.9
Ovis dalli	84	91 ± 10	−7.7	80 ± 12	5.0	79 ± 8	6.3	79 ± 6	6.3	74 ± 11	13.5	75 ± 10	12.0
		95	−11.6	85	−1.2	70	20.0	91	−7.7	112	−25.0	73	15.1
Gazella thomsoni	23	21 ± 3	9.5	28 ± 3	−17.9	27 ± 2	−14.8	26 ± 3	−11.5	28 ± 3	−17.9	31 ± 5	−25.8
		13	76.9	32	−28.1	62	−62.9	23	0.0	31	−25.8	49	−53.1
Damaliscus	123	130 ± 10	−5.4	137 ± 4	−10.2	136 ± 10	−9.6	129 ± 18	−4.7	133 ± 10	−7.5	132 ± 12	−6.8
korrigum		119	3.4	208	−40.9	167	−26.3	110	11.8	143	−14.0	148	−16.9
Syncerus caffer	820	714 ± 85	14.8	553 ± 53	48.3	667 ± 181	22.9	762 ± 113	7.6	670 ± 122	22.4	693 ± 146	18.3
		421	94.8	217	277.9	80	925.0	442	94.3	304	169.7	93	781.7
Mazama	20	19 ± 2	5.3	19 ± 2	5.3	20 ± 2	0.0	21 ± 3	−4.8	23 ± 3	−13.0	22 ± 2	−9.1
americana		25	−20.0	19	5.3	27	−25.9	22	−9.1	21	−4.8	34	−41.2
Axis porcinus	47	39 ± 5	20.5	45 ± 9	4.4	40 ± 4	17.5	50 ± 11	−6.0	50 ± 5	−6.0	43 ± 3	9.3
		23	104.3	26	80.8	29	62.1	41	14.6	35	34.3	31	51.6
Alces alces	450	501 ± 78	−10.2	502 ± 68	−10.4	482 ± 50	−6.6	588 ± 161	−23.5	589 ± 85	−23.6	579 ± 136	−22.3
		800	−43.8	549	−18.0	338	33.1	588	−23.5	691	−34.9	554	−18.8

Table 14.2. (*cont.*)

	Mass	HM ± SD	%PE	RD ± SD	%PE	MC ± SD	%PE	FM ± SD	%PE	TB ± SD	%PE	MT ± SD	%PE
Odocoileus	45	43 ± 3	4.7	43 ± 3	4.7	47 ± 7	−4.3	43 ± 7	4.7	47 ± 4	−4.3	49 ± 7	−8.2
virginianus		53	−15.1	50	−10.0	66	−31.8	52	−13.5	53	−15.1	84	−46.4
Moschus	11	12 ± 1	−8.3	13 ± 2	−15.4	14 ± 2.2	−21.4	12 ± 2	−8.3	12 ± 2	−8.3	13 ± 4	−15.4
moschiferous		24	−54.2	15	−26.7	15	−26.7	17	−35.3	21	−47.6	22	−50.0
Antilocapra	50	59 ± 5	−15.3	54 ± 3	−7.4	56 ± 5.6	−10.7	49 ± 5	2.0	51 ± 4	−2.0	54 ± 5	−7.4
americana		63	−20.6	67	−25.4	102	−51.0	57	−12.3	69	−27.5	96	−47.9
Giraffa	1150	1533 ± 219	−25.0	1875 ± 881	−38.7	1796 ± 419	−36.0	1254 ± 369	−8.3	1543 ± 325	−25.5	1705 ± 388	−32.6
camelopardalis		2152	−46.6	3990	−71.2	2519	−54.3	1163	−1.1	1697	−32.2	3848	−70.1
Okapia johnstoni	200	357 ± 108	−44.0	352 ± 32	−43.2	337 ± 43	−40.7	256 ± 63	−21.9	292 ± 86	−31.5	339 ± 52	−41.0
		324	−38.3	345	−42.0	267	−25.1	205	−2.4	160	25.0	295	−32.2
Vicugna vicugna	50	53 ± 9	−5.7	49 ± 7	2.0	63 ± 15	−20.6	44 ± 17	13.6	44 ± 16	13.6	54 ± 19	−7.4
		61	−18.0	80	−37.5	70	−28.6	60	−16.7	38	31.6	52	−3.8
Camelus	750	751 ± 275	−.1	859 ± 180	−12.7	813 ± 178	−7.7	762 ± 272	−1.6	705 ± 245	6.4	753 ± 200	−.4
dromedarius		1257	−40.3	1181	−36.5	360	108.3	1488	−49.6	622	20.6	399	88.0

Note: The top line for each species gives the mean (± SD) of all measurements for that bone, and the percent predictive error (%PE) for that estimate. The lower value for each bone is the mass predicted by length of that bone, followed by %PE of the prediction. Abbreviations as in Table 14.1.

this species (and possibly for other giraffids). Results for the okapi similarly give large overestimates, although they are not as large as the giraffe estimates. Again, it is often a single measurement that increases the estimates, although it is not always the same dimension. The non-length predictions for camelids are mostly within 15% of actual mass, although the standard deviations are higher. However, most long-bone lengths, including femur and humerus, consistently overestimate body mass. As was pointed out above, in camelids the humerus and femur are long relative to body mass.

Artiodactyl predictions. Table 14.3 gives mass predictions for various artiodactyl taxa based on all-artiodactyl regressions. Since ruminants dominate these regressions, the results are very similar to those for the bovid, cervid, and ruminant regressions. Reliable mass estimates are given for all of the ruminant taxa, with the exception of the giraffids. As might be expected, masses predicted for nonruminant taxa are less accurate. However, in most cases masses predicted by non-length measurements are within 25–30% of actual body masses. Results are generally poorer for large suids than for tayassuids or smaller suids. Metapodials of suoids give the poorest estimates and have large standard deviations. Although this is not surprising given the morphological differences between these bones in groups with fused and unfused metapodials, the increased variance and erroneous estimates largely come from midshaft diameters. In suids the metapodials are laterally expanded and flattened anteroposteriorly, a trend that becomes more pronounced in larger suids. This may partly explain the poor predictions.

Ungulate predictions. The mass estimates for artiodactyls based on the all-ungulate equations do not differ greatly from those of the artiodactyl-only regressions, again since artiodactyls dominate these regressions, but the standard deviations within a bone are generally higher than for any other estimates (see Table 14.4). For equids, the mass predictions are reliable, and in fact are nearly as good as those based on equids only. Again, the standard deviations of the estimates from a single bone are much higher, suggesting that the similar mass predictions result from an averaging out of a series of larger over- and underestimates.

Table 14.3. *Body mass predictions for living artiodactyl genera, based on artiodactyl regressions*

	Mass	HM ± SD	%PE	RD ± SD	%PE	MC ± SD	%PE	FM ± SD	%PE	TB ± SD	%PE	MT ± SD	%PE
Cephalophus	5.5	5 ± .4	10.0	5 ± .3	10.0	5.1 ± .4	7.8	5.2 ± .5	5.8	5.2 ± .7	5.8	4.1 ± .7	34.1
monticola		6	−8.3	6	−8.3	8	−31.2	5.5	0.0	4.8	14.6	6.8	−19.1
Madoqua kirki	5	5.6 ± .7	−10.7	5.7 ± .4	−12.3	5.3 ± .4	−5.7	5.5 ± .5	−9.1	6.4 ± 1	−21.9	5.5 ± .6	−9.1
		6	−16.7	8.6	−41.9	14	−64.3	5.2	−3.8	8.4	−40.5	10.6	−52.8
Redunca	32	36 ± 5	−11.1	35 ± 3	−8.6	35 ± 5	−8.6	32 ± 2	0.0	37 ± 7	−13.5	35 ± 4	−8.6
fulvorufula		31	3.2	45	−28.9	67	−52.2	41	−22.0	53	−39.6	56	−42.9
Ovis dalli	84	92 ± 11	−8.7	81 ± 12	3.7	79 ± 8	6.3	80 ± 5	5.0	76 ± 11	10.5	75 ± 10	12.0
		95	−11.6	89	−5.6	70	20.0	94	−10.6	116	−27.6	74	13.5
Gazella thomsoni	23	29 ± 3	−20.7	28.2 ± 3	−18.4	28 ± 2	−17.9	26 ± 3	−11.5	28 ± 4	−17.9	31 ± 5	−25.8
		19	21.1	36	−36.1	62	−62.9	24	−4.2	36	−36.1	49	−53.1
Damaliscus	123	131 ± 9	−6.1	138 ± 4	−10.9	137 ± 10	−10.2	130 ± 18	−5.4	135 ± 10	−8.9	132 ± 12	−6.8
korrigum		120	2.5	208	−40.9	168	−26.8	113	8.8	145	−15.2	148	−16.9
Syncerus caffer	820	716 ± 86	14.5	750 ± 44	9.3	668 ± 180	22.8	765 ± 110	7.2	670 ± 119	22.4	694 ± 146	18.2
		423	93.9	240	241.7	80	925.0	426	92.5	288	184.7	93	781.7
Mazama	20	19 ± 2	5.3	20 ± 2	0.0	20 ± 2	0.0	22 ± 3	−9.1	24 ± 3	−16.7	22 ± 1	−9.1
americana		25	−20.0	21	−4.8	27	−25.9	23	−13.0	25	−20.0	34	−41.2
Axis porcinus	47	44 ± 6	6.8	45 ± 9	4.4	40 ± 3	17.5	51 ± 11	−7.8	51 ± 5	−7.8	42 ± 10	11.9
		44	6.8	28	67.9	29	62.1	43	9.3	40	17.5	91	−48.4
Alces alces	450	503 ± 83	−10.5	506 ± 69	−11.1	482 ± 51	−6.6	591 ± 162	−23.9	590 ± 85	−23.7	580 ± 135	−22.4
		803	−44.0	525	−14.3	337	33.5	591	−23.9	607	−25.9	554	−18.8

Odocoileus	45	43 ± 3	4.7	43 ± 3	4.7	47 ± 7	−4.3	44 ± 7	2.3	49 ± 4	−8.2	49 ± 7	−8.2
virginianus		53	−15.1	54	−16.7	65	−30.8	54	−16.7	59	−23.7	84	−46.4
Giraffa	1150	1533 ± 227	−25.0	1883 ± 898	−38.9	1797 ± 420	−36.0	1259 ± 372	−8.7	1529 ± 314	−24.8	1704 ± 387	−32.5
camelopardalis		2162	−46.8	3447	−66.6	2517	−54.3	1160	−.9	1375	−16.4	3849	−70.1
Vicugna vicugna	50	53 ± 8	−5.7	50 ± 7	0.0	63 ± 15	−20.6	44 ± 18	13.6	46 ± 17	8.7	54 ± 19	−7.4
		61	−18.0	84	−40.5	70	−28.6	62	−19.4	43	16.3	52	−3.8
Camelus	750	754 ± 279	−.5	863 ± 182	−13.1	814 ± 177	−7.9	765 ± 274	−2.0	704 ± 240	6.5	734 ± 198	2.2
dromedarius		1263	−40.6	1085	−30.9	360	108.3	1480	−49.3	552	35.9	399	88.0
Phacochoerus	80	78 ± 21	2.6	73 ± 19	9.6	44 ± 22	81.8	64 ± 10	25.0	68 ± 24	17.6	93 ± 111	−14.0
aethiopicus		86	−7.0	38	110.5	6	1233.3	48	66.7	25	220.0	5	1500.0
Hylochoerus	215	157 ± 49	36.9	119 ± 24	80.7	116 ± 62	85.3	136 ± 26	58.1	125 ± 40	72.0	222 ± 290	−3.2
meinertzhageni		142	51.4	48	347.9	9	2288.9	97	121.6	28	667.9	4	5275.0
Tayassu pecari	30	34 ± 9	−11.8	28 ± 11	7.1	30 ± 9	0.0	25 ± 5	20.0	25 ± 5	20.0	44 ± 50	−31.8
		32	−6.2	14	114.3	4	650.0	22	36.4	10	200.0	3	900.0

Note: The top line for each species gives the mean (± SD) of all measurements for that bone, and the percent predictive error (%PE) for that estimate. The lower value for each bone is the mass predicted by length of that bone, followed by %PE of the prediction. Abbreviations as in Table 14.1

Table 14.4. *Body mass predictions for living ungulate genera, based on ungulate regressions*

	Mass	HM ± SD	%PE	RD ±SD	%PE	MC ± SD	%PE	FM ± SD	%PE	TB ± SD	%PE	MT ± SD	%PE
Cephalophus	5.5	5.1 ± .4	7.8	5.2 ± .2	5.8	5.1 ± .3	7.8	5.4 ± .7	1.9	5.3 ± .8	3.8	4.5 ± .8	22.2
monticola		6	−8.3	5.6	−1.8	15	−63.3	5.4	1.9	4.9	12.2	15	−63.3
Madoqua kirki	5	5.6 ± .6	−10.7	5.8 ± .5	−13.8	5.4 ± .3	−7.4	5.7 ± .5	−12.3	6.5 ± 1	−23.1	6 ± .8	−16.7
		5.6	−10.7	8.4	−40.5	25	−80.0	5.1	−2.0	8.7	−42.5	21	−76.2
Redunca	32	36 ± 5	−11.1	35 ± 3	−8.6	36 ± 6	−11.1	31 ± 3	3.2	38 ± 3	−15.8	36 ± 5	−11.1
fulvorufula		32	0.0	45	−28.9	82	−61.0	37	−13.5	62	−48.4	72	−55.6
Ovis dalli	84	92 ± 11	−8.7	80 ± 12	5.0	82 ± 8	2.4	79 ± 5	6.3	76 ± 12	10.5	76 ± 9	10.5
		98	−14.3	93	−9.7	85	−1.2	97	−13.4	129	−34.9	88	−4.5
Gazella thomsoni	23	29 ± 3	−20.7	28 ± 3	−17.9	29 ± 3	−20.7	26 ± 3	−11.5	29 ± 4	−20.7	32 ± 7	−28.1
		19	21.1	36	−36.1	78	−70.5	24	−4.2	39	−41.0	66	−65.2
Damaliscus	123	132 ± 11	−6.8	137 ± 6	−10.2	142 ± 12	−13.4	128 ± 21	−3.9	133 ± 12	−7.5	133 ± 16	−7.5
korrigum		124	−.8	221	−44.3	169	−27.2	116	6.0	161	−23.6	146	−15.8
Syncerus caffer	820	719 ± 78	14.0	733 ± 42	11.9	709 ± 94	15.7	733 ± 139	11.9	642 ± 114	27.7	665 ± 123	23.3
		449	82.6	256	220.3	778	5.4	450	82.2	325	152.3	104	688.5
Mazama	20	19 ± 2	5.3	20 ± 2	0.0	20 ± 3	0.0	22 ± 2	−9.1	24 ± 3	−16.7	23 ± 2	−13.0
americana		25	−20.0	21	−4.8	40	−50.0	23	−13.0	27	−25.9	49	−59.2
Axis porcinus	47	44 ± 6	6.8	46 ± 9	2.2	41 ± 4	14.6	51 ± 11	−7.8	51 ± 5	−7.8	44 ± 3	6.8
		45	4.4	29	62.1	42	11.9	44	6.8	43	9.3	46	2.2
Alces alces	450	505 ± 82	−10.9	499 ± 84	−9.8	509 ± 50	−11.6	565 ± 167	−20.4	571 ± 108	−21.2	564 ± 148	−20.2
		864	−47.9	569	−20.9	292	54.1	628	−28.3	697	−35.4	385	16.9
Odocoileus	45	43 ± 3	4.7	43 ± 4	4.7	49 ± 9	−8.2	44 ± 7	2.3	49 ± 4	−8.2	51 ± 9	−11.8
virginianus		54	−16.7	55	−18.2	81	−44.4	55	−18.2	64	−29.7	96	−53.1

Giraffa	1150	1545 ± 241	−25.6	1831 ± 889	−37.2	1906 ± 400	−39.7	1163 ± 267	−1.1	1458 ± 338	−21.1	1593 ± 318	−27.8
camelopardalis		2367	−51.4	3930	−70.7	1404	−18.1	1247	−7.8	1610	−28.6	1592	−27.8
Vicugna vicugna	50	53 ± 8	−5.7	50 ± 7	0.0	65 ± 16	−23.1	44 ± 17	13.6	45 ± 16	11.1	55 ± 19	−9.1
		63	−20.6	87	−42.5	86	−41.9	63	−20.6	47	6.4	68	−26.5
Camelus	750	767 ± 310	−2.2	847 ± 191	−11.5	870 ± 216	−13.8	729 ± 279	2.9	665 ± 199	12.8	715 ± 232	4.9
dromedarius		1369	−45.2	1200	−37.5	307	144.3	1598	−53.1	632	18.7	302	148.3
Phacochoerus	80	79 ± 22	1.3	73 ± 19	9.6	62 ± 28	29.0	63 ± 10	27.0	67 ± 24	19.4	87 ± 95	−8.0
aethiopicus		89	−10.1	38	110.5	16	400.0	49	63.3	27	196.3	12	566.7
Hylochoerus	215	159 ± 53	35.2	118 ± 24	82.2	120 ± 64	79.2	133 ± 24	61.7	123 ± 40	74.8	198 ± 239	8.6
meinertzhageni		148	45.3	49	338.8	17	1164.7	99	117.2	30	616.7	11	1854.5
Tayassu pecari	30	34 ± 9	−11.8	28 ± 11	7.1	31 ± 9	−3.2	25 ± 5	20.0	25 ± 5	20.0	42 ± 45	−28.6
		33	−9.1	14	114.3	10	200.0	23	30.4	11	172.7	7	328.6
Equus asinus	245	226 ± 65	8.4	263 ± 38	−6.8	213 ± 34	15.0	292 ± 73	−16.1	274 ± 68	−10.6	245 ± 48	0.0
		200	22.5	225	8.9	109	124.8	203	20.7	121	102.5	127	92.9
Equus kiang	230	244 ± 49	−5.7	253 ± 34	−9.1	196 ± 40	17.3	256 ± 81	−10.2	257 ± 66	−10.5	224 ± 50	2.7
		173	32.9	207	11.1	125	84.0	179	28.5	102	125.5	152	51.3
Equus burchelli	230	260 ± 56	−11.5	292 ± 29	−21.2	225 ± 30	2.2	341 ± 64	−32.6	315 ± 68	−27.0	259 ± 53	−11.2
		169	36.1	168	36.9	101	127.7	199	15.6	94	144.7	111	107.2
Equus grevyi	390	353 ± 83	10.5	396 ± 52	−1.5	296 ± 46	31.8	446 ± 94	−12.6	436 ± 90	−10.6	331 ± 51	17.8
		280	39.3	275	41.8	135	188.9	318	22.6	134	191.0	161	142.2
Equus caballus	490	555 ± 108	−11.7	591 ± 101	−17.1	443 ± 53	10.6	718 ± 138	−31.8	661 ± 119	−25.9	595 ± 105	−17.6
(th)		594	−17.5	470	4.3	180	172.2	547	−10.4	288	70.1	222	120.7

Note: The top line for each species gives the mean (± SD) of all measurements for that bone, and the percent predictive error (%PE) for that estimate. The lower value for each bone is the mass predicted by length of that bone, followed by %PE of the prediction. Abbreviations as in Table 14.1

Table 14.5. *Predicted masses for fossil antilocaprids, dromomerycids, and moschids, comparing masses predicted by bovid, cervid, and ruminant regressions*

	HM ± SD	RD ± SD	MC ± SD	FM ± SD	TB ± SD	MT ± SD
Bovid based						
Aletomeryx	15 ± 3	23 ± 2	23 ± 4	15 ± 22	20 ± 3	21 ± 3
	15	25	36	14	15	32
Bouromeryx	43 ± 8	50 ± 8	49 ± 4	65 ± 14	66 ± 15	60 ± 8
	50	45	52	65	39	67
Cranioceras	57 ± 12	108 ± 13	90 ± 11	87 ± 22	133 ± 51	100 ± 10
	48	92	77	89	56	102
Blastomeryx	—	11 ± 1	12 ± 2	12 ± 1.4	11 ± 1.3	15 ± 1.4
		14	26	12	11	21
Merycodus	12 ± 1.5	14 ± 1.5	15 ± 3	—	15 ± 6	17 ± 1
	17	19	25		12	20
Ramoceros	12 ± 2	14 ± .5	16 ± 2	13 ± 1.4	13 ± 2	16 ± 2
	15	17	33	15	15	31
Stockoceros	51 ± 6	43 ± 3	48 ± 6	46 ± 3	56 ± 6	55 ± 8
	57	50	78	55	73	76
Cervid based						
Aletomeryx	15 ± 3	23 ± 2	23 ± 4	15 ± 2	19 ± 3	21 ± 3
	14	26	35	14	16	31
Bouromeryx	43 ± 7	50 ± 7	48 ± 3	64 ± 13	65 ± 14	59 ± 7
	47	46	51	61	39	63
Cranioceras	57 ± 12	108 ± 12	89 ± 10	85 ± 21	131 ± 51	97 ± 8
	45	92	76	85	55	94
Blastomeryx	—	11 ± 1	12 ± 2	12 ± 1	11 ± 1	15 ± 2
		16	26	11	11	20
Merycodus	12 ± 1.3	14 ± 1.4	15 ± 3	—	15 ± 6	17 ± 1.4
	16	20	25		12	19
Ramoceros	12 ± 2	14 ± .4	16 ± 2	13 ± 1	13 ± 2	17 ± 3
	14	18	32	14	15	29
Stockoceros	51 ± 6	43 ± 2	47 ± 5	46 ± 4	55 ± 6	54 ± 8
	54	51	77	53	71	71

Mass estimates on fossils

Artiodactyls. Body mass estimates (Table 14.5) for the fossil pecoran genera tested are in general consistent within a bone and among the long bones tested. Results for genera in both subfamilies of antilocaprids are consistently uniform, with similar estimates from bone to bone and low standard deviations. Predictive equations based on bovids, cervids, and all ruminants give very similar and equally uniform results. The two genera of fossil moschids included give consistent results from bone to

Table 14.5. (*cont.*)

Ruminant based						
Aletomeryx	15 ± 3	23 ± 2	23 ± 4	15 ± 2	19 ± 3	21 ± 3
	15	26	35	14	16	31
Bouromeryx	43 ± 7	49 ± 7	48 ± 3	65 ± 13	65 ± 14	59 ± 7
	46	46	51	61	39	62
Cranioceras	57 ± 13	107 ± 11	88 ± 9	86 ± 21	130 ± 50	97 ± 8
	45	88	76	83	54	93
Blastomeryx	—	11 ± 1	12 ± 2	12 ± 1	11 ± 1	15 ± 2
		16	26	12	11	20
Merycodus	12 ± 1.3	14 ± 1.4	15 ± 3	—	15 ± 6	17 ± 1
	16	21	25		12	20
Ramoceros	12 ± 2	14 ± .5	16 ± 2	13 ± 1	13 ± 2	17 ± 3
	15	19	33	15	15	30
Stockoceros	51 ± 6	43 ± 2	47 ± 5	46 ± 4	55 ± 6	54 ± 7
	53	51	76	52	71	70

Note: The top line for each species gives the mean (± SD) for the predicted mass of all measurements for that bone. The lower value for each bone is the predicted mass based on length of that bone.
Abbreviations: HM, humerus; RD, radius; MC, metacarpal; FM, femur; TB, tibia; MT, metatarsal. AMNH, American Museum of Natural History; F:AM, Frick Collection, AMNH; LACM, Natural History Museum of Los Angeles County.
Specimens measured: *Aletomeryx*, F:AM42883; *Bouromeryx* and *Blastomeryx*, AMNH, various, Trinity River; *Cranioceras*, AMNH, various, Xmas Quarry Zone; *Merycodus*, AMNH 51128; *Ramoceros*, F:AM 31512; *Stockoceros*, LACM, various, San Josecito Cave.

bone regardless of the predictive equations used. There are also few differences among the three sets of predicted masses for dromomerycids, but there are consistent differences between bones. Forelimb bones generally give lower mass estimates than hindlimb bones, with the humerus being consistently lower. This trend was also noted by Scott (1983), who attributed it to the cervidlike skeleton of this family; however, the cervid-based predictive masses make it clear that this group is no more cervidlike than bovidlike. In some cases the pool of fossil bones that formed the basis for the estimates may have included more than one species. This may be the case for *Aletomeryx,* since there is less difference when estimates are made on a single associated skeleton (see Scott 1983). However, some of the forelimb/hindlimb differences in mass predictions are probably real.

Predicted body masses for camelids are relatively uniform among proximal bones for ruminant-, artiodactyl-, and ungulate-based predictions (Table 14.6). Standard deviations are high for these bones, as

Table 14.6. *Predicted masses for fossil camelids, comparing predictions by ruminant, artiodactyl, and ungulate regressions*

	HM ± SD	RD ± SD	MC ± SD	FM ± SD	TB ± SD	MT ± SD
Ruminant based						
Aepycamelus	473 ± 143	322 ± 100	445 ± 116	376 ± 151	381 ± 110	391 ± 100
	935	928	698	873	937	794
Michenia	61 ± 20	68 ± 24	92 ± 19	51 ± 21	64 ± 16	67 ± 16
	115	156	159	77	102	142
Megatylopus	704 ± 227	589 ± 134	1072 ± 199	739 ± 189	856 ± 330	1155 ± 309
	1154	918	715	1483	1269	913
Stenomylus	34 ± 11	32 ± 9	42 ± 6	26 ± 2	24 ± 5	26 ± 12
	68	69	60	30	29	51
Artiodactyl based						
Aepycamelus	472 ± 134	325 ± 100	445 ± 117	450 ± 233	383 ± 111	392 ± 242
	939	864	697	874	801	795
Michenia	61 ± 19	68 ± 23	92 ± 19	55 ± 22	65 ± 16	67 ± 16
	116	159	159	79	107	142
Megatylopus	702 ± 217	593 ± 135	1073 ± 201	847 ± 327	854 ± 324	1155 ± 307
	1160	854	714	1476	1056	913
Stenomylus	35 ± 11	32 ± 9	41 ± 6	27 ± 3	25 ± 5	26 ± 11
	68	74	60	31	34	51
Ungulate based						
Aepycamelus	482 ± 158	321 ± 103	468 ± 120	445 ± 256	365 ± 88	383 ± 104
	1013	949	515	934	926	501
Michenia	62 ± 20	67 ± 23	96 ± 21	55 ± 22	64 ± 16	68 ± 17
	120	167	162	81	118	141
Megatylopus	717 ± 264	585 ± 147	1139 ± 205	827 ± 370	787 ± 246	1099 ± 312
	1255	939	525	1593	1229	555
Stenomylus	35 ± 11	33 ± 9	43 ± 7	27 ± 3	25 ± 5	27 ± 11
	70	76	76	32	36	67

Note: The top line for each species gives the mean (± SD) for the predicted mass of all nonlength measurements for that bone. The lower value for each bone is the predicted mass based on length of that bone. Abbreviations as in Table 14.5. Masses are predicted from the average measurements of a number of specimens from the following collections: *Aepycamelus*, AMNH, various, Olcott Formation; *Michenia*, AMNH, various, Dunlop Camel Quarry; *Megatylopus*, AMNH, various, Edson Quarry; *Stenomylus*, AMNH, various, Harrison Formation.

might be expected since no attempt was made to sort out measurements of characters that are obviously modified in camelids. Lengths generally give much larger estimates than nonlength dimensions, suggesting that proximal long bones were elongated in fossil camels as they are in the living genera. Metapodials give widely aberrant results for most genera in both length and nonlength predictions. Again, this is not surprising in light of the secondarily digitigrade posture of this family. Discrepancies among predictions are greater in large species than in small and medium-sized taxa.

The oreodont mass predictions (Table 14.7) are based on a small sample size and are thus somewhat difficult to interpret. The variance among predicted masses for different measurements of the same bone are highest for the suid-based regressions, and approximately the same when artiodactyl and ruminant regressions are used. The masses predicted by the suid regressions are generally higher than for the artiodactyl or ruminant regressions. For the single associated specimen measured, *Promerycochoerus carrikeri,* the predicted masses are reasonably consistent from bone to bone, although the radius gives a lower estimate than the other long bones. Both the radius and tibia of *Merycochoerus proprius* also give lower estimates than the femur, but the association of this specimen is not as certain.

Perissodactyls. The body mass estimates (Table 14.8) based on equid regressions give consistent results from bone to bone for most genera, and again, some of the variation may be related to mixing of taxa. The standard deviations for within-bone estimates are of about the same magnitude as those for living taxa, except for some of the estimates based on the humerus. Most of the larger variance of these estimates comes from a single measurement, in most cases H4. The reason for high estimates from this measurement is unclear. In general, standard deviations within a bone are lower for the ungulate-based predictions, and mass estimates are also somewhat lower.

Discussion

Most of the skeletal dimensions analyzed here change in some regular way with body mass, although some dimensions are more tightly controlled by body mass than others. All of these relationships are thus potential bases for predicting body masses of fossil ungulates. Although all skeletal dimensions show significant correlations with body mass, lengths of humerus and femur and nonlength dimensions are the most highly correlated, and consistently give the lowest predictive errors. All of these dimensions can be used individually to predict body masses

Table 14.7. *Predicted masses for oreodonts, comparing predictions by suid, artiodactyl, and ungulate regressions*

	Regressions			
	HM ± SD	RD ± SD	FM ± SD	TB ± SD
Suid				
Merycochoerus proprius	—	85 ± 34	138 ± 43	88 ± 29
	—	54	107	22
Merycochoerus proprius	—	—	127 ± 54	—
	—	—	95	—
Promerycochoerus	—	—	205 ± 42	—
	—	—	178	—
Promerycochoerus carrikeri	176 ± 57	151 ± 44	178 ± 38	180 ± 61
	113	92	120	76
Artiodactyl				
Merycochoerus proprius	—	63 ± 15	115 ± 36	63 ± 19
	—	22	68	6
Merycochoerus proprius	—	—	120 ± 46	—
	—	—	60	—
Promerycochoerus	—	—	167 ± 28	—
	—	—	111	—
Promerycochoerus carrikeri	135 ± 41	101 ± 25	142 ± 19	130 ± 21
	104	38	76	23
Ruminant				
Merycochoerus proprius	—	62 ± 15	114 ± 35	62 ± 19
	—	19	66	4
Merycochoerus proprius	—	—	119 ± 46	—
	—	—	58	—
Promerycochoerus	—	—	165 ± 28	—
	—	—	109	—
Promerycochoerus carrikeri	134 ± 42	100 ± 25	140 ± 20	128 ± 22
	103	19	74	19

Note: The top line for each species gives the mean (± SD) for the predicted mass of all nonlength measurements for that bone. The lower value for each bone is the predicted mass based on the length of that bone. Abbreviations as in Table 14.5. Specimens measured are as follows: *Merycochoerus proprius*, AMNH, 42469A-E; *Promerycochoerus*, AMNH, uncataloged, Harrison Formation; *Promerychochoerus carrikeri*, F:AM33353.

within about 20% for living species within the same families. However, more accurate body mass predictions can be obtained by averaging predicted masses over a number of measurements. In all families, lengths of distal long bones are poor predictors of body mass and should not be used for this purpose. Distal bones vary significantly in length, and

Table 14.8. *Body mass predictions for fossil equid genera*

	Regressions					
	HM ± SD	RD ± SD	MC ± SD	FM ± SD	TB ± SD	MT ± SD
Equid						
Neohipparion	—	131 ± 40	123 ± 21	149 ± 54	151 ± 25	129 ± 40
whitneyi		179	322	166	298	372
Neohipparion	197 ± 116	145 ± 34	146 ± 25	183 ± 75	161 ± 32	155 ± 25
trampasense	176	158	322	271	370	373
Neohipparion	193 ± 104	144 ± 29	135 ± 40	—	194 ± 30	151 ± 22
trampasense	192	156	314		288	336
Neohipparion	—	157 ± 38	—	—	—	157 ± 25
trampasense		160				372
Pliohippus sp.	149 ± 80	123 ± 35	—	122 ± 41	116 ± 26	107 ± 15
	139	152		126	239	222
Pliohippus sp.	—	123 ± 35	—	—	114 ± 22	—
		152			235	
Dinohippus	224 ± 126	220 ± 25	179 ± 14	—	235 ± 98	174 ± 17
leidyanus	184	222	283		266	255
Dinohippus	—	189 ± 14	193 ± 10	—	212 ± 27	151 ± 26
leidyanus		214	257		270	225
Dinohippus	—	207 ± 40	180 ± 22	—	173 ± 12	173 ± 17
leidyanus		184	271		275	228
Dinohippus	—	—	—	—	207 ± 29	165 ± 21
leidyanus					275	243
Parahippus	94 ± 31	81 ± 33	—	75 ± 44	—	—
cognatus	96	146		99		

(*continued*)

sometimes in other dimensions, depending on the habitat-specific adaptations of a species.

Examination of the predicted masses for individual species indicates that there are systematic patterns of bias in the mass predictions, and that the overall predictive errors are in some cases misleading. The pattern of biases in these erroneous predictions is instructive in determining how to apply predictive equations. For example, in bovids, cervids, and suoids mass estimates are poorer for large species, a trend that is most marked in bovids and suids. Individual predictive errors for small to medium-sized species are much smaller, so that weight predictions for these species are likely to be more accurate than the average predictive errors indicate. Conversely, mass estimates in the upper size range of a family are likely to be severely underestimated and less

Table 14.8. (*cont.*)

	Regressions					
	HM ± SD	RD ± SD	MC ± SD	FM ± SD	TB ± SD	MT ± SD
Ungulate						
Neohipparion	—	139 ± 26	—	186 ± 36	149 ± 30	102 ± 31
whitneyi		140		121	117	159
Niohipparion	148 ± 39	153 ± 13	103 ± 21	197 ± 74	169 ± 32	130 ± 29
trampasense	119	124	120	209	123	159
Neohipparion	147 ± 34	153 ± 15	111 ± 37	—	198 ± 34	126 ± 23
trampasense	132	123	117		111	148
Neohipparion	—	166 ± 10	120 ± 25	—	—	132 ± 31
trampasense		126	120			159
Pliohippus sp.	113 ± 28	131 ± 26	89 ± 15	137 ± 28	109 ± 18	78 ± 13
	90	120	105	89	83	108
Pliohippus sp.	—	131 ± 25	—	—	107 ± 15	—
		120			80	
Dinohippus	171 ± 46	233 ± 28	143 ± 14	—	255 ± 123	153 ± 38
leidyanus	125	171	109		98	120
Dinohippus	—	202 ± 38	159 ± 15	—	222 ± 46	128 ± 26
leidyanus		165	102		101	109
Dinohippus	—	219 ± 19	149 ± 23	—	175 ± 32	146 ± 26
leidyanus		143	106		103	110
Dinohippus	—	—	—	—	216 ± 49	139 ± 29
leidyanus					103	116
Parahippus	74 ± 10	84 ± 17	—	61 ± 23	79 ± 9	—
cognatus	58	115		68	97	

Note: The first set of predictions is derived from the equid regressions and the second from the combine ungulate regressions. The top line for each species gives the mean (± SD) for the predicted mass base on all nonlength measurements of each long bone. The lower value for each bone is the predicted ma based on the length of that bone. Abbreviations as in Table 14.5. *Neohipparion whitneyi*, AMNH 981 9815A: *Neohipparion trampasense*, various, Arens Quarry; *Pliohippus*, F:AM60810; *Dinohippu* AMNH, various, Edson Quarry; *Parahippus*, F:AM71700.

accurate than the average predictive error suggests. The seriousness of this problem varies among families. In bovids, scaling relationships of most dimensions tend to level out at high body masses, although the tendency is less marked for proximal bones and articular dimensions (Scott 1985). Although mass estimates are poorer for large suids, there is no obvious tendency for plots of scaling relationships to flatten (Karp 1987). For cervids the problem is minor; throughout the size range of living cervids scaling relationships are constant and mass estimates for *Alces* are uniform within and between bones (Scott 1983). The problem is difficult to evaluate in equids since the only available large equids are domestic breeds. These results suggest that mass estimate errors are likely to be even more severe if attempts are made to extrapolate beyond the size range of living members of a family.

Comparison of the mass predictions based on "own" versus combined regressions shows that for artiodactyls there is little difference among the predictive methods. The lower standard deviations for "own" taxon mass predictions suggest that it would be best to use the family predictive equations where a fossil species is known to be a bovid or a cervid. However, these two groups scale so similarly, their skeletons are so similar even to casual examination, and they so completely dominate the ruminant and artiodactyl regressions numerically that the choice of family or combined equations is not critical. Despite the fact that the ruminant and artiodactyl regressions are dominated by bovids and cervids, they appear to work well for other pecoran taxa with the possible exception of giraffids. Mass estimates for living and fossil taxa of antilocaprids and moschids are consistent from bone to bone, and consistent with recorded masses for the living taxa. These groups are clearly very similar skeletally to other ruminants, and there would seem to be little problem with using the ruminant regressions to predict body masses of these taxa with a reasonable degree of confidence.

Giraffids are problematical in that none of the methods gives highly reliable results, although body masses are closer to actual and less variable with ruminant or artiodactyl regressions. Because there are only two living genera, and *Giraffa* is so highly modified, these results may not be typical for the entire family. Without results for any fossil taxa it is somewhat difficult to judge how accurately masses can be predicted for this family.

The most puzzling results among the fossil pecoran families are those for dromomerycids. Although the skeleton of this group is not obviously modified in any way, for many taxa the forelimb gives consistently lower mass estimates than does the hindlimb. Some of the discrepancies, such as the very low estimates for the humerus of *Cranioceras,* might be attributable to misreferred material or mixing of more than one species. However, at least to some degree the results are probably real. Functionally, the lower estimated masses from forelimb bones suggest that they are more slender on average than those of other ruminants. This in turn suggests that more of the muscle mass is concentrated in the hindquarters, and that more emphasis may be placed on the hindlimb in locomotion. Cervids, which utilize longer hindlimb suspensions in galloping than bovids, also show slightly heavier hindlimbs and slightly lighter forelimbs (Scott 1987). It has been suggested that this distinction is related to the generally closed habitat of cervids as opposed to bovids (Gambaryan 1974; Scott 1987). However, although many dromomerycids appear to have been woodland or forest species, *Aletomeryx* was probably found in relatively open habitat, suggesting that the hindlimb emphasis was not related to habitat differences. Whatever the explanation in functional terms, the distinction is probably not critical for mass estimation. The true mass probably lies somewhere between, and the range of estimated masses for most dromomerycids (e.g., 15–23 kg for *Aletomeryx*) is still less than the mass ranges commonly given for actual field masses of living adult wild ungulates.

Camelids would seem to present the greatest difficulties for mass estimation since they are the most highly modified of the ruminant families. Besides the modifications clearly associated with the secondary assumption of digitigrade posture (i.e, widening of the distal ends of the metapodial), the proximal bones are clearly elongate compared to other ruminants of similar size. This elongation is presumably related to the adoption of a pacing gait, in which additional elongation of the limbs is possible since interference between limbs on one side is reduced. (For a discussion of use of the pace in camelids, see Webb 1972.) However, despite these problems, the mass estimates for living camelids based on ruminant or artiodactyl regressions are relatively reliable if one ignores all long-bone lengths. The estimates for most fossil camelids were also uniform within a bone, and among bones of the same taxon. This suggests that the ruminant and artiodactyl regressions can be used to predict masses of fossil camelids with reasonable confidence. For some bones it is only a single estimate that increases the variance, and better specific predictive equations might be developed by eliminating these dimensions.

For suoids, the choice of appropriate predictive equations is much more critical. Suoid-only predictive equations give satisfactory mass predictions, but the ruminant or combined artiodactyl predictive equations give many wildly aberrant masses. Suids differ significantly from other artiodactyls in postcranial morphology and may even deviate more widely from ruminant curves than do equids. Tayassuids are somewhat more similar to other artiodactyls, perhaps because they have partially fused metapodials. Nevertheless, for any fossil suid or tayassuid the suoid regressions would clearly be the method of choice.

Of all of the single family groups, equids are the most problematical, both at high and low body masses. Scaling relationships generally have lower r^2 values than in any of the artiodactyl families. It is difficult to determine whether this reflects something inherent in equids or the inclusion of domestic species in the baseline regressions, or is a consequence of poorer body mass data. I have chosen to include domestic horses mainly because without them, the size range of living equids is quite small, and may give biased regressions. Like all domestic animals, horses have been selectively bred for a variety of purposes that may be reflected in variation in skeletal dimensions. However, all breeds of horses (except perhaps the miniature horses) are working breeds, and their bone dimensions must still be controlled by body mass to a fairly high degree, so that there is some mechanical justification for including them in the mass predictive equations. Nevertheless, the breed mass data are poor and probably contribute to the variablity, compounding problems with the masses of wild ungulates. I believe that it is primarily poor body mass data that contribute to the greater variability within equids, and that when better mass data become available, these predictive equations can be refined. Despite these problems, %PE for the equid regressions are similar to those for artiodactyls and give reasonably consistent predictions for living species. For both fossil and living genera, standard deviations are lower for the equid regressions than for the combined ungulate regressions, suggesting that the equid-only regressions are the better choice for fossil mass predictions.

Despite the fact that same family regressions consistently give "better" mass predictions, there is surprisingly little scatter in the all-ungulate plots for femur length and many nonlength dimensions. This suggests that many skeletal dimensions are generally similar enough that these results could be extrapolated to other fossil ungulate groups of at least basically similar morphology, for example, Litopterna. Whether these similarities would extend to "archaic" ungulate taxa, such as the Pantodonta, which appear to be more heavily built, is uncertain. The distinctions between suids and other ungulates certainly suggest caution in applying these estimates to heavily built taxa. However, the fact that general equations give acceptable estimates for smaller suids also suggests that these problems may be minor at smaller body sizes. Inclusion of tapirs, rhinos, and hippos in the ungulate equations will give a better indication of the extent to which these results can be generalized. Nevertheless, the general ungulate regressions, especially for lengths of proximal bones and articular surfaces, would seem to provide a viable method of predicting ungulate body masses for a variety of ungulate taxa.

Summary

A series of equations for predicting body mass of ungulates based on postcranial dimensions were calculated for artiodactyl and perissodactyl families and combinations of families. For all taxa, nonlength dimensions of long bones are the best predictors of body weight; lengths of humerus and femur give acceptable predictions in most cases, but distal long bones are highly unreliable. In general, equations based on single families give the most accurate predictions for members of those families, but for ruminants there is little difference between the accuracy of single-family or all-ruminant regressions. Although regressions based on all living ungulate taxa are less accurate predictors of mass for living species than "own taxon" regressions, these equations give sufficiently accurate results to suggest that all-ungulate equations can be used for fossil ungulate taxa with no living relatives. When the predictive equations are used to estimate body mass of fossil ungulates belonging to six families, the estimates are consistent from bone to bone and the standard deviations are equivalent to those obtained for living species of known body mass.

Acknowledgments

I thank the participants of the symposium, especially V. L. Roth, J. Damuth, and C. Janis, for many helpful discussions on this topic. I am grateful to the many curators and staff members of the museum collections I visited who provided access to the living and fossil specimens, to P. Jones for preparing the figures, and to L. Karp and A. Hayes-Grillo for the data on suoids and fossil camelids. This work was supported by grants from the Rutgers University Research Council and by N.S.F. Grant No. BSR–8418349.

References

Alexander, R. McN. 1977. Allometry of the limbs of antelopes (Bovidae). *J. Zool., Lond. 183*:125–146.

Best, G. A., Edmond-Blanc, F., & Whitting, R. C. 1962. *Rowland Ward's Records of Big Game.* London: Rowland Ward.

Clark, J. L. 1963. *The Great Arc of Wild Sheep.* Norman: University of Oklahoma Press.

Gambaryan, P. P. 1974. *How Mammals Run.* New York: Halstead Press.

Karp, L. 1987. Allometric effects and habitat influences on the postcranial skeleton of suids and tayassuids. Master's dissertation, Rutgers University, New Brunswick.

McMahon, T. 1975a. Allometry and biomechanics: limb bones in adult ungulates. *Am. Nat., 109*(969):547–563.

McMahon, T. 1975b. Using body size to understand the structural design of animals: quadrupedal locomotion. *J. Appl. Physiol. 39*(4):619–627.

Meinertzhagen, R. 1938. Some weights and measurements of large mammals. *Proc. Zool. Soc. Lond. 108*:433–439.

Sachs, R. 1967. Liveweights and body measurements of Serengeti game animals. *East Afr. Wildlife J. 5*:24–36.

Scott, K. M. 1983. Prediction of body weight of fossil Artiodactyla. *Zool. J. Linn. Soc. 77*:199–215.

Scott, K. M. 1985. Allometric trends and locomotor adaptations in the Bovidae. *Bull. Am. Mus. Nat. Hist. 179* (2):197–288.

Scott, K. M. 1987. Allometry and habitat-related adaptations in the postcranial skeleton of Cervidae. In *The Biology and Management of the Cervidae,* ed. C. Wemmer, pp. 65–80. Washington, D.C.: Smithsonian University Press.

Von la Chevallerie, M. 1970. Meat production from wild ungulates. *Proc. S. Afr. Soc. Animal Prod. 9*:73–87.

Walker, E. P. 1968. *Mammals of the World.* Baltimore: The Johns Hopkins University Press.

Webb, S. D. 1972. Locomotor evolution in camels. *Forma Functio* 5:99–112.

15

Body size estimates and size distribution of ungulate mammals from the Late Miocene Love Bone Bed of Florida

BRUCE J. MACFADDEN AND
RICHARD C. HULBERT, JR.

Body size is among the most revealing characters from which inferences can be made about the adaptation of a given species to its environment and how it functions in a biotic community. Although body size data are frequently presented and analyzed for extant taxa and communities, until recently little has been done with this character in the fossil record. The reason for this is simple; among the best measures of body size is mass, and, of course, this is impossible to determine directly from paleontological data. Any body size analysis of extinct mammalian species must first estimate mass of the species in question by assuming similar (isometric) scaling relative to modern analogues. The further back in time one ventures, the more tenuous this assumption becomes. As is clear from this book, there are numerous character systems (e.g., cranial, dental, and postcranial) from which body sizes of mammals have been estimated, but, in practice, the choice of estimating parameters depends upon available material and the group in question.

Once the hurdle of estimating body mass is adequately addressed, then a new spectrum of paleontological analysis is opened for study at several levels of biological organization. The data used for these types of investigations are inferential relative to modern species and communities; however, their appeal is that they provide a temporal perspective over an evolutionary time frame. Although in its infancy, body size studies (prior to those presented in this book) in the fossil record have examined questions of microevolution of rodent species (Martin 1986), macroevolution of horses (MacFadden 1987), primate locomotory

In *Body Size in Mammalian Paleobiology: Estimation and Biological Implications,* John Damuth and Bruce J. MacFadden, eds.

adaptations (Jungers 1985), mammalian predator guilds (Van Valkenburgh 1982, 1985, 1988), ungulate community evolution (Janis 1982), and general community patterns (Damuth 1981, 1982).

The purpose of this paper is to present new data on body size estimates and distribution for 21 species of ungulates from the Late Miocene (Late Clarendonian) Love Bone Bed of northern Florida. As will be discussed in the next section, this locality seems to demonstrate many of the prerequisites of a paleocommunity from which inferences using modern analogues can be made. In addition, our results also lead us to recommend (1) some cautions to be noted when body size regressions equations are used, and (2) the types of characters that can be used in paleocommunity analysis.

The Love Bone Bed: setting and previous investigations

In this section we briefly summarize the context of the Love Bone Bed; the geology and a preliminary overview of the vertebrate fauna was presented by Webb, MacFadden, and Baskin (1981).

The Love Bone Bed is located 1.5 km north of the town of Archer in western Alachua County, Florida, about 20 km southwest of Gainesville. The site was discovered in 1974, and was worked extensively by field crews from the Florida Museum of Natural History until 1981. During this time, approximately 15,000 specimens of identifiable, medium- to large-sized mammals were recovered, as well as numerous other vertebrates. The rich microfauna has only been partially curated. Webb et al. (1981) determined that the Love Site was produced during a single, rapid depositional (cut and fill) cycle representing no discernible interval of geological time.

Fossil material from the Love Site represents all modern classes of vertebrates. Sharks and bony fishes, although extremely abundant and diverse, are at present unstudied. Based on an analysis of the emydid turtles, Jackson (1976, 1978) reconstructed a habitat of quiet water like a sinkhole pond or shallow stream, and an equable, warm climate. The avifauna from this site is among the richest of pre-Pleistocene age in North America (Becker 1987). Based on this segment of the fauna, Becker (1985: 200) stated: "It would seem likely, therefore, that the environments around the Love Bone Bed during the time of deposition would indicate freshwater ponds and streams, but probably also with wet marshes, streams, estuaries, and mudflats nearby." The terrestrial mammals consist of at least 45 taxa and are the best-studied part of the

Table 15.1. *The 21 species of ungulate mammals currently recognized in the Love Bone Bed local fauna*

Order Proboscidea, Perissodactyla	Order Artiodactyla
Order Proboscidea	**Order Artiodactyla**
Family Gomphotheriidae	Family Tayassuidae
Amebelodon cf. *A. barbourensis*	*"Prosthennops"* sp. (small)
Order Perissodactyla	*"Prosthennops"* sp. (large)
Family Equidae	Family Camelidae
Neohipparion trampasense	*Procamelus grandis*
Pseudhipparion skinneri	*Aepycamelus major*
Hipparion cf. *H. tehonense*	*"Hemiauchenia" minima*
Nannippus westoni	Family Dromomerycidae
Cormohipparion ingenuum	*Pediomeryx hamiltoni*
Cormohipparion plicatile	Family Paleomerycidae
Protohippus gidleyi	*Pseudoceras* sp.
Calippus elachistus	Family Antilocapridae
Calippus cerasinus	(genus & species undetermined)
Family Tapiridae	
Tapirus simpsoni	
Family Rhinocerotidae	
Aphelops malachorhinus	
Teleoceras proterum	

Love Bone Bed local fauna. Primary systematic descriptions have been presented for two of the rodents (Baskin 1980a, 1986), the Felidae (Baskin 1981), some of the Canidae (Baskin 1980b), Procyonidae (Baskin 1982), Nimravidae (Baskin 1981; Bryant 1988), the proboscidean (Lambert 1989), Dugongidae (Domning 1988), Equidae (Hulbert 1987; 1988a,b; Webb & Hulbert 1986), Tapiridae (Yarnell 1980), and Dromomerycidae (Webb 1983a). Other taxa, including the tayassuids, camelids, rhinoceratids, gomphotherid, small carnivores, and microfauna are currently unstudied or under study by various workers.

Of relevance to this study, 21 species of terrestrial ungulates (Proboscidea, Perissodactyla, and Artiodactyla) are currently recognized from the Love Site (Table 15.1). This number is equaled in modern faunas only in African savannas (Bourlière 1963), and probably approaches the actual number of ungulate taxa that lived in the surrounding area during the Late Miocene in northern Florida. The only major herbivore clades not represented at the site that are known from contemporaneous faunas in North America are merycoidodontids, blastomerycids (=moschids), protoceratids, anchitheriine equids, and mammutids. The absence of these browsers, in conjunction with the numerical dominance of grazing taxa (Table 15.1), suggests a preponderance of open grasslands, with only limited availability of browse. Thus, the relatively minor taxonomic bias recognizable in the fauna seems to reflect natural ecological factors. Given its limited period of deposition, taxonomic richness, and great diversity of represented body sizes (literally from mouse to elephant), the Love Site's mammalian fauna seems appropriate for paleocommunity analysis.

The age of the Love Bone Bed is latest Clarendonian (Late Miocene, about 9.0 myr), based on the stage of evolution of the mammalian fauna (Webb & Hulbert 1986; Webb et al. 1981). It is similar in age to (although slightly younger than) the well-known Late Miocene faunas of the midcontinent (e.g., Clarendon of Texas, Minnechaduza and Upper Snake Creek of Nebraska [Webb 1969; Skinner & Johnson 1984]), and represents a typical mammalian assemblage from the Clarendonian Chronofauna (Webb 1977, 1983b).

Materials and methods

Dental and skull measurements

The following measurements (allowing for incompleteness and breakage) were taken for 480 ungulate specimens from the Love Site (also see Figure 15.1):

1. UM1APL; greatest occlusal anteroposterior length of M1 (excluding cement in equids for this and all relevant characters below). For this character and characters 2, 5, and 6 below, in *Amebelodon* these were measured on either M1 or M2 because these teeth are difficult to distinguish (the same was done for the corresponding lower m1's and m2's). This character and/or characters 2, 5, and 6 below have been used to estimate body mass in previous studies (e.g., Creighton 1980; Gingerich, Smith, & Rosenberg 1982; MacFadden 1987).
2. UM1TRW; greatest transverse width of M1 (occlusal surface width for equids).
3. UP4M3; greatest occlusal length of P4 through M3. In advanced artiodactyls, this and character 7 constitute the functional chewing surfacc.
4. UTRL; greatest occlusal length of functional upper cheektooth row (P2 through M3 in most forms).

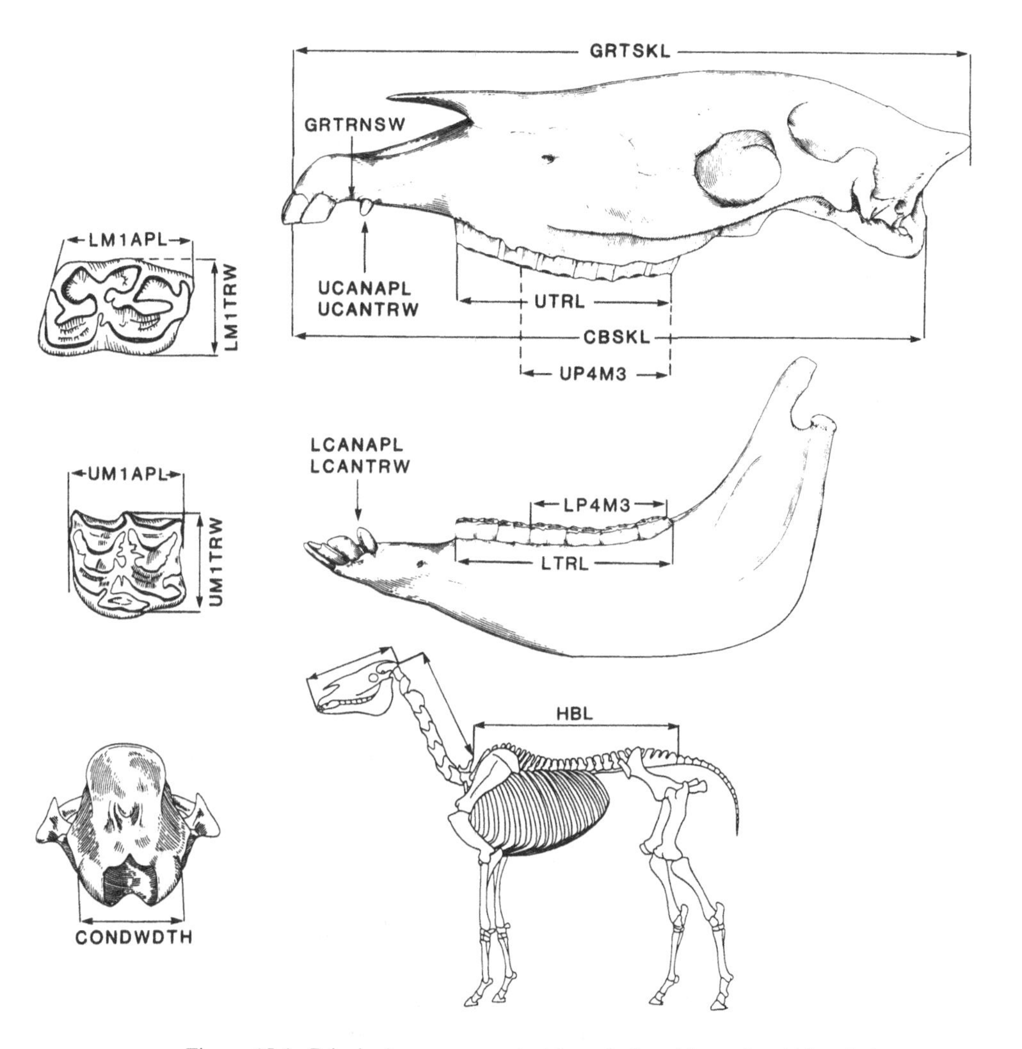

Figure 15.1. Principal measurements taken during this study. Abbreviations are given in text.

5. LM1APL; greatest occlusal anteroposterior length of m1. This equals FLML of Janis (this volume).
6. LM1TRW; greatest transverse width of m1 (occlusal surface width for equids from the metaconid to protoconid).
7. LP4M3; greatest occlusal length of lower p4 through m3.
8. LTRL; greatest occlusal length of functional lower cheektooth row (p2 through m3 in most forms).

9. GRTRNSW; greatest transverse width across symphysis (either I3s or Cs when the former are functionally lost). This was included to test the hypothesis that muzzle morphology can be used to distinguish browsers from grazers (Janis & Ehrhardt 1988; Owen-Smith 1985). However, our results indicate that there were insufficient specimens that preserve this character (also see discussion below).
10. UCANAPL; greatest anteroposterior length of upper canine (C).
11. UCANTRW; greatest transverse width of upper C. The canine characters were included to determine if sexual dimorphism could be discerned for the Love Site ungulates.
12. LCANAPL; greatest anteroposterior length of lower canine (c), or i2 in rhinoceratids.
13. LCANTRW; greatest transverse width of c, or i2 in rhinoceratids.
14. CONDWDTH; greatest condylar width. Martin (1980) used this as an estimator of body mass.
15. GRTSKL; greatest skull length from anteriormost tip of prostion to posteriormost part of nuchal crest.
16. CBSKL; condylarbasal skull length, greatest length from anteriormost tip of prostion to posteriormost part of occipital condyles.
17. HBL; head–body length, greatest length from tip of prostion to anterior part of tail. This character is highly correlated with body mass (Creighton 1980), and is a frequently cited size parameter for modern mammals (e.g., Walker 1975).

In addition to the characters listed above, first lower molar area was calculated by LM1APL × LM1TRW, which equals FLMA of Janis (this volume), and first upper molar area was calculated by UM1APL × UM1TRW, which equals M1AREA of MacFadden (1987).

Measurements less than 200 mm were taken using Fowler Maxcal digital calipers interfaced to an IBM-PC with the INCAL software program. Larger specimens were measured with large manual calipers ("anthropometers") to the nearest millimeter. When all measurements of the 480 specimens were completed, the data set was transferred to the database management program dBASE III+ for editing, and then exported, using a modem, to a CMS file on an IBM 4341 mainframe computer. SAS programs were used to compute statistics.

Twenty of the 21 ungulate species from the Love Site could be measured by use of this character matrix, although some sample sizes were very small (see Discussion below). The small sample of the gelocid *Pseudoceras* from the Love Site lacked any of the characters listed above. Therefore, a morphologically similar but much more numerous sample from the Early Hemphillian Withlacoochee River 4A Site was substituted for the Love Site *Pseudoceras*. Likewise, the characters represented in the Love Site material of *Aepycamelus* were insufficient to accurately estimate the body size for this giraffe-camel, so a specimen from the Early Hemphillian Mixson's Bone Bed was substituted.

As will be clear from the discussion below, for some taxa, the measurements listed above were not sufficient to estimate body mass. Accordingly, the following were also measured on selected taxa:

18. HUMCIR; minimum circumference of the humeral shaft (Anderson, Hall-Martin, & Russell 1985; Roth, this volume).
19. FEMCIR; minimum circumference of the femoral shaft (Anderson et al. 1985; Roth, this volume).
20. MC2, transverse width distal articular surface of metacarpal (Scott, this volume).
21. MC3, anteroposterior length of distal articular surface of metacarpal (Scott, this volume).
22. MC6, transverse diameter of metacarpal at midshaft (Scott, this volume).
23. MC7, anteroposterior length of metacarpal at midshaft (Scott, this volume).

Relative abundances

The minimum number of individuals (MNI) was calculated for each of the 21 taxa (Table 15.2). MNIs were obtained from both dental and postcranial elements. For the former, MNI was determined as the sum of the maximum number of right or left M3s or m3s (MNI of adults), and of right or left DP4s or dp4s (MNI of juveniles). For taxa whose isolated DP4s cannot always be distinguished from DP3s (or dp4s from dp3s), juvenile MNI was obtained from DP2s (or dp2s), or by adding the number of DP4s and DP3s (and dp4s plus dp3s) per side and dividing the result by two. At the Love Site, the most common postcranial elements are usually either astragali or calcanea. For all taxa except equids, the MNI for each species reported in Table 15.2 was the maximum value based on either of these two elements, or the MNI obtained with the

Table 15.2. *Minimum number of individuals (MNI) represented by terrestrial ungulate species at the Love Site*

	MNI	%Total	%By family
Amebelodon cf. *A. barbourensis*	25	4.4	4.4
Neohipparion trampasense	86	15.0	
Pseudhipparion skinneri	10	1.7	
Hipparion cf. *H. tehonense*	3	0.5	
Nannippus westoni	23	4.0	
Cormohipparion ingenuum	62	10.8	
Cormohipparion plicatile	83	14.5	
Protohippus gidleyi	31	5.4	
Calippus cerasinus	28	4.9	
Calippus elachistus	9	1.6	58.6
Tapirus simpsoni	24	4.2	4.2
Aphelops malacorhinus	22	3.8	
Teleoceras proterum	42	7.3	11.2
"*Prosthennops*" sp. (small)	1	0.2	
"*Prosthennops*" sp. (large)	17	3.0	3.1
Aepycamelus major	1	0.2	
Procamelus grandis	35	6.1	
"*Hemiauchenia*" *minima*	40	7.0	13.3
Pediomeryx hamiltoni	23	4.0	4.0
Pseudoceras skinneri	6	1.0	1.0
Antilocapridae (genus and species undetermined)	1	0.2	0.2

dentitions. For individual species of equids, computation of MNIs based on postcranial elements was not attempted, since they cannot be unambiguously assigned to the proper species. The total MNI of Love Site equids based on dental remains is 334. This is only slightly less than the overall MNI for equids of 345, based on right astragali. The differences in MNIs for Love Site ungulates between those listed in Table 3 of Webb et al. (1981) and those reported here reflect additional collecting and further curation of most groups and some differences in systematics.

There has been some question about the use of MNIs in paleoecological studies (e.g., Damuth 1982). However, Badgley (1986), reviewed the different procedures for obtaining relative abundances of vertebrate taxa at fluvial sites, including the MNI and number of identifiable specimens per taxon (NISP) methods. She concluded that the appropriate index depended on the taphonomic history of the site. The Love Site is taphonomically complex, and does not correspond to any of Badgley's (1986:332) simpler models. Rather it probably represents multiple sources of bone accumulation, and possibly several modes of transport. As there are some associated individuals at the Love Site, the MNI method was chosen to estimate relative abundance (Badgley 1986).

Results and discussion

Characters for body size estimation

What types of characters should be used to estimate the body mass of an extinct mammal? For example, it is well known that HBL is highly correlated to body mass in extant mammals (Creighton 1980). However, for most fossil taxa, this character is lacking. In the current study, only one of 480 specimens measured had this character; a mounted, composite skeleton of *Teleoceras proterum* (on exhibit at the Florida Museum of Natural History). Other workers have suggested various skull or postcranial characters as good estimators of body mass (e.g., Martin 1980). By looking at modern taxa of known skeletal dimensions and weight, Scott (this volume) provides convincing data to show that postcranial characters, which are for the most part unrelated to feeding adaptations, but are more related to the weight-bearing function of the species, are good estimators of body size. The theoretical presentation of this argument cannot be faulted; however, there are practical problems to be considered with the use of limbs in fossil samples. Perhaps paramount in this regard is the fact that in many quarry samples, particularly those containing morphologically similar, coexisting species, limb elements are usually disarticulated and disassociated from the taxonomically diagnostic elements (usually teeth). For the Miocene of North America, this problem is particularly acute for equids. This numerically dominant group (e.g., Table 15.2) frequently has a species richness of six to ten, and because of overlap in size among the group, isolated postcranial elements cannot always be unambiguously assigned to a particular species.

For the present study, if characters other than single or isolated tooth dimensions are used (see below), then only 25% of the ungulate species from the Love Site have sufficient samples (arbitrarily taken at $N \geq 3$ for a given character) to estimate body size (Figure 15.2). For samples of $N < 3$, there is no way to control for sampling error or to determine the extent of acceptable character variation (e.g., CVs < 10; Simpson,

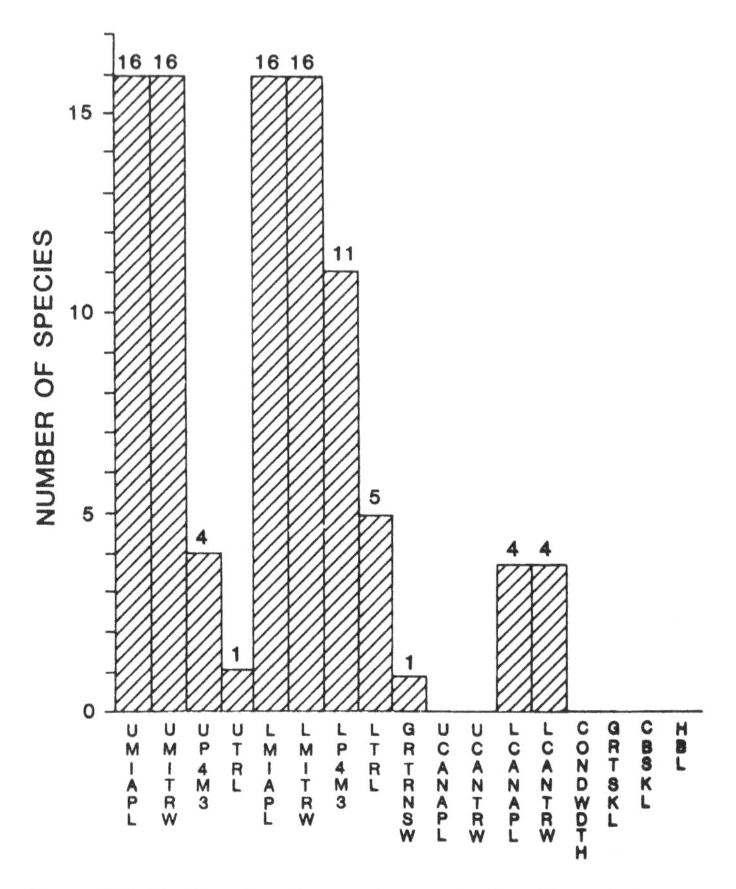

Figure 15.2. Number of ungulate species from the Love Site with three or more specimens exhibiting the measurable character listed on the histogram (see text and Figure 15.1 for abbreviations).

Roe, & Lewontin 1960). These problems in using characters other than those of dentitions arise from several factors:

1. Other structures (e.g., skulls and skeletons) usually require extraordinary preservation.
2. Fragmented skull parts (e.g., occipital condyles) or, as mentioned above, postcranial elements from fossil samples in which most specimens are disarticulated, are difficult, if not impossible, to assign to the proper taxon unless they are directly associated with dental material.

On the other hand, if tooth dimensions are used, then many more fossil sites will provide sufficient samples for body size estimates because these elements are usually both numerically abundant and taxonomically diagnostic. The caution with these characters is that they can be highly prone to both taxonomic specialization and ontogenetic variation. However, if only mature adult specimens (i.e., excluding juvenile and late wear stages) are measured, then CVs typically lie within acceptable limits (<10 for horses; MacFadden 1984, 1990). Likewise, body size estimates of taxa with highly specialized dentitions (e.g., mylagaulids and proboscideans) should be based only on modern analogues with similar specializations or on postcranial characters. The benefits of the use of dental characters far outweigh the drawback of the need for ontogenetic filtering. In the case of the Love Site, 19 of the 21 ungulate taxa preserved either or both M1s and m1s, and 16 (75%) had these characters represented by three or more specimens (Figure 15.2).

In summary, without the use of dental characters, comparative estimates of body masses at the Love Site would be impossible. For long-term research in paleocommunity reconstruction, it must be realized that the Love Site represents an *extraordinarily* well-preserved fossil assemblage. If one imposes the requirements of using only nondental characters in body size estimates, then there are only about a dozen known faunas of pre-Pleistocene age that would satisfy these criteria in the entire North American sequence.

Body size estimates

Two characters were used to estimate body mass: first lower molar length (LM1APL) and first lower molar area (LM1APL × LM1TRW). These were chosen because (1) they are represented in 18 of the 21 ungulate mammals from the Love site; (2) Janis (this volume) has computed regression equations using these characters for various combinations of ungulate groups; and (3) she has found that they are relatively good predictors of body mass. Of the equations available in Janis (this volume; Table 16.8, FLML and FLMA), the all-ungulate regressions were used for each taxon. When estimating body mass within a monophyletic group, the regressions derived for that clade (e.g., "perissodactyls and hyracoids" of Janis [this volume] for horses, see below) are more appropriate. However, in the current study we chose the all-ungulate regressions of Janis (this volume) because our principal purpose was to estimate and then compare body sizes across taxonomic categories. In this way, probably less error would be introduced from intertaxon sources than if different equations specific for each monophyletic group had been employed.

With the use of mean values for relevant characters that we measured (Table 15.3), the estimated body masses seemed reasonable for 17 of the 21 Love Site ungulates – that is, the horses, tapir, camels (except the aberrant *Aepycamelus*), *Pediomeryx,* peccaries, *Pseudoceras,* and antilocaprid (Table 15.4). By "reasonable," we mean similar to previously reported estimates (see Comparisons with other body mass estimates for similar taxa, below) and not dissimilar to extant forms of known body mass (as taken from Walker 1975).

Using the same regressions as above, we found that the remaining four taxa of Love Site ungulates had estimated body masses that were clearly absurd. With the use of the FLML and FLMA, respectively, these estimates were 44,361 and 63,710 kg for *Amebelodon* cf. *A. barbourensis,* 2385 and 4310 kg for *Aphelops malacorhinus,* 3605 and 2831 kg for *Teleoceras proterum,* and 2153 and 2940 kg for *Aepycamelus major.* These taxa are all of large size and the source of the error in the estimates is probably related to the fact that each of these is greater than the majority of taxa used to calculate the original regression equations (Janis, this volume). It is clear that large-scale errors in body size estimation can occur when taxa must be estimated by extrapolation beyond the original range of body sizes used to calculate the regression. It may also be possible that dental specializations of the fossil taxa may introduce additional errors in body size estimates.

Initially, we were not optimistic about being able to accept the results for these large taxa, and the absurd results merely underscored the need not to use these kinds of estimates. In an attempt to estimate body sizes that seemed reasonable, the following solutions were tried, using characters other than those measured for the small and medium-sized taxa. Our basic method here was to use postcranial characters, particularly cortical circumference of humeri (HUMCIR) and femora (FEMCIR), and the regression equations of Anderson et al. (1985) for large mammals (>300 kg, ranging from the kudu to African elephant).

Amebelodon cf. *A. barbourensis.* Measurements of HUMCIR ($\bar{x}$ = 321.2, N = 5, s [sample SD] = 35.0, OR [observed range] = 278–350 mm) and FEMCIR ($\bar{x}$ = 300.2, N = 5, s = 32.3, OR = 264–340) yielded body mass estimates of, respectively, 3095 and 3534 kg for this gomphothere. We also used the equations of Roth (this volume) which are specific to proboscideans, and the same characters yielded very similar estimated body masses of 3310 and 3822 kg, respectively. We therefore report an estimated mass of 3440 kg for this taxon (Table 15.4), based on the mean of these four determinations. *Amebelodon* certainly was smaller than modern-day proboscideans, which weigh about 5000 kg for the Indian elephant (*Elephas maximus*) and 5000–7500 kg for the African elephant (*Loxodonta africana;* weights in Walker 1975). A body mass of about 3440 kg therefore seems a reasonable estimate for *Amebelodon* cf. *A. barbourensis.*

Aphelops malacorhinus and *Teleoceras proterum.* Occipital condyle width (CONDWDTH) was first tried as a body mass estimator because (1) the relevant measurements were available from the Love Site sample (Table 15.3) and (2) regression equations for this character were published by Martin (1980). In particular, we used the regression for "large mammals," which produced estimated body masses of, respectively 1729 and 1906 kg, for *Aphelops malachorhinus* and *Teleoceras proterum.* Both of these estimates appeared high, as body masses of modern-day rhinocerotids range from 1000 to 3500 kg (Walker 1975), and the species from the Love Site were probably smaller than extant forms. The reason for the overestimate again seems to be that the distributions of body sizes used to produce the original regression equations were smaller than those that we were trying to estimate from our fossil sample.

We then used the regressions for HUMCIR and FEMCIR (equations 5 and 6 in Anderson et al. 1985), and the results obtained seem more reasonable. For *Aphelops,* HUMCIR ($\bar{x}$ = 185.4 mm, N = 13, s = 9.2, OR = 170–196 mm) and FEMCIR ($\bar{x}$ = 181.4 mm, N = 11, s = 11.5, OR = 168–199 mm) yielded estimated body masses of 737 and 1059 kg; the mean of these two determinations is used here (Table 15.4). Likewise, for *Teleoceras,* HUMCIR ($\bar{x}$ = 167.5 mm, N = 27, s = 9.34, OR = 148–189 mm) and FEMCIR ($\bar{x}$ = 163.5, N = 14, s = 8.58, OR = 144–178 mm) produced estimated body masses of 568 and 702 kg; a mean of these two determinations is used here (Table 15.4).

Aepycamelus major. The large "giraffe-camel" posed the greatest problem to estimating body mass because we had so few specimens available from the Love Site, all of which could potentially belong to the same individual (Table 15.2). The only available part of the postcranial skeleton that could be used in previously calculated regressions (Scott, this volume) was a single metacarpal (UF 39049). Four nonlength characters (MC2, MC3, MC6, and MC7 of Scott, this volume) yielded

Table 15.3 *Summary of univariate statistics for the Love Site ungulate mammals*

Taxon	Sample[a]	UM1APL	UM1TRW	UP4M3	UTRL	LM1APL	LM1TRW
Amebelodon cf. *A. barbourensis*	M	124.0, 5.68, 6 115.7–130.5,4.6	75.8, 2,68, 6 72.0–79.8, 3.5	—	—	115.5, 3.94, 4 112.2–121.2, 3.4	67.3, 3.26, 4 64.8–72.1, 4.8
Neohipparion trampasense	M	19.1, 0.98, 26 17.3–21.6, 5.1	17.5, 0.63, 25 16.6–18.5, 3.6	76.0, 1.53, 2 74.9–77.1, 2.0	121.0, – , 1	18.7, 0.60, 24 17.4–19.8, 3.2	9.1, 0.40, 25 8.5–9.9, 4.4
Pseudhipparion skinneri	M	13.7, 0.40, 7 13.1–14.2, 2.9	12.4, 0.22, 7 12.2–12.8, 1.8	—	—	14.1, 0.47, 4 13.6–14.7, 3.3	7.4, 0.09, 4 7.3–7.5, 1.2
Hipparion cf. *H. tehonense*	M	—	—	—	—	19.5, 1.51, 2 18.4–20.5, 7.7	8.5, 0.88, 2 7.9–9.1, 10.4
Nannippus westoni	M	16.0, 0.69, 9 14.7–16.8, 4.3	15.1, 0.82, 9 13.4–16.2, 5.5	—	—	16.4, 1.47, 3 15.5–18.1, 9.0	8.0, 0.67, 3 7.3–8.6, 8.4
Cormohipparion ingenuum	M	18.2, 1.11, 21 16.9–20.5, 6.1	16.7, 0.72, 21 15.3–18.2, 4.3	75.1, – , 1	119.8, – , 1	18.3, 1.06, 24 15.5–20.0, 5.8	9.6, 0.71, 24 8.6–11.4, 7.5
Cormohipparion plicatile	M	20.4, 1.06, 27 18.3–22.7, 5.2	19.6, 0.78, 27 18.2–21.4, 4.0	—	—	20.9, 2.0, 26 19.4–23.8, 5.7	11.1, 0.66, 26 9.5–12.1, 6.0
Protohippus gidleyi	M	21.2, 0.96, 19 19.4–22.8, 4.5	20.8, 0.87, 19 18.8–22.7, 4.2	—	—	20.7, 1.08, 24 18.8–23.9, 5.2	10.8, 0.44, 24 10.0–11.5, 4.1
Calippus elachistus	M	13.0, 0.8, 5 11.9–13.9, 6.2	12.6, 1.00, 5 11.2–14.0, 8.0	—	—	12.9, 1.00, 6 12.0–14.8, 7.8	7.4, 0.41, 6 6.8–7.8, 5.5
Calippus cerasinus	M	16.7, 0.86, 15 15.7–18.7, 5.1	17.2, 0.75, 15 15.7–18.4, 4.4	—	—	16.6, 1.01, 13 15.2–18.9, 6.4	8.7, 0.67, 13 7.6–10.0, 7.7
Tapirus simpsoni	A	23.3, 1.38, 10 21.1–25.4, 5.9	25.3, 1.50, 10 22.8–27.5, 59	93.0, 3.56, 3 89.6–96.7, 3.8	135.8, 14.3, 2 125.7–146.0, 10.6	23.2, 1.12, 17 21.4–25.3, 4.8	17.0, 1.42, 17 14.1–19.3, 8.
Alphelops malachorhinus	A	50.1, 9.50, 4 36.8–58.4,19.0	55.3, 3.36, 4 50.5–57.8, 6.1	188.1, 5.73, 3 181.6–192.3,3.05	271.3, 1.06, 2 270.5–272.0, 0.4	42.2, 8.55, 25 26.6–55.3,20.3	41.6, 7.8, 25 30.6–55.3, 18
Teleoceras proterum	A	56.2, 5.69, 9 46.6–64.0,10.13	56.9, 5.49, 9 49.8–66.8, 9.6	193.5, 16.8, 5 171.0–216.0, 8.7	241.5, 14.7, 4 221.6–255.6, 6.1	47.9, 2.69, 20 43.2–53.6, 5.6	28.0, 2.21, 2[illegible] 23.3–32.9, 7.
"*Prosthennops*" sp. (small)	A	12.5, – , 1	8.5, – , 1	—	—	—	—
"*Prosthennops*" sp. (large)	A	14.3, 0.64, 10 13.5–15.8, 4.5	11.7, 1.05, 10 10.7–13.4, 8.9	63.2, – , 1	84.3, – , 1	14.5, 0.98, 18 13.1–16.3, 6.8	10.9, 0.64, 1[illegible] 9.7–11.9, 5.[illegible]
Procamelus grandis	A	26.8, 1.01, 4 26.0–28.3, 3.8	13.9, 1.41, 4 12.3–15.3, 10.2	—	—	27.1, 2.82, 10 23.6–33.3,10.4	14.8, 1.42, 1 12.2–16.8, 9
Aepycamelus major	A	40.9, – , 1	33.6, – , 1	152.1, – , 1	—	—	—
Hemiauchenia minima	A	18.5, 1.21, 12 16.6–20.8, 6.5	12.1, 2.08, 12 8.7–15.9, 17.3	—	—	17.1, 1.47, 10 14.2–19.6, 8.6	8.8, 1.56, 1[illegible] 7.3–112.2,1[illegible]
Pediomeryx hamiltoni	M	21.4, 0.99, 7 20.1–22.6, 4.6	18.7, 1.79, 7 16.3–21.1, 9.6	81.2, 4.93, 2 77.7–84.7,6.07	116.3, – , 1	20.8, 1.16, 12 18.5–22.0, 5.6	12.9, 1.18, [illegible] 11.2–14.7, [illegible]
Pseudoceras sp.[b]	A	7.2, 1.28, 5 5.7–8.9, 18.0	7.1, 0.66, 5 6.4–8.0, 9.2	29.0, 1.56, 5 27.3–30.6, 5.4	39.9, 1.72, 4 37.8–41.8, 4.3	7.6, 1.15, 18 6.1–10.6,15.1	4.5, 0.26, [illegible] 4.2–5.0, 5.
Antilocapridae	—	12.9, – , 1	8.7, – , 1	—	—	—	—

[a]Refers to whether sample represents all (A) or only mature (M) wear classes.

[b]Sample taken from Withlacoochee River 4A site (see discussion in text).

LP4M3	LTRL	GRTRNSW	UCANAPL	UCANTRNW	LCANAPL	LCANTRNW	CONDWDTH
—	—	—	—	—	—	—	—
76.4, 1.74, 5 74.1–78.5, 2.27	117.0, 3.71, 4 112.4–121.5, 3.2	—	—	—	8.6, 0.58, 2 8.1–9.0, 6.8	7.6, 0.47, 2 7.2–7.9, 6.2	—
—	—	—	—	—	—	—	—
—	—	—	—	—	—	—	—
—	—	—	—	—	—	—	—
'6.6, 1.27, 8 '5.2–78.4, 166	115.6, 0.98, 5 114.1–116.6, 0.8	—	9.8, – , 1	6.9, – , 1	—	—	—
5.5, 2.98, 4 3.9–90.0, 3.5	130.5, 1.00, 2 129.7–131.2, 0.8	—	—	—	—	—	—
3.9, 2.23, 3 .5–85.8, 2.7	127.0, – ,1	—	—	—	—	—	—
—	—	—	—	—	—	—	—
.8, – , 1	—	—	—	—	—	—	—
.8, 9.4, 5 .0–99.1, 10.1	140.7, 2.27, 4 137.7–143.1, 1.6	35.7, 2.66, 4 33.4–39.2, 7.5	9.8, – , 1	9.7, – , 1	14.2, 1.84, 6 11.9–17.4, 13.0	12.0, 1.72, 6 10.0–14.8, 14.3	—
4.4, 8.63, 14 4.8–207.4, 4.4	276.0, 11.02, 5 263.5–287.4, 4.0	77.0, – , 1	—	—	21.1, – , 1	25.9, – , 1	115.8, – , 1
5.5, 10.13, 9 .3–200.3, 5.5	242.3, 13.28, 3 227.0–250.0, 5.5	103.8, 6.79, 2 99.0–108.6, 6.5	45.2, – , 1	23.0, – , 1	27.7, 6.28, 6 16.6–33.7, 22.7	36.3, 4.16, 6 28.1–39.7, 11.5	118.9, 2.63, 2 117.1–120.8, 2.2
—	—	—	—	—	10.3, – , 1	8.0, – , 1	—
5, 2.63, 5 0–67.5, 4.02	88.5, 2.56, 4 84.9–90.4, 2.9	29.4, 5.34, 2 25.6–32.1, 18.2	13.6, – , 1	8.0, – , 1	12.9, 2.86, 3 10.8–16.1, 22.3	9.6, 1.61, 3 8.4–11.4, 16.8	52.2, – , 1
.7, 7.18, 3 .5–119.8, 6.3	141.6, – , 1	—	—	—	12.2, 4.65, 2 8.9–15.5, 38.2	7.4, 3.39, 2 5.0–9.8, 45.6	—
—	—	—	—	—	—	—	—
5, – , 1	79.0, – , 1	—	—	—	—	—	—
, 4.25, 4 –88.7, 5.0	113.6, – , 1	—	—	—	—	—	—
, 2.69, 16 –35.2, 8.4	42.6, 2.55, 10 39.1–46.2, 6.0	—	5.7, – , 1	4.2, – , 1	3.8, 0.38, 4 3.4–4.1, 9.9	3.6, 0.86, 4 3.0–4.9, 23.8	25.4, 0.53, 2 24.5–25.7, 2.1
—	—	—	—	—	—	—	—

: For each measured character, data are arranged as follows: first line – mean, standard deviation, and number of observations; nd line – observed range and coefficient of variation. For description of characters see text and Figure 15.1. Because of the very l number of available specimens, GRTSKL, CBSKL, and HBL are not included in this chart.

Table 15.4. *Characters used to estimate body mass in the 21 species of ungulates represented at the Love Site*

Taxon	Estimates derived for individual characters[a]				Mean body mass estimate (kg)
	FLML (Janis, this volume) (kg)	LMAREA (Janis, this volume) (kg)	HUMCIRC (Anderson et al., 1985) (kg)	FEMCIRC (Anderson et al., 1985) (kg)	
Amebelodon cf. *A. barbourensis*	(63,710)[b]	(44,361)	3,095	3,534	
			3,310[c]	3,822[c]	3,440
Neohipparion trampasense	168	115	—	—	142
Pseudhipparion skinneri	67	54	—	—	61
Hipparion cf. *H. tehonense*	192	110	—	—	151
Nannippus westoni	109	77	—	—	93
Cormohipparion ingenuum	156	121	—	—	139
Cormohipparion plicatile	241	186	—	—	214
Protohippus gidleyi	223	175	—	—	204
Calippus elachistus	50	47	—	—	49
Calippus cerasinus	114	89	—	—	102
Tapirus simpsoni	339	423	—	—	381
Aphelops malacorhinus	(2,385)	(4,310)	737	1,059	889
Teleoceras proterum	(3,605)	(2,831)	568	702	635

"*Prosthennops*" sp., small	45	55	—	—	50
"*Prosthennops*" sp., large	73	102	—	—	88
Procamelus grandis	562	434	—	—	498
Aepycamelus major	(2,153)	(2,940)	1,026	—	1,026[d]
Hemiauchenia minima	125	95	—	—	110
Pediomeryx hamiltoni	237	233	—	—	235
Pseudoceras sp.	9	10	—	—	10
Antilocapridae	50	60	—	—	55

[a]The regression equations used to calculate body mass for a given linear dimension are as follows: (1) $\log W = (3.263 \times \log \text{FLML}) + 1.337$; (2) $\log W = (1.553 \times \log \text{LMAREA}) + 1.701$; (3) $\log W = (\log \text{HUMCIR} - \log 1.09) \times 10.38$; (4) $\log W = (\log \text{FEMCIR} - \log 1.47) \times 10.35$. Equations 1 and 2 are taken from Janis (this volume); FLML (first lower molar length) and LMAREA (first lower molar area) are in cm, *W* is in kg. Equations 3 and 4 are from Anderson et al. (1985); HUMCIR (minimum humeral circumference) and FEMCIR (minimum femoral circumference) are in mm, *W* is in g.
[b]Entries in parentheses are considered inaccurate for reasons given in the text.
[c]In addition to using the equations in Anderson et al. (1985), those in Roth (this volume) were also used for comparison of mean body weight estimates, i.e., $W(\text{kg}) = 9.448 \times 10^{-4} \times \text{HUMCIR (mm)}^{2.611}$; and $W(\text{kg}) = 3.790 \times 10^{-4} \times \text{FEMCIR (mm)}^{2.827}$.
[d]This determination is based on HUMCIR of a single specimen of *Aepycamelus* from the early Hemphillian Mixson's Bone Bed.

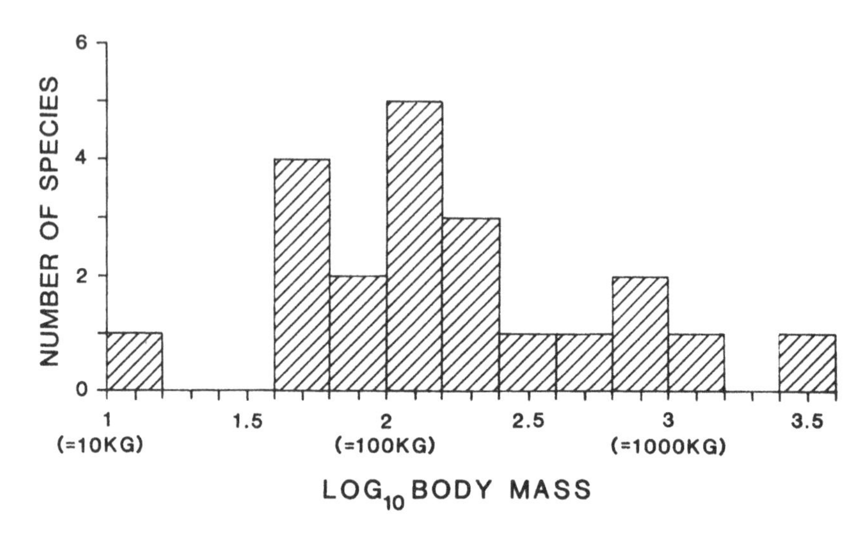

Figure 15.3. Histogram of estimated body sizes for the 21 species of ungulate mammals represented at the Love Site.

a mean body mass estimate of 2080 kg. Modern *Giraffa,* which is probably the best functional analogue of *Aepycamelus,* has a body mass of about 550–1180 kg for females and 800–1900 kg for males (Kingdon 1979). The Love Site giraffe-camel was of approximately similar size to modern giraffes; thus this estimate is probably the least accurate of the 21 ungulates from the Love Site. Another reason for this skepticism is that Scott (this volume) has already concluded that elongated artiodactyl metapodials, particularly of camelids, are prone to error in body size estimates.

On the basis of these questionable results, we measured the humeral circumference of one available specimen of *Aepycamelus major* (UF/FGS V322) from the early Hemphillian Mixson's Bone Bed, also located in northcentral Florida. The regressions of Anderson et al. (1985) produced an estimated body mass of 1026 kg, which seems more consistent with modern giraffids; hence this is the value we report in Table 15.4. The distribution of final estimates of body mass for the 21 species of Love Site ungulates is shown in Figure 15.3.

Comparisons with other body mass estimates for similar taxa

There is a close correspondence between the body masses estimated here for the Love Site ungulates and those produced for the same taxa in other studies, but using a different combination of regression equations. For example, using the mean determinations of many more characters and with the all-ungulate regressions, Janis (this volume) has estimated *Pediomeryx hamiltoni* to be 190 kg; our estimate is 235 kg (Table 15.4). The horses seem even closer; Janis (this volume) has estimated *Cormohipparion plicatile, Neohipparion trampasense,* and *Pseudhipparion skinneri* to be, respectively, 201, 155, and 65 kg, whereas our estimates are 214, 142, and 61 kg (Table 15.4). This consistency perhaps is not surprising because the same samples were measured; only the mix of character regressions is different.

MacFadden (1987) published regression equations to estimate the body masses of fossil horses that are scaled to different extant samples and different dental characters than those of Janis (this volume). MacFadden (1987) based his regressions on characters such as UM1APL (=M1APL), UM1TRW (=M1TRNW), and UM1AREA (=M1AREA). Using those regressions and the data from the nine species of fossil horses from the Love Site (Table 15.3), we calculated the estimated body masses presented in Table 15.5. For the sake of comparison, body mass estimates for the same nine species of Love Site horses are presented in Table 15.5 using the "perissodactyl–hyracoid" regressions in Janis (this volume) – that is, LM1APL (=FLML), LM1TRW (=FLMW), and LM1AREA (=FLMA). A Wilcoxon ranked-signed test indicates that there is no significant difference (T = 9, $p > .05$) between the masses estimated from the regressions in MacFadden (1987) versus those derived from the equations reported in Janis (this volume). These estimates, however, which are based on clade-specific regressions, are all lower than those produced by the all-ungulate regressions above, and the former are probably more accurate.

Patterns of body mass distribution

Ecologists have long been interested in the distribution of body size among coexisting organisms and any possible causal mechanism to explain observed patterns. Hutchinson (1959) noted that sympatric, ecologically similar species (often, but not always, limited to congeners) differed in size by at least a factor of about 1.3 for linear structures related to feeding, or about 2.0–2.2 for body mass. He interpreted these ratios as the minimum amount of similarity sufficient to decrease competition for limiting resources between individuals of sympatric species to a degree where coexistence was possible. Since that time, considerable

Table 15.5. *Body mass estimates for the nine species of horses represented at the Love Site*

	Masses (kg) based on MacFadden (1987)				Masses (kg) based on Janis (this volume)			
Taxon	UM1APL	UM1TRW	UM1AREA	$\bar{x}$	LM1APL	LM1TRW	LM1AREA	$\bar{x}$
Neohipparion trampasense	137	101	85	108	135	67	92	98
Pseudhipparion skinneri	63	50	44	52	55	37	44	45
Hipparion cf. *H. tehonense*	—	—	—	—	154	55	89	99
Nannippus westoni	88	72	59	73	89	46	62	66
Cormohipparion ingenuum	120	90	75	95	126	78	97	100
Cormohipparion plicatile	165	135	110	137	192	112	148	152
Protohippus gidleyi	185	160	129	158	187	109	140	145
Calippus elachistus	57	51	43	50	41	37	38	39
Calippus cerasinus	97	97	70	88	92	59	72	74

Note: For the regression equations in MacFadden (1987), the characters UM1APL, UM1TRW, UM1AREA were used. For those in Janis (this volume), the "perissodactyl–hyracoid" regression equations were used with LM1APL (=FLML), LM1TRW (=FLMW), and LM1AREA (=FLMA).

Table 15.6. *Expected and observed size distribution ratios for the 13 Love Site grazing ungulates*

Expected	1.22	1.13	1.10	1.08	1.07	1.07	1.07	1.07	1.08	1.10	1.13	1.30
Observed	1.12	1.11	1.52	1.10	1.08	1.26	1.02	1.06	1.35	1.05	2.33	1.28

Note: These ungulates are *Calippus elachistus* (49 kg), antilocaprid (55 kg), *Pseudhipparion skinneri* (61 kg), *Nannippus westoni* (93 kg), *Calippus cerasinus* (102 kg), "*Hemiauchenia*" *minima* (110 kg), *Cormohipparion ingenuum* (139 kg), *Neohipparion trampasense* (142 kg), *Hipparion* cf. *H. tehonense* (151 kg), *Protohippus gidleyi* (204 kg), *Cormohipparion plicatile* (214 kg), *Procamelus grandis* (498 kg), and *Teleoceras proterum* (635 kg). Expected values were calculated by linear extrapolation from Table 1 of Eadie et al. (1987) for a log-transformed population standard deviation of 0.34. The two series are not significantly different based on a Wilcoxon matched-pair signed-ranks test (T = 30.5).

evidence has been marshaled for and against the "reality" of so-called Hutchinsonian ratios, and their putative biological cause has been repeatedly investigated (e.g., Eadie et al. 1987 and references therein; Horn & May 1977; Maiorana 1978; Roth 1981; Strong et al. 1979). In particular, the Eadie et al. study surveyed size distributions across a broad array of vertebrate and invertebrate groups, and presented evidence that (1) animal sizes are distributed log-normally; (2) the variance of these distributions falls within a limited range; and (3) Hutchinsonian ratios are a direct mathematical artifact of (1) and (2), and not necessarily the product of competition or any other biological factor. These results also explain how Hutchinsonian ratios can be found in groups of nonbiological origin, such as musical instruments and bicycles (Horn & May 1977; Maiorana 1978).

We analyzed the distribution of body size for the grazing ungulate guild at the Love Site (N = 13, Table 15.6) to determine if it followed the pattern predicted by Eadie et al. (1987; their Table 1). Species were classified as grazers on the basis of hypsodont cheekteeth, previous interpretations of their diets (Webb 1983), and also by analogy with closely related or ecomorphologically similar modern mammals. The size distribution of these 13 species does not differ significantly from either a log-normal, normal, or log-uniform distribution, using the Kolmogorov–Smirnoff goodness-of-fit test ($p > .05$). However, the D statistic was least (.125) when the observed pattern was compared to the log-normal distribution (D vs. normal distribution = 0.214; D vs. log-uniform distribution = 0.192). Eadie et al. (1987) found that the D statistic was consistently least when compared to a log-normal distri-

Table 15.7. *Ratios of upper and lower molar lengths and estimated body masses of ecologically and/or taxonomically "similar" species represented from the Love Site*

Taxa	UM1APL	LM1APL	Estimated body mass
Neohipparion trampasense / *Pseudhipparion skinneri*	$\frac{19.1}{13.7} = 1.39$	$\frac{18.7}{14.1} = 1.33$	$\frac{142}{61} = 2.38$
Cormohipparion plicatile / *Cormohipparion ingenuum*	$\frac{20.4}{18.2} = 1.12$	$\frac{20.9}{18.3} = 1.14$	$\frac{214}{139} = 1.54$
Protohippus gidleyi / *Calippus cerasinus*	$\frac{21.2}{16.7} = 1.27$	$\frac{20.7}{16.6} = 1.25$	$\frac{204}{102} = 2.00$
Calippus cerasinus / *Calippus elachistus*	$\frac{16.7}{13.0} = 1.28$	$\frac{16.6}{12.9} = 1.29$	$\frac{102}{49} = 2.08$
Procamelus grandis / *Hemiauchenia minor*	$\frac{26.8}{18.6} = 1.44$	$\frac{26.4}{17.0} = 1.55$	$\frac{498}{110} = 4.53$
Hemiauchenia minor / Antilocaprid	$\frac{18.6}{12.9} = 1.44$	17.0 / ----	$\frac{110}{55} = 2.00$
Peccary #1 / Peccary #2	$\frac{21.4}{11.4} = 1.49$	$\frac{14.9}{12.5} = 1.19$	$\frac{88}{50} = 1.76$
	$\bar{x} = 1.35$	$\bar{x} = 1.29$	$\bar{x} = 2.33$

bution of body sizes, although it was not always the only nonsignificant statistic. Eadie et al. (1987; Table 1) calculated expected size ratios for groups of 2–15 species when $\sigma = .25$ and $\sigma = .75$, and determined that linear extrapolation would produce expected values for standard deviations of intermediate magnitude. The population standard deviation of the log-transformed body masses of the 13 Love Site grazers is .34. This value was used to produce expected values for size ratios (Table 15.6). Although there are some differences between the observed and expected values (Table 15.6), these are not significant according to a Wilcoxon matched-pair signed-ranks test ($N = 12$, $T = 30.5$). Therefore, the size distributions of the grazing ungulates from the Love Site conform to the model of Eadie et al. (1987).

Roth (1981) noted that many workers uncritically accepted size ratios of about 1.3 (or near 2.0 for mass) as the result of competitive pressure, although ratios of differing magnitude were justified with post hoc explanations. Similar conclusions could be deduced from the Love Site body size data (Table 15.7). For example, the two species of *Calippus*

– *C. cerasinus* and *C. elachistus* – have a ratio of body mass very close to 2.0, an apparent example of a Hutchinsonian ratio. However, the site's other congeneric pair of equids, *Cormohipparion plicatile* and *C. ingenuum,* differ only by a factor of 1.5. Furthermore, this pair coexisted in Florida for at least 3 my without any significant change in size in either species (Hulbert 1988b). If body size is influenced by competition, then their similarity requires a post hoc explanation for this exception to the 2.0 rule. The model of Eadie et al. (1987) requires no such explanations, predicting instead a range of ratios that includes both 1.5 and 2.0. Competition should be hypothesized as a cause for size distributions only in situations where character displacement is actually observed. Given this fact, we therefore report the linear and body size ratios for similar ungulates from the Love Site (Table 15.7), but make no claims about the possible causal relationship between competition and these observed ratios.

Conclusions

The Late Miocene Love Site is an extraordinarily rich deposit containing an inferred paleocommunity of at least 21 terrestrial ungulate mammals. Our results indicate that despite the richness and basically excellent preservation, the "best" (i.e., the most widely preserved and taxonomically distributed) characters that can be used to estimate body size are those of the dentitions. We emphasize that workers using these characters, particularly in hypsodont taxa, must carefully deal with ontogenetic changes in the teeth. On the other hand, if postcranial or skull characters are used, then the problems of unambiguous taxonomic recognition result in major problems in estimating body size for the various faunal components. We agree that in situations where there is unambiguous and taxonomically diagnostic association of postcranial elements, then many of the characters discussed in Scott (this volume) are probably better estimators of body mass. However, to use these requires exceedingly rare taphonomic situations not normally encountered in terrestrial vertebrate deposits.

Our results also underscore the importance of choosing the appropriate regression equations for estimating body size. A corollary of this conclusion is the empirical observation that extrapolations of regression equations, particularly for animals that lie far outside the original size ranges used to determine the regression, oftentimes yield absurd results

– for example, about 55 metric tons for *Amebelodon,* which is an order of magnitude greater than extant proboscideans.

Our data result in nonrandom patterns of body sizes for the 13 species of grazing ungulates; that is, they best conform to a log-normal distribution. As Eadie et al. (1987) have shown, this distribution best fits many animal assemblages and guilds. Similarly, although some of our size ratios for linear characters and body masses are similar to previously reported "Hutchinsonian ratios," there is no available evidence for an underlying biological causal explanation for these results.

Acknowledgments

We thank Mr. Joe Russo of the Smithsonian Institution for making the INCAL program available for our use and Mr. Minh Ngo for integrating the hardware configuration and software. The collecting and curation of the Love Site specimens were supported by NSF Grants DEB 78-10672, DEB 8022779, and BSR 8412748. The current research was supported by grants from the University of Florida (UF) Graduate School and NSF BSR–8515003. Statistical analysis was performed using the UF Faculty Support Center for Computing. This is UF Contribution to Paleobiology No. 314.

References

Anderson, J. F., Hall-Martin, A., & Russell, D. A. 1985. Long-bone circumference and weight in mammals, birds and dinosaurs. *J. Zool., Lond.* [*A*] *207*:53–61.

Badgley, C. 1986. Counting individuals in mammalian fossil assemblages from fluvial environments. *Palaios. 1*:328–338.

Baskin, J. A. 1980a. Evolutionary reversal in *Mylagaulus* (Mammalia, Rodentia) from the Late Miocene of Florida. *Am. Midl. Nat. 104*:155–162.

Baskin, J. A. 1980b. The generic status of *Aelurodon* and *Epicyon* (Carnivora, Canidae). *J. Paleontol. 54*:1349–1351.

Baskin, J. A. 1981. *Barbourofelis* (Nimravidae) and *Nimravides* (Felidae), with a description of two new species from the Late Miocene of Florida. *J. Mammal. 62*:122–139.

Baskin, J. A. 1982. Tertiary Procyonidae (Mammalia: Carnivora) of North America. *J. Vert. Paleontol. 2*:71–93.

Baskin, J. A. 1986. The Late Miocene radiation of neotropical sigmodontine rodents in North America. In *Vertebrates, Phylogeny, and Philosophy,* ed. K. M. Flanagan & J. A. Lillegraven, pp. 287–303. Contrib. Geol., Univ. Wyoming, Spec. Pap.

Becker, J. J. 1985. The fossil birds of the Late Miocene and Early Pliocene of Florida. Unpublished Ph.D. dissertation, University of Florida, Gainesville.

Becker, J. J. 1987. *Neogene Avian Localities of North America.* Washington, D.C.: Smithsonian Institution Press.

Bourlière, F. 1963. Observations on the ecology of some large African mammals. In *African Ecology and Human Evolution,* ed. F. C. Howell & F. Bourlière, pp. 43–54. Chicago: Aldine Publ.

Bryant, H. N. 1988. Delayed eruption of the deciduous upper canine in the sabertoothed carnivore *Barbourofelis lovei* (Carnivora, Nimravidae). *J. Vert. Paleontol. 8*:295–306.

Creighton, G. K. 1980. Static allometry of mammalian teeth and the correlation of tooth size and body size in contemporary mammals. *J. Zool., Lond. 191*:435–443.

Damuth J. D. 1981. Population density and body size in mammals. *Nature 290*:699–700.

Damuth, J. 1982. Analysis of the preservation of community structure in assemblages of fossil mammals. *Paleobiology 8*:434–446.

Domning, D. P. 1988. Fossil Sirenia of the West Atlantic and Caribbean region. I. *Metaxytherium floridanum* Hay 1922. *J. Vert. Paleontol. 8*:395–426.

Eadie, J. McA., Broekhoven, L., & Colgan, P. 1987. Size ratios and artifacts: Hutchinson's rule revisited. *Am. Nat. 129*:1–17.

Gingerich, P. D., Smith, B. H., & Rosenberg, K. 1982. Allometric scaling in the dentitions of primates and prediction of body weight from tooth size in fossils. *Am. J. Phys. Anthropol. 58*:81–100.

Horn, H. S., & May, R. M. 1977. Limits to similarity among coexisting competitors. *Nature 270*:660–661.

Hulbert, R. C., Jr. 1987. Late Neogene *Neohipparion* (Mammalia, Equidae) from the Gulf Coastal Plain of Florida and Texas. *J. Paleontol. 61*: 809–830.

Hulbert, R. C., Jr. 1988a. *Calippus* and *Protohippus* (Mammalia, Perissodactyla, Equidae) from the Miocene (Barstovian–Early Hemphillian) of the Gulf Coastal Plain. *Bull. Florida State Mus., Biol. Sci. 32*:221–340.

Hulbert, R. C., Jr. 1988b. *Cormohipparion* and *Hipparion* (Mammalia, Perissodactyla, Equidae) from the Late Neogene of Florida. *Bull. Florida State Mus., Biol. Sci. 33*:229–338.

Hutchinson, G. E. 1959. Homage to Santa Rosalia, or why are there so many kinds of animals? *Am. Nat. 93*:145–159.

Jackson, D. 1976. The status of the Pliocene turtles *Pseudemys caelata* Hay and *Chrysemys carri* Rose and Weaver. *Copeia 1976*:655–659.

Jackson, D. 1978. Evolution and fossil record of the chicken turtle *Deirochelys,* with a re-evaluation of the genus. *Tulane Studies Zool. Botany 20*:35–55.

Janis, C. 1982. Evolution of horns in ungulates: ecology and palaeoecology. *Biol. Rev. 57*:261–318.

Janis, C. M., & Ehrhardt, D. 1988. Correlation of relative muzzle width and relative incisor width with dietary preference in ungulates. *Zool. J. Linn. Soc. 92*:267–284.

Jungers, W. L. 1985. Body size and scaling of limb proportions in primates. In

Size and Scaling in Primate Biology, ed. W. L. Jungers, pp. 345–381. New York: Plenum Press.

Kingdon, J. 1979. *East African Mammals,* Vol. 3, Part B: *Large Mammals.* London: Academic Press.

Lambert, W. D. 1989. The morphology and ecology of feeding in the shovel-tusked gomphotheres (Proboscidea, Gomphotheriidae, Amebelodontidae). Unpublished Master's thesis, Gainesville, University of Florida.

MacFadden, B. J. 1984. Systematics and phylogeny of *Hipparion, Neohipparion, Nannippus,* and *Cormohipparion* (Mammalia, Equidae) from the Miocene and Pliocene of the New World. *Bull. Am. Mus. Nat. Hist. 179*:1–196.

MacFadden, B. J. 1987. Fossil horses from "Eohippus" (*Hyracotherium*) to *Equus*: scaling, Cope's law, and the evolution of body size. *Paleobiology 12*:355–369.

MacFadden, B. J. 1990. Dental character variation in paleopopulations and morphospecies of fossil horses and extant analogs. In *The Evolution of Perissodactyls,* ed. D. R. Prothero & R. M. Shoch, pp. 128–141. Oxford: Oxford University Press.

Maiorana, V. C. 1978. An explanation of ecological and developmental constants. *Nature 273*:375–377.

Martin, R. A. 1980. Body mass and basal metabolism of extinct mammals. *Comp. Biochem. Physiol. 66*:307–314.

Martin, R. A. 1986. Energy, ecology, and cotton rat evolution. *Paleobiology 12*:370–382.

Owen-Smith, N. 1985. Niche separation among African ungulates. In *Species and Speciation,* ed. E. S. Vrba, pp. 167–171. Transvaal Museum Monograph No. 4.

Roth, V. L. 1981. Constancy in the size ratios of sympatric species. *Am. Nat. 118*:394–404.

Simpson, G. G., Roe, A., & Lewontin, R. C. 1960. *Quantitative Zoology*. New York: Harcourt, Brace.

Skinner, M. F., & Johnson, F. W. 1984. Tertiary stratigraphy and the Frick Collection of fossil vertebrates from north-central Nebraska. *Bull. Am. Mus. Nat. Hist. 178*:215–368.

Strong, D. R., L. A. Szyska, & Simberloff, D. S. 1979. Tests of community-wide characters displacement against null hypotheses. *Evolution 33*: 897–913.

Van Valkenberg, B. 1982. Evolutionary dynamics of terrestrial large predator guilds. *Proc. 3rd N. Am. Paleontol. Conv. 2*:557–562.

Van Valkenberg, B. 1985. Locomotory diversity within past and present guilds of large predatory mammals. *Paleobiology 11*:406–428.

Van Valkenburgh, B. 1988. Trophic diversity within past and present guilds of large predatory mammals. *Paleobiology 14*:156–173.

Walker, E. P. 1975. *Mammals of the World.* Baltimore: Johns Hopkins University Press.

Webb, S. D. 1969. The Burge and Minnechaduza Clarendonian mammalian faunas of northcentral Nebraska. *Univ. California Publ. Geol. Sci. 78*: 1–191.

Webb, S. D. 1977. A history of savanna vertebrates in the New World. Part I. North America. *Annu. Rev. Ecol. Syst.* 8:355–380.

Webb, S. D. 1983a. A new species of *Pediomeryx* from the Late Miocene of Florida, and its relationships within the subfamily Cranioceratinae (Ruminantia: Dromomerycidae). *J. Mammal. 64*:261–276.

Webb, S. D. 1983b. The rise and fall of the Late Miocene ungulate fauna in North America. In *Coevolution,* ed. M. H. Nitecki, pp. 267–306. Chicago: University of Chicago Press.

Webb, S. D., & Hulbert, R. C., Jr. 1986. Systematics and evolution of *Pseudhipparion* (Mammalia, Equidae) from the Late Neogene of the Gulf Coastal Plain and the Great Plains. In *Vertebrates, Phylogeny, and Philosophy,* K. M. Flanagan & J. A. Lillegraven, pp. 237–272. Contrib. Geology Univ. Wyoming, Special Paper No. 3.

Webb, S. D., MacFadden, B. J., and Baskin, J. A. 1981. Geology and paleontology of the Love Bone Bed from the of Florida. *Am. J. Sci. 281*:513–544.

Yarnell, K. 1980. Systematics of Late Miocene Tapiridae (Mammalia, Perissodactyla) from Florida and Nebraska. Unpublished Master's thesis, University of Florida, Gainesville.

16

Appendix: Prediction equations

Many of the chapters in this book are based on a considerable number of empirical relationships between body mass and various morphological variables. Estimation of the body masses of fossil species from regression equations requires the choice of a proper reference population of extant species. Of course, there must also be a regression equation at hand that involves the skeletal part(s) available for the fossil species. The editors felt that it would be a service to the reader to collect in one appendix the large number of regression equations provided by the authors. Using this chapter, one may more easily discover whether there exists a known relationship that is appropriate for a given body mass estimation task.

Not all of the equations included here have been individually discussed in the text; however, readers are cautioned to refer to the source chapters for details of measurement and data manipulation. Although some indication of probable error is presented for each equation, we do not guarantee that any of the equations will yield adequate body mass estimates for a particular research question; that determination remains the responsibility of the reader.

For the most part, the equations are presented as originally provided by the authors. All values are expressed as logarithms to the base ten. However, note that the *units* vary from one table to the next. Abbreviations for particular measurements follow the original authors. Citations should be to the source chapter and author.

Abbreviations used in the Appendix tables include r^2, the square of

In *Body Size in Mammalian Paleobiology: Estimation and Biological Implications,* John Damuth and Bruce J. MacFadden, eds.

the correlation coefficient; %SEE, percent standard error of the estimate (Van Valkenburgh, this volume); %PE, mean percent prediction error (Van Valkenburgh, this volume); NR, not reported.

All mammals: postcranial dimensions

Table 16.1. *Mammalian postcranial dimensions*

Measure	Slope	Intercept	r^2	%SEE	%PE
Humerus length	2.675	−4.558	>.90	82	53
Femur length	2.654	−4.751	>.90	89	56
Humerus circumference	2.611	−3.024	.987	34	25
Femur circumference	2.827	−3.421	.984	39	26
Sum of humerus and femur circumferences[a]	2.733	−4.075	.990	29	21

Note: Values in kg and mm.
[a]Recalculated from Anderson et al. (1983); reference in Chapter 9.
Source: Roth (this volume).

Proboscidea: postcranial dimensions

Table 16.2. *Proboscidean postcranial dimensions*

Regressions of body mass on shoulder height	Slope	Intercept	%SEE	%PE
a	3.11	−3.9914	NR	NR
b	2.631	−2.8972	NR	NR
c	2.803	−3.2950	NR	NR
d	2.917	−3.5884	NR	NR
e	2.890	−3.5143	NR	NR
f	2.97	−3.7423	NR	NR
g	2.711	−3.0844	4	NR
h	2.934	−3.6819	16	11
i	2.917	−3.5127	13	NR
j	3.263	−4.3296	19	NR
k	3.356	−4.4895	18	12
l	3.387	−4.5638	20	NR

Note: Values in kg and cm.
Source: Roth (this volume).

Primates: postcranial dimensions

Table 16.3. *Hindlimb bone cortical area*

Section	Group	r^2	Intercept	Slope	%SEE
Tib. 20%	Total	.984	−1.373	1.419	16.2
	Nonhuman	.988	−1.409	1.440	17.5
	Hominoid	.876	−1.252	1.366	17.6
	Af. ape + human	.880	−1.534	1.484	17.6
	Pongid	.912	−1.452	1.460	20.6
Tib. 35%	Total	.966	−1.396	1.380	24.8
	Hominoid	.762	−0.806	1.129	25.0
Tib. 50%	Total	.956	−1.375	1.349	28.9
	Nonhuman	.984	−1.574	1.466	20.6
	Hominoid	.696	−0.665	1.052	28.8
	Af. ape + human	.726	−1.226	1.276	27.7
	Pongid	.893	−1.342	1.367	23.0
Tib. 65%	Total	.955	−1.398	1.354	29.4
	Hominoid	.672	−0.704	1.063	30.0
Tib. 80%	Total	.960	−1.661	1.477	27.5
	Hominoid	.696	−0.987	1.192	29.8
Fem. 20%	Total	.984	−1.677	1.434	16.2
	Tot. w/o *Pongo*	.985	−1.669	1.431	17.8
	Hominoid	.880	−1.330	1.294	17.2
	Af. ape + human	.861	−1.330	1.294	19.0
Fem. 35%	Total	.990	−1.496	1.331	12.8
	Tot. w/o *Pongo*	.993	−1.489	1.323	11.6
	Hominoid	.933	−1.385	1.287	12.5
	Af. ape + human	.949	−1.519	1.336	11.0
Fem. 50%	Total	.988	−1.453	1.291	13.8
	Tot. w/o *Pongo*	.996	−1.454	1.284	9.1
	Nonhuman	.992	−1.488	1.312	14.8
	Hominoid	.920	−1.181	1.184	13.9
	Af. ape + human	.966	−1.447	1.282	9.0
	Pongid	.947	−1.128	1.222	15.7
Fem. 65%	Total	.990	−1.521	1.306	13.4
	Tot. w/o *Pongo*	.997	−1.526	1.300	7.4
	Hominoid	.933	−1.075	1.133	12.6
	Af. ape + human	.976	−1.288	1.208	7.3
Fem. 80%	Total	.990	−1.451	1.281	13.4
	Tot. w/o *Pongo*	.996	−1.452	1.274	8.5
	Nonhuman	.990	−1.468	1.289	15.6
	Hominoid	.935	−1.024	1.115	12.5
	Af. ape + human	.972	−1.179	1.170	8.0
	Pongid	.943	−1.088	1.142	16.1

Table 16.3. (*cont.*)

Section	Group	r^2	Intercept	Slope	%SEE
Fem. Neck	Total	.953	−1.631	1.571	30.1
	Nonhuman	.992	−1.704	1.582	14.6
	Hominoid	.734	−0.634	1.120	26.6
	Af. ape + human	.696	−0.442	1.037	29.4
	Pongid	.929	−1.680	1.571	18.1

Note: Values in mm^2 and kg.
Source: Ruff (this volume).

Table 16.4. *Second polar moment of area of femoral midshaft*

Section	Group	r^2	Intercept	Slope	%SEE
Tib. 20%	Total	.843	−0.804	.633	26.1
	Nonhuman	.986	−0.920	.678	19.2
	Hominoid	.712	−0.376	.531	27.9
	Af. ape + human	.803	−1.177	.715	23.0
	Pongid	.887	−0.792	.646	23.6
Tib. 35%	Total	.949	−0.877	.648	31.7
	Hominoid	.634	−0.196	.485	32.0
Tib. 50%	Total	.941	−0.906	.639	34.2
	Nonhuman	.982	−1.116	.711	22.0
	Hominoid	.604	−0.128	.458	33.5
	Af. ape + human	.643	−0.765	.599	32.2
	Pongid	.870	−0.859	.650	25.4
Tib. 65%	Total	.951	−0.971	.634	30.8
	Hominoid	.663	−0.267	.475	30.6
Tib. 80%	Total	.949	−1.033	.624	31.3
	Hominoid	.677	−0.799	.567	29.8
Fem. 20%	Total	.994	−1.133	.626	9.1
	Tot. w/o *Pongo*	.997	−1.131	.623	7.4
	Hominoid	.958	−0.985	.595	9.9
	Af. ape + human	.968	−1.084	.613	8.5
Fem. 35%	Total	.988	−1.101	.636	13.8
	Tot. w/o *Pongo*	.997	−1.107	.633	7.5
	Hominoid	.918	−0.790	.569	14.1
	Af. ape + human	.970	−1.032	.618	8.5
Fem. 50%	Total	.984	−1.099	.638	16.4
	Tot. w/o *Pongo*	.996	−1.114	.636	8.4
	Nonhuman	.984	−1.113	.643	20.3
	Hominoid	.889	−0.698	.551	16.5
	Af. ape + human	.960	−1.113	.643	9.6
	Pongid	.897	−0.721	.558	22.5

Table 16.4. (*cont.*)

Section	Group	r^2	Intercept	Slope	%SEE
Fem. 65%	Total	.980	−1.100	.635	18.7
	Tot. w/o *Pongo*	.996	−1.125	.635	8.6
	Hominoid	.870	−0.599	.527	17.9
	Af. ape + human	.960	−1.025	.614	9.6
Fem. 80%	Total	.988	−1.060	.619	14.7
	Tot. w/o *Pongo*	.997	−1.107	.616	7.4
	Nonhuman	.988	−1.089	.629	17.0
	Hominoid	.902	−0.777	.559	15.5
	Af. ape + human	.970	−1.111	.626	8.4
	Pongid	.920	−0.820	.572	19.6
Fem. Neck	Total	.972	−1.036	.666	22.5
	Nonhuman	.988	−1.149	.703	18.1
	Hominoid	.819	−0.471	.535	26.6
	Af. ape + human	.861	−0.771	.600	19.0
	Pongid	.918	−0.853	.634	19.7

Note: Values in mm^4 and kg.
Source: Ruff (this volume).

Table 16.5. *Hindlimb lengths and breadth*

Measure and group	r^2	Intercept	Slope	%SEE
Tibial length				
Total	.864	−4.512	2.586	83.7
Nonhuman	.981	−7.914	4.123	22.7
Hominoid	.108	0.234	0.649	54.2
Af. ape + human	.040	0.863	0.400	58.0
Pongid	.921	−11.581	5.674	19.3
Femoral length				
Total	.810	−4.380	2.450	69.3
Nonhuman	.987	−6.750	3.498	18.4
Hominoid	.214	−0.421	0.888	50.2
Af. ape + human	.099	0.125	0.673	55.7
Pongid	.899	−7.343	3.740	22.1
Average femoral midshaft external breadth				
Total	.981	−1.696	2.492	18.2
Nonhuman	.982	−1.737	2.533	22.1
Hominoid	.865	−1.168	2.120	18.4
Af. ape + human	.969	−1.793	2.541	8.6
Pongid	.877	−1.227	2.174	24.7

Note: Values in mm and kg.
Source: Ruff (this volume).

Carnivorous mammals: craniodental and postcranial dimensions

Table 16.6. *Craniodental and postcranial dimensions of carnivorous mammals*

Group	Measure[a]	Slope	Intercept	r^2	%SEE	%PE
Total sample	HBL	2.88	−7.24	.92	53	36
	M_1L	2.97	−2.27	.69	138	97
	OOL	3.44	−5.74	.90	61	42
	SKL	3.13	−5.59	.90	66	47
Canidae	HBL	2.30	−5.58	.92	24	17
	M_1L	1.82	−1.22	.76	44	27
	OOL	3.08	−5.03	.88	30	22
	SKL	2.86	−5.21	.86	31	21
Ursidae	HBL	2.98	−7.43	.90	21	15
	M_1L	0.49	1.26	.18	78	46
	OOL	1.98	−2.38	.41	61	42
	SKL	2.02	−2.8	.49	56	39
Mustelidae	HBL	2.81	−7.08	.85	73	48
	M_1L	3.48	−3.04	.86	66	45
	OOL	3.29	−5.53	.86	66	47
	SKL	3.39	−6.03	.90	58	40
Felidae	HBL	2.72	−6.83	.85	54	36
	M_1L	3.05	−2.15	.90	41	28
	OOL	3.54	−5.86	.85	57	37
	SKL	3.11	−5.38	.85	57	38

Note: Values in kg and mm.
[a]Abbreviations as in Table 10.2.
Source: Van Valkenburgh (this volume).

Ungulates: postcranial dimensions

Table 16.7. *Limb measures for ungulates*

Measure	Slope	Intercept	r^2	%SEE	%PE
All ungulates					
H1	3.4026	−2.3707	.9209	43	28
H2	3.3951	−2.5130	.9196	44	29
H3	2.7146	0.2594	.9415	36	24
H4	2.4815	0.4516	.9516	33	22
H5	2.5752	0.2863	.9575	30	20
H6	2.7884	1.3696	.9224	43	29
H7	2.4850	1.0934	.9546	31	20
H8	2.4937	0.8760	.9438	35	21

Table 16.7. (*cont.*)

Measure	Slope	Intercept	r^2	%SEE	%PE
U1	2.9762	−2.3087	.8825	55	38
U2	2.7611	−0.1556	.9331	39	27
R1	2.8455	−1.8223	.8319	69	47
R2	2.4304	0.4856	.9494	33	22
R3	2.5360	1.0605	.9381	37	23
R4	2.3884	0.4011	.9594	30	20
R5	2.4054	0.4987	.9586	30	20
R6	2.5894	0.9092	.9444	35	24
R7	2.5038	1.4661	.9378	38	25
MC1	1.9313	−0.4760	.4455	159	112
MC2	2.6485	0.6080	.9509	33	21
MC3	2.7928	1.0784	.9404	37	24
MC4	2.3780	0.7366	.8746	59	29
MC5	2.6061	1.2757	.9094	47	31
MC6	2.5266	1.1921	.9127	47	26
MC7	2.6193	1.4156	.9057	48	33
F1	3.4855	−2.9112	.9241	42	28
F2	2.6886	−0.2471	.9438	97	23
F3	2.9405	−0.0870	.9334	99	26
F4	3.0284	0.7196	.9429	36	25
F5	2.7820	−0.0107	.9439	35	24
F6	2.8210	0.9062	.9402	37	23
F7	2.6016	0.9119	.9373	38	26
T1	3.5808	−3.1732	.7567	88	65
T2	2.8491	−0.2495	.9534	32	21
T3	3.1568	0.1370	.9153	45	31
T4	2.6075	0.4247	.9269	41	28
T5	2.8949	0.6420	.9450	35	24
T6	2.7382	0.8761	.9469	34	24
T7	2.9060	0.9909	.9243	42	27
MT1	2.1344	−0.8149	.4632	156	113
MT2	2.8269	0.6408	.9414	36	24
MT3	2.8481	0.6681	.8907	53	29
MT4	2.7634	0.5401	.9348	39	26
MT5	2.8246	1.1872	.9063	48	33
MT6	2.7335	1.1904	.8962	51	30
MT7	2.7417	1.2078	.8756	57	40
Equids					
H1	2.8794	−1.5178	.8804	24	16
H2	2.7142	−1.3571	.6912	41	27
H3	2.6562	0.3506	.8841	23	16
H4	2.8535	0.0471	.8612	26	16
H5	2.5392	0.2196	.8761	24	15
H6	2.9047	1.4284	.7970	32	21
H7	2.2001	1.3111	.8646	25	16
H8	2.3416	1.0596	.8744	24	15
U1	3.1121	−2.4111	.8250	29	21
U2	2.8706	−0.0754	.8746	24	16

Table 16.7. (*cont.*)

Measure	Slope	Intercept	r^2	%SEE	%PE
R1	3.0770	−2.0373	.7940	32	24
R2	2.7991	0.0890	.8570	26	16
R3	2.5513	1.0488	.8488	27	19
R4	2.7703	−0.0067	.8385	28	18
R5	2.5996	0.2520	.8828	23	16
R6	2.2382	1.1988	.7907	32	20
R7	2.1409	1.6146	.8764	24	16
MC1	2.6838	−1.0402	.6792	41	28
MC2	2.7422	0.6135	.8395	28	18
MC3	2.7860	1.1468	.8210	30	20
MC4	2.7086	0.7046	.8554	26	16
MC5	2.3308	1.4910	.8213	30	20
MC6	2.4313	1.2990	.8307	29	18
MC7	2.4850	1.5380	.8694	25	16
F1	3.1274	−2.2619	.8615	26	18
F2	2.9812	−2.1543	.8773	24	17
F3	2.8419	−1.4078	.7996	32	22
F4	2.8685	0.8680	.7936	32	20
F5	2.5856	0.0445	.8690	25	27
F6	2.0469	1.3123	.8622	26	15
F7	2.1794	1.0246	.8628	25	17
T1	2.2529	−0.8241	.4825	55	37
T2	2.6798	−0.0915	.8889	23	14
T3	2.5490	0.6566	.8703	25	17
T4	2.5026	0.3548	.8497	27	18
T5	2.5712	0.7648	.8374	28	18
T6	2.3573	1.0746	.8164	30	20
T7	2.2549	1.3171	.8647	25	16
MT1	2.8417	−1.4453	.6873	41	27
MT2	2.4076	0.8452	.8061	30	19
MT3	1.9467	1.2419	.6632	41	23
MT4	2.4247	0.8911	.8526	27	16
MT5	2.4560	1.3638	.8404	28	20
MT6	2.3802	1.3643	.8303	29	17
MT7	2.3286	1.4629	.8470	27	16
Artiodactyls					
H1	3.3414	−2.3068	.9182	43	28
H2	3.3142	−2.4273	.9286	40	26
H3	2.6909	0.2689	.9360	37	25
H4	2.5332	0.4312	.9506	32	21
H5	2.6454	0.2538	.9580	29	19
H6	2.7340	1.3692	.9191	43	28
H7	2.4600	1.0947	.9526	31	20
H8	2.4548	0.8824	.9413	35	21
U1	2.9111	−2.2284	.8736	56	38
U2	2.6853	−0.1174	.9378	37	25
R1	2.7722	−1.7397	.8176	71	48
R2	2.4768	0.4677	.9478	33	22
R3	2.5374	1.0603	.9321	39	24

Table 16.7. (*cont.*)

Measure	Slope	Intercept	r^2	%SEE	%PE
R4	2.4226	0.3872	.9588	32	19
R5	2.4522	0.4805	.9577	50	19
R6	2.5550	0.9131	.9438	46	23
R7	2.5133	1.4647	.9317	41	26
MC1	2.4720	−1.2370	.5791	156	150
MC2	2.6115	0.6185	.9489	33	22
MC3	2.7570	1.0819	.9369	37	24
MC4	2.3014	0.7536	.8727	58	28
MC5	2.6075	1.2497	.9189	47	33
MC6	2.4965	1.1937	.9047	48	28
MC7	2.6556	1.3819	.9231	50	36
F1	3.4263	−2.8404	.9202	42	28
F2	2.6886	−0.2471	.9438	34	23
F3	2.9405	−0.0870	.9334	38	26
F4	2.9962	0.7259	.9401	36	25
F5	2.8940	−0.0719	.9458	34	22
F6	2.8771	0.8915	.9376	37	23
F7	2.7585	0.8737	.9449	34	23
T1	3.4985	−3.0950	.7914	77	55
T2	2.8560	−0.2540	.9489	33	22
T3	3.1232	0.1473	.9091	46	31
T4	2.7425	0.3677	.9305	39	26
T5	2.9777	0.6171	.9444	34	23
T6	2.7672	0.8689	.9437	35	24
T7	2.9356	0.9834	.9176	43	27
MT1	2.9135	−1.8977	.6219	156	185
MT2	2.8912	0.6210	.9401	36	24
MT3	2.8712	0.6581	.8850	53	29
MT4	2.7095	0.5527	.9349	38	25
MT5	2.9239	1.1406	.9249	50	36
MT6	2.8721	1.1860	.9336	53	28
MT7	2.8727	1.1410	.9177	59	45
Suids					
H1	3.4122	−2.3639	.9524	22	17
H2	3.3910	−2.5183	.9458	23	18
H3	3.0904	0.2716	.9138	30	21
H4	2.7528	0.4996	.8741	38	27
H5	3.0861	0.0216	.9150	30	21
H6	2.2222	1.3837	.7766	53	33
H7	2.6530	1.1157	.9114	31	20
H8	2.6672	0.7875	.9353	26	16
U1	2.8975	−1.9239	.9017	33	22
U2	2.8708	−0.3834	.9160	30	20
R1	2.6673	−1.2336	.8883	35	22
R2	2.8435	0.5580	.9243	28	20
R3	2.5962	1.1392	.9405	25	18
R5	2.4855	1.1374	.9718	19	16

Table 16.7. (*cont.*)

Measure	Slope	Intercept	r^2	%SEE	%PE
R6	3.0290	1.3634	.9375	30	13
R7	2.3860	1.2913	.9489	25	20
MC1	2.7639	−0.4650	.8236	46	30
MC2	2.9534	1.0052	.9577	20	16
MC3	2.9362	1.1772	.9463	23	16
MC4	2.6377	1.3056	.9160	31	21
MC5	2.6955	1.5010	.8525	42	22
MC6	2.2955	1.5468	.9189	31	21
MC7	3.0858	1.7350	.8312	47	29
F1	3.5007	−2.7428	.9585	20	14
F2	2.6845	−0.0411	.9150	30	20
F3	3.0588	−0.0858	.9443	24	14
F4	2.9139	0.7925	.8865	36	22
F5	3.0451	−0.0643	.9398	25	16
F6	2.6940	1.0774	.9270	28	17
F7	2.8226	0.8736	.9029	33	21
T1	3.3916	−2.4421	.8668	39	24
T2	2.9746	−0.1783	.9215	29	19
T3	2.9538	0.4142	.9165	30	20
T4	3.0297	0.5102	.8976	34	21
T5	3.0289	0.6660	.9351	26	19
T6	2.5831	1.0961	.9146	30	20
T7	2.6749	1.0588	.9117	31	20
MT1	3.2270	−0.9333	.7976	53	34
MT2	2.9934	1.2300	.9173	31	19
MT3	2.8820	1.6360	.7335	63	36
MT4	2.6821	1.3219	.9542	25	18
MT5	3.2878	1.5054	.9202	31	20
MT6	2.7074	1.5614	.9192	34	24
MT7	3.4254	1.7129	.8715	42	29
Ruminants					
H1	3.3412	−2.3085	.9173	44	28
H2	3.3123	−2.4252	.9281	41	26
H3	2.7156	0.2411	.9465	34	23
H4	2.5518	0.4093	.9600	29	19
H5	2.6372	0.2574	.9599	29	19
H6	2.7556	1.3708	.9260	41	28
H7	2.4603	1.0899	.9551	31	20
H8	2.4492	0.8856	.9417	36	21
U1	2.9900	−2.3584	.8928	52	35
U2	2.6873	−0.1103	.9423	36	24
R1	2.9204	−1.9583	.8532	63	43
R2	2.5149	0.4297	.9638	27	18
R3	2.5443	1.0519	.9339	39	24
R4	2.4226	0.3872	.9587	29	19
R5	2.4520	0.4758	.9594	29	19
R6	2.5784	0.8948	.9502	33	21

Table 16.7. (*cont.*)

Measure	Slope	Intercept	r^2	%SEE	%PE
R7	2.5041	1.4672	.9324	39	26
MC1	2.4722	−1.2370	.5791	128	93
MC2	2.6091	0.6225	.9498	33	22
MC3	2.7633	1.0742	.9401	37	23
MC4	2.3014	0.7536	.8727	58	28
MC5	2.6075	1.2497	.9189	44	29
MC6	2.4965	1.1938	.9047	48	28
MC7	2.6556	1.3819	.9231	42	29
F1	3.4661	−2.9080	.9281	41	27
F2	2.7237	−0.2882	.9546	31	20
F3	2.9439	−0.0941	.9343	39	26
F4	3.0012	0.7215	.9421	36	25
F5	2.9006	−0.0830	.9484	34	22
F6	2.9022	0.8739	.9426	36	22
F7	2.7577	0.8725	.9462	34	23
T1	3.8468	−3.6242	.8602	61	43
T2	2.8751	−0.2758	.9554	31	20
T3	3.1659	0.1114	.9175	44	30
T4	2.7940	0.3222	.9484	34	23
T5	2.9821	0.6107	.9457	35	23
T6	2.8018	0.8446	.9523	32	22
T7	2.9451	0.9804	.9179	44	27
MT1	2.9135	−1.8978	.6219	119	87
MT2	2.8871	0.6262	.9404	36	24
MT3	2.9054	0.6300	.8946	51	27
MT4	2.7095	0.5527	.9349	38	25
MT5	2.9239	1.1406	.9249	42	29
MT6	2.8721	1.1860	.9336	39	26
MT7	2.8727	1.1410	.9177	44	31
Cervids only					
H1	3.4337	−2.4141	.9183	40	27
H2	3.3580	−2.4788	.9260	37	24
H3	2.7080	0.2432	.9463	27	17
H4	2.5568	0.4084	.9590	27	20
H5	2.6389	0.2582	.9596	27	19
H6	2.7230	1.3730	.9241	39	31
H7	2.4930	1.0845	.9531	29	22
H8	2.4630	0.8836	.9390	26	20
U1	3.0910	−2.4906	.8919	38	30
U2	2.6396	−0.0825	.9461	25	22
R1	3.0795	−2.1515	.8549	45	36
R2	2.5150	0.4304	.9630	25	18
R3	2.5588	1.0498	.9306	33	24
R4	2.4301	0.3842	.9560	23	19
R5	2.4956	0.4581	.9600	26	20
R6	2.5970	0.8916	.9494	31	21
R7	2.5681	1.4651	.9347	42	28

Table 16.7. (*cont.*)

Measure	Slope	Intercept	r^2	%SEE	%PE
MC1	2.5036	−1.2742	.5404	57	37
MC2	2.6568	0.6070	.9500	32	26
MC3	2.8040	1.0687	.9393	35	24
MC4	2.3300	0.7466	.8664	98	28
MC5	2.6352	1.2473	.9184	34	23
MC6	2.4787	1.1964	.8971	39	28
MC7	2.6919	1.3805	.9219	31	23
F1	3.5335	−2.9926	.9299	34	24
F2	2.6928	−0.2670	.9526	27	17
F3	2.9500	−0.1030	.9348	33	23
F4	2.9714	0.7268	.9433	30	21
F5	2.9100	−0.0883	.9509	32	22
F6	2.8949	0.8767	.9380	29	23
F7	2.7547	0.8739	.9450	33	32
T1	3.8551	−3.6352	.8513	46	21
T2	2.8861	−0.2841	.9548	30	29
T3	3.1432	0.1174	.9193	42	23
T4	2.8486	0.3000	.9494	31	23
T5	2.9628	0.6159	.9429	36	18
T6	2.8333	0.8378	.9512	25	26
T7	2.9694	0.9733	.9163	47	35
MT1	3.0039	−2.0105	.5936	55	24
MT2	2.9334	0.6132	.9398	36	24
MT3	2.8848	0.6369	.8879	82	27
MT4	2.7521	0.5397	.9356	38	25
MT5	2.9391	1.1375	.9232	36	24
MT6	2.8760	1.1850	.9312	34	23
MT7	2.8693	1.1399	.9128	35	24
Bovids only					
H1	3.4591	−2.9190	.9286	40	25
H2	3.3633	−2.4637	.9317	39	24
H3	2.7311	0.2334	.9434	35	23
H4	2.5499	0.4078	.9590	29	18
H5	2.6246	0.2756	.9604	28	18
H6	2.7630	1.3617	.9274	40	27
H7	2.5013	1.0794	.9529	31	19
H8	2.5424	0.8843	.9339	38	21
U1	3.1886	−2.6320	.8872	52	36
U2	2.6310	−0.0844	.9436	35	23
R1	3.2052	−2.3203	.8487	63	43
R2	2.5069	0.4311	.9629	28	17
R3	2.5472	1.0525	.9279	39	23
R4	2.4305	0.3736	.9543	31	19
R5	2.4824	0.4635	.9596	29	18
R6	2.6208	0.8854	.9497	33	21
R7	2.5679	1.4742	.9409	36	23

Table 16.7. (*cont.*)

Measure	Slope	Intercept	r^2	%SEE	%PE
MC1	2.5109	−1.2773	.4703	149	109
MC2	2.6495	0.6016	.9529	31	20
MC3	2.8291	1.0620	.9402	36	22
MC4	2.3765	0.7443	.9203	43	25
MC5	2.6469	1.2458	.9127	45	29
MC6	2.4354	1.2059	.8922	51	28
MC7	s.7444	1.3815	.9187	43	30
F1	3.5526	−2.9997	.9326	39	26
F2	2.6934	−0.2555	.9529	31	20
F3	2.9531	−0.0808	.9420	35	24
F4	2.9573	0.7271	.9415	35	24
F5	2.9053	−0.0768	.9530	31	20
F6	2.9170	0.8627	.9347	38	22
F7	2.7671	0.8731	.9456	34	22
T1	3.9842	−3.8078	.8453	64	46
T2	2.8850	−0.2758	.9567	30	19
T3	3.2248	0.0753	.9231	42	28
T4	2.8409	0.3222	.9545	31	20
T5	2.9720	0.6222	.9464	34	22
T6	2.8562	0.8318	.9479	33	22
T7	2.9469	0.9879	.9224	42	26
MT1	3.0701	−2.0622	.5434	134	96
MT2	2.9220	0.6162	.9405	36	24
MT3	3.0306	0.5755	.9307	39	26
MT4	2.7421	0.5614	.9418	35	22
MT5	2.9763	1.1416	.9239	41	28
MT6	2.8710	1.1939	.9294	40	26
MT7	2.9361	1.1375	.9118	45	32

Note: Values in cm and kg.
Source: Scott (this volume).

Craniodental measures

Table 16.8. *Cranial and occlusal dental measures*

Measure[a]	r^2	Intercept	Slope	%SEE	%PE	Dietary type differences	Isometry
All ungulates (137 species)							
SLPL	.633	1.957	2.185	164	101.1	O*,B>I	–
SLPW	.687	2.636	1.990	145	89.8	B>I	X
TLPL	.776	1.686	2.714	113	84.9	O*,B>I	–
TLPW	.802	2.389	2.224	104	72.7		X
FLPL	.869	1.533	3.023	78.6	75.0	O,B>I	–
FLPW	.867	2.226	2.486	79.5	57.5	G>I	X

Table 16.8. (*cont.*)

Measure[a]	r^2	Intercept	Slope	%SEE	%PE	Dietary type differences	Isometry
FLPA	.890	1.913	1.398	70.2	59.2	B>I*	–
FLML	.933	1.337	3.263	51.4	34.6		–
FLMW	.918	2.030	2.909	58.1	38.4	B,O>I*	–
FLMA	.934	1.701	1.553	51.0	33.2	B>I*	–
SLML	.944	1.130	3.201	46.6	31.9	O>G,I,B	–
SLMW	.900	1.932	2.967	66.3	42.1	O>B; B*>G,I	–
SLMA	.935	1.541	1.563	50.7	33.5	O>G,I,B	–
TLML	.905	0.801	3.183	64.1	41.7	I,O>G,B; O>I	–
TLMW	.889	1.991	2.933	71.0	42.9	B>I,G; O>G,I,B	–
TLMA	.927	1.404	1.580	54.2	33.1	I,B*>G; O>G,I,B	–
SUML	.932	1.091	3.184	51.7	34.7	O>G,I*,B; I*>G	–
SUMW	.919	1.469	3.004	57.8	38.9	O*>I	–
SUMA	.939	1.277	1.568	48.6	32.7	O>I*	–
LPRL	.799	0.438	2.673	105	171.0	B,G*>I	–
LMRL	.940	−0.552	3.285	47.9	32.8	O>G,I,B	–
LMRL[a]	.941	−0.536	3.265	45.9	31.9	I>G,B	–
MZW	.871	0.640	2.313	77.4	50.4	O*>G*>I,B	+
PAW	.858	−0.036	3.070	83.2	60.2	G,I,B>O; I*>G	–
PAW[a]	.917	−0.196	3.270	56.3	38.2		–
MFL	.938	−1.289	2.950	48.9	35.0	G*>I>B; G,I*>O	–
OCH	.905	−0.420	2.783	63.7	34.7	B>I; O>G,I,B	X
OCH[a]	.948	−0.457	2.873	42.5	28.1	B>G*,I	X
PSL	.942	−0.973	2.758	47.2	33.4	G>I*,B	X
BCL	.901	−1.062	3.137	65.6	51.8		–
TSL	.950	−2.344	2.975	43.2	30.5	O*>I,B*	X
AJL	.877	−0.902	2.806	75.4	51.9		X
PJL	.931	0.031	2.412	52.8	36.5	G*>I*>B; O>B	–
DMA	.879	−0.331	2.448	75.0	45.2	G,O>I,B	X
WMA	.916	−0.352	2.803	59.2	40.5	G,O>I,B	X
TJL	.942	−1.952	2.884	47.2	33.4	O>I,B*	X
Perissodactyls and hyracoids only (19 species)							
SLPL	.839	1.527	2.637	128	88.6	G>B*	X
SLPW	.916	2.275	2.739	81.1	52.7	G>B*	X
TLPL	.957	1.383	2.965	52.8	37.6	G>I,B*	X
TLPW	.951	2.090	2.753	57.4	38.4		X
FLPL	.986	1.290	3.090	27.9	21.0	G>I*,B	X
FLPW	.973	1.965	2.854	39.6	25.9		X
FLPA	.984	1.637	1.491	29.7	19.7		X
FLML	.985	1.264	3.187	28.5	16.5		X
FLMW	.970	1.940	2.856	42.9	29.5		X
FLMA	.983	1.616	1.516	30.6	20.4		X
SLML	.986	1.216	3.010	27.4	16.9		X
SLMW	.959	1.873	2.910	51.4	33.4		X
SLMA	.983	1.542	1.495	30.9	21.1	B>G*	X
TLML	.987	1.162	2.999	26.2	18.0		X
TLMW	.937	2.001	2.941	67.1	38.4	B>G	X
TLMA	.980	1.572	1.513	33.4	20.9		X

Table 16.8. (*cont.*)

Measure[a]	r^2	Intercept	Slope	%SEE	%PE	Dietary type differences	Isometry
SUML	.964	1.209	2.900	47.6	30.5		X
SUMW	.993	1.345	2.887	19.1	12.9		X
SUMA	.985	1.270	1.456	28.5	18.6		X
LPRL	.932	0.026	2.865	70.6	50.0	G>B*	X
LMRL	.978	−0.246	3.061	35.5	21.6		X
MZW	.924	0.419	2.952	75.8	50.8		X
PAW	.940	−0.027	3.243	64.8	45.5		−
MFL	.918	−1.106	2.815	79.5	51.4	G>B	X
OCH	.971	−0.569	2.914	41.6	30.1	B>G	X
PSL	.959	−0.988	2.763	51.7	33.6	G>B*	X
BCL	.928	−0.724	2.830	73.4	50.8		X
TSL	.945	−1.931	2.751	62.2	42.1		X
AJL	.835	−0.240	2.361	130	95.0		X
PJL	.945	0.283	2.202	61.8	41.9	G>I*,B	+
DMA	.858	−0.979	2.824	116	75.6	G>B	X
WMA	.984	−1.152	3.385	30.0	20.8		−
TJL	.940	−1.384	2.564	65.6	44.1		+
Ruminants only (103 species)							
SLPL	.600	2.067	2.723	151	92.3	B>G,I	−
SLPW	.696	2.888	2.409	123	80.0		X
TLPL	.755	1.715	3.147	106	74.7	B>G,I*	−
TLPW	.818	2.491	2.466	85.8	55.7		X
FLPL	.877	1.553	3.433	66.3	45.1	B>G	−
FLPW	.890	2.299	2.664	61.8	40.4		X
FLPA	.914	1.979	1.550	53.1	36.3		−
FLML	.930	1.344	3.469	46.9	34.0		−
FLMW	.918	2.086	3.105	51.7	34.9		−
FLMA	.937	1.735	1.662	43.9	30.7		−
SLML	.939	1.118	3.337	43.5	31.1		−
SLMW	.912	1.993	3.250	53.8	36.6	B,I>G*	−
SLMA	.941	1.557	1.674	42.6	30.4		−
TLML	.939	0.749	3.264	43.2	30.1	B>I>G*	−
TLMW	.900	2.036	3.089	58.5	37.8	B>G	−
TLMA	.947	1.396	1.635	40.0	27.9	I>G*	−
SUML	.931	1.068	3.360	46.6	32.3		−
SUMW	.922	1.480	3.366	50.3	34.2		−
SUMA	.941	1.266	1.707	42.6	30.8		−
LPRL	.783	0.285	3.107	93.6	80.1	B>G,I	−
LMRL	.945	−0.604	3.352	40.6	29.9		−
MZW	.894	0.690	2.190	60.7	36.9	G*>I,B	+
PAW	.933	−0.196	3.191	45.5	32.2	I,B>G*	−
MFL	.953	−1.337	2.973	33.0	26.9	G>B	X
OCH	.944	−0.481	2.913	41.3	26.6	B>I*	−
PSL	.943	−0.925	2.705	41.6	28.7	G*>I,B	X
BCL	.913	−1.209	3.281	53.5	35.6		−
TSL	.960	−2.348	2.969	33.4	24.5		X
AJL	.919	−1.005	2.871	51.4	35.9		X

Table 16.8. (*cont.*)

Measure[a]	r^2	Intercept	Slope	%SEE	%PE	Dietary type differences	Isometry
PJL	.931	0.003	2.435	46.6	31.2	G>B	+
DMA	.920	−0.427	2.631	51.0	36.1	G>I,B	X
WMA	.933	−0.419	2.973	45.9	31.7	G>I,B	–
TJL	.955	−2.018	2.923	36.1	26.6		X
Cervids only (23 species)							
SLPL	.873	1.885	3.619	51.4	33.0		–
SLPW	.752	2.827	2.855	78.2	50.5		X
TLPL	.905	1.556	3.399	43.2	40.5		–
TLPW	.794	2.405	2.819	69.4	48.1		X
FLPL	.938	1.471	3.130	33.7	27.2		–*
FLPW	.869	2.139	2.761	52.1	35.9		X
FLPA	.919	1.827	1.495	39.3	29.8		X
FLML	.934	1.270	3.334	34.9	23.9		–
FLMW	.910	1.953	3.149	41.6	30.7		–
FLMA	.952	1.616	1.673	28.8	21.7		–
SLML	.951	1.119	3.106	29.1	20.4	I>B*	–
SLMW	.907	1.865	3.279	42.6	31.4		–
SLMA	.955	1.474	1.638	27.9	23.0		–
TLML	.957	0.799	3.143	27.4	19.1		–
TLMW	.880	1.877	2.875	49.6	35.9		
TLMA	.953	1.346	1.561	28.8	19.9		–
SUML	.959	1.073	3.218	26.8	18.3	I>B*	–
SUMW	.921	1.375	3.286	38.7	23.8		–
SUMA	.954	1.214	1.651	28.2	19.2	I>B*	–
LPRL	.943	−0.051	3.324	32.1	28.5		–
LMRL	.956	−0.524	3.209	27.6	19.9	I>B*	–
MZW	.874	0.499	2.374	51.0	30.0	I>B*	X
PAW	.903	−0.502	3.551	43.5	26.6		–
MFL	.946	−1.482	3.093	31.2	20.1		–
OCH	.947	−0.578	3.035	30.6	20.7		–
PSL	.956	−0.893	2.718	27.6	16.7		X
BCL	.967	−1.308	3.403	23.6	15.6	I>B*	–
TSL	.955	−2.176	2.828	27.9	19.4		X
AJL	.907	−0.872	2.634	42.2	30.8		X
PJL	.937	0.029	2.431	33.7	20.3		+*
DMA	.964	−0.325	2.547	24.5	16.2		X
WMA	.949	−0.530	3.181	30.0	19.4		–
TJL	.943	−1.831	2.745	31.8	23.1		X
Bovids only (72 species)							
SLPL	.662	2.200	2.720	130	79.7	B>G*	–
SLPW	.666	2.981	2.433	129	79.0		–
TLPL	.751	1.797	2.988	105	67.3	B>G*	–
TLPW	.815	2.570	2.570	85.4	53.9		X
FLPL	.865	1.606	3.434	69.4	43.9		–
FLPW	.925	2.389	2.916	48.3	34.2		X
FLPA	.925	2.035	1.626	47.9	32.7		–

Table 16.8. (*cont.*)

Measure[a]	r^2	Intercept	Slope	%SEE	%PE	Dietary type differences	Isometry
FLML	.931	1.372	3.520	45.5	34.0		–
FLMW	.927	2.140	3.157	47.2	32.0		–
FLMA	.942	1.776	1.689	41.3	29.2		–
SLML	.928	1.119	3.375	47.2	33.6		–
SLMW	.917	2.046	3.273	51.4	33.2		–
SLMA	.935	1.586	1.684	44.5	30.8		–
TLML	.929	0.745	3.236	46.9	33.3	I>G,B*	–
TLMW	.919	2.103	3.129	50.3	33.4		–
TLMA	.943	1.427	1.624	40.9	29.1	I>G*	–
SUML	.912	1.061	3.395	53.1	36.5		–
SUMW	.922	1.517	3.436	49.3	32.7		–
SUMA	.931	1.279	1.734	45.9	32.4		–
LPRL	.817	0.398	3.092	84.9	69.5	B>G*,I*	–
LMRL	.934	−0.581	3.335	44.5	32.8		–
MZW	.916	0.807	2.058	51.7	33.9	G>I	+
PAW	.938	−0.156	3.174	42.9	29.4	B*>G	–
MFL	.952	−1.253	2.902	36.7	26.9	G>B*	X
OCH	.944	−0.356	2.775	40.6	26.7	B>I	X
PSL	.933	−1.006	2.760	45.2	31.1	G>I*	X
BCL	.894	−1.215	3.281	59.6	40.3		–
TSL	.958	−2.450	3.047	34.3	25.6	G>I*	X
AJL	.932	−1.102	3.023	45.2	33.3	G,B*>I	X
PJL	.922	−0.006	2.429	49.6	34.0	G>B*	+
DMA	.895	−0.472	2.663	59.2	42.9	G>I,B*	X
WMA	.920	−0.458	2.994	50.0	35.1	G>I,B	–
TJL	.956	−2.085	2.988	35.2	26.8		X

Note: Values in cm and kg. Abbreviations and symbols as in Table 13.2 and Figure 13.1.
[a]Variable considered without suines in data set.
Source: Janis (this volume).

Table 16.9. *Maximal tooth dimensions, bivariate regression equations*

Variable	Slope	Intercept	r^2	%SEE	%PE	*N*
All ungulates						
HBL	3.16	−5.12	.95	45.54	31.42	91
M^3 length	2.81	1.29	.86	81.61	48.60	93
M^2 length	3.03	1.06	.91	64.23	42.21	94
M^1 length	3.09	1.21	.92	57.76	39.89	94
P^4 length	2.80	1.89	.84	93.43	58.62	94
P^3 length	2.72	1.96	.82	101.55	65.27	94
P^2 length	2.51	2.23	.72	143.66	93.54	88
P^1 length	2.89	2.01	.89	125.15	80.35	10
M_3 length	2.99	0.80	.88	78.57	50.29	94
M_2 length	3.07	1.07	.92	59.09	38.78	94

Table 16.9. (*cont.*)

Variable	Slope	Intercept	r^2	%SEE	%PE	*N*
M_1 length	3.11	1.24	.92	60.73	42.34	93
P_4 length	2.94	1.59	.85	89.91	59.73	94
P_3 length	2.76	1.90	.80	110.69	73.67	89
P_2 length	2.38	2.56	.72	146.95	90.62	86
M^3 width	2.77	1.58	.82	100.57	63.17	93
M^2 width	2.78	1.51	.84	94.10	61.02	94
M^1 width	2.83	1.52	.87	79.29	52.03	94
P^4 width	2.62	1.91	.83	94.95	61.54	94
P^3 width	2.24	2.48	.77	121.21	83.29	94
P^2 width	2.30	2.58	.75	132.93	92.14	88
P^1 width	2.54	2.67	.73	253.51	210.05	10
M_3 width	2.73	2.04	.83	97.52	57.82	94
M_2 width	2.75	1.99	.85	87.79	50.89	94
M_1 width	2.83	2.00	.89	73.37	45.65	93
P_4 width	2.47	2.55	.83	97.81	61.01	94
P_3 width	2.45	2.72	.84	95.53	62.21	89
P_2 width	2.39	3.03	.80	115.34	75.70	86
M^3 area	1.47	1.26	.88	73.69	44.92	93
M^2 area	1.48	1.23	.90	69.57	46.17	95
M^1 area	1.50	1.33	.92	60.07	41.20	95
P^4 area	1.35	1.91	.85	90.40	58.17	95
M_3 area	1.48	1.32	.89	75.15	45.55	95
M_2 area	1.48	1.49	.91	66.70	41.08	95
M_1 area	1.50	1.60	.92	61.94	41.07	94
P_4 area	1.36	2.07	.86	88.33	57.38	95
M^{1-3} area	1.52	0.45	.91	63.28	40.79	93
M_{1-3} area	1.51	0.69	.91	64.02	39.37	94
M^{1-2} area	1.49	0.81	.91	63.64	42.97	95
M_{1-2} area	1.50	1.07	.92	62.66	39.45	94
M^{1-3} length	3.05	−0.37	.92	60.30	39.54	93
M_{1-3} length	3.14	−0.60	.93	55.82	36.34	94
P^4–M^3 area	1.51	0.36	.91	64.25	40.98	93
P_4–M_3 area	1.50	0.59	.91	63.13	38.39	94
P^4–M^3 length	3.07	−0.67	.92	58.41	38.05	93
P_4–M_3 length	3.17	−0.94	.94	53.02	34.51	94
P^4–M^2 area	1.51	0.60	.90	65.64	43.73	93
P^4–M^2 length	3.00	−0.28	.90	66.54	43.17	93
P_4–M_2 area	1.48	0.93	.91	64.39	41.21	94
P_4–M_2 length	3.13	−0.31	.93	57.04	38.81	94
P^3–M^3 area	1.50	0.31	.90	65.07	41.95	93
P_3–M_3 area	1.50	0.52	.92	61.68	39.06	89
PCRU	3.07	−1.10	.91	62.87	41.60	93
PCRL	3.15	−1.28	.92	59.43	40.39	94
UPCTA	1.48	0.29	.90	68.28	44.85	93
LPCTA	1.48	0.51	.91	64.27	40.73	94

Table 16.9. (*cont.*)

Variable	Slope	Intercept	r^2	%SEE	%PE	*N*
Nonselenodonts						
HBL	3.17	−5.06	.99	30.30	22.41	19
M^3 length	2.63	1.49	.88	119.14	64.05	18
M^2 length	2.96	1.09	.95	64.63	42.15	19
M^1 length	3.11	1.11	.97	50.53	34.61	19
P^4 length	3.13	1.16	.94	74.09	52.76	19
P^3 length	2.93	1.43	.98	37.25	23.29	19
P^2 length	2.75	1.72	.92	96.03	52.24	18
P^1 length	2.89	2.01	.89	125.15	80.35	10
M_3 length	2.81	1.18	.89	114.01	60.54	19
M_2 length	2.98	1.11	.97	49.86	30.61	19
M_1 length	3.17	1.04	.98	39.69	26.52	18
P_4 length	3.11	1.15	.97	49.91	32.26	19
P_3 length	2.94	1.47	.98	42.65	27.88	18
P_2 length	2.71	1.89	.90	113.36	70.61	18
M^3 width	2.88	1.27	.92	91.38	58.00	18
M^2 width	2.90	1.14	.96	61.83	39.82	19
M^1 width	2.86	1.30	.97	52.25	32.79	19
P^4 width	2.81	1.41	.95	66.17	40.02	19
P^3 width	2.72	1.66	.95	68.20	44.31	19
P^2 width	2.62	2.03	.93	85.05	59.45	18
P^1 width	2.54	2.67	.73	253.51	210.05	10
M_3 width	2.71	1.90	.84	153.70	86.15	19
M_2 width	2.74	1.82	.89	111.35	62.26	19
M_1 width	2.79	1.88	.95	72.05	44.69	18
P_4 width	2.98	1.70	.96	61.57	36.57	19
P_3 width	2.86	2.03	.98	43.51	29.98	18
P_2 width	2.75	2.41	.94	80.72	53.18	18
M^3 area	1.40	1.32	.92	94.05	53.68	18
M^2 area	1.45	1.15	.96	58.90	39.23	20
M^1 area	1.48	1.22	.98	43.10	29.72	20
P^4 area	1.46	1.33	.95	65.86	43.75	20
M_3 area	1.41	1.48	.88	121.97	65.09	20
M_2 area	1.46	1.42	.94	72.90	42.33	20
M_1 area	1.51	1.44	.97	51.86	33.36	19
P_4 area	1.54	1.38	.97	48.76	29.32	20
M^{1-3} area	1.48	0.43	.96	62.12	38.21	18
M_{1-3} area	1.47	0.72	.93	84.37	48.16	19
M^{1-2} area	1.47	0.72	.97	48.63	33.70	20
M_{1-2} area	1.49	0.97	.96	62.91	38.03	19
M^{1-3} length	2.95	−0.24	.94	72.91	45.24	18
M_{1-3} length	3.03	−0.39	.96	63.78	37.19	19
P^4–M^3 area	1.50	0.22	.96	53.52	34.11	18
P_4–M_3 area	1.49	0.49	.95	72.93	42.93	19
P^4–M^3 length	3.04	−0.73	.96	59.77	39.35	18

Table 16.9. (*cont.*)

Variable	Slope	Intercept	r^2	%SEE	%PE	N
P_4–M_3 length	3.06	−0.80	.97	55.33	32.34	19
P^4–M^2 area	1.52	0.37	.97	44.69	29.08	18
P^2–M^2 length	3.00	−0.35	.93	81.34	51.42	18
P_4–M_2 area	1.51	0.67	.97	55.02	33.64	19
P_4–M_2 length	3.12	−0.43	.98	38.71	24.02	19
P^3–M^3 area	1.50	0.11	.97	46.21	29.88	18
P_3–M_3 area	1.49	0.39	.97	51.64	32.11	18
PCRU	3.13	−1.45	.99	28.94	20.72	18
PCRL	3.17	−1.47	.99	31.40	21.87	19
UPCTA	1.51	−0.01	.98	36.40	23.66	18
LPCTA	1.52	0.20	.97	48.04	30.57	19
All selenodonts						
HBL	3.13	−5.04	.93	46.79	32.12	70
M^3 length	3.12	0.94	.90	60.71	39.46	73
M^2 length	3.15	0.94	.89	64.24	41.25	73
M^1 length	3.13	1.17	.90	58.32	39.78	73
P^4 length	3.12	1.64	.88	67.45	42.71	73
P^3 length	3.02	1.72	.80	92.85	58.32	73
P^2 length	2.65	2.15	.64	144.56	93.50	70
M_3 length	3.19	0.51	.90	59.50	39.42	73
M_2 length	3.21	0.92	.90	60.30	39.14	73
M_1 length	3.20	1.17	.89	61.08	41.58	73
P_4 length	3.18	1.38	.84	81.24	52.93	73
P_3 length	2.92	1.79	.73	114.83	73.81	71
P_2 length	2.62	2.42	.68	134.01	82.85	68
M^3 width	2.87	1.51	.78	98.59	58.74	73
M^2 width	2.96	1.36	.80	92.47	57.05	73
M^1 width	3.06	1.31	.86	74.11	47.47	73
P^4 width	2.87	1.70	.84	79.90	50.61	73
P^3 width	2.28	2.49	.73	113.04	78.52	73
P^2 width	2.31	2.62	.67	135.65	89.86	70
M_3 width	2.93	1.89	.84	78.52	49.11	73
M_2 width	2.95	1.84	.85	75.78	46.21	73
M_1 width	3.02	1.86	.88	66.76	43.11	73
P_4 width	2.58	2.52	.84	79.61	52.61	73
P_3 width	2.53	2.72	.82	87.01	56.09	71
P_2 width	2.55	2.99	.78	101.89	66.68	68
M^3 area	1.59	1.01	.89	62.67	39.18	73
M^2 area	1.60	0.99	.88	65.86	42.47	73
M^1 area	1.60	1.13	.91	56.16	38.15	73
P^4 area	1.52	1.61	.87	68.22	43.00	73
M_3 area	1.58	1.11	.90	59.91	38.88	73
M_2 area	1.57	1.32	.89	61.29	39.06	73
M_1 area	1.58	1.47	.90	58.39	38.72	73
P_4 area	1.45	1.95	.86	74.64	49.96	73
M^{1-3} area	1.61	0.23	.90	59.02	37.94	73
M_{1-3} area	1.59	0.49	.91	56.64	36.52	73

Table 16.9. (*cont.*)

Variable	Slope	Intercept	r^2	%SEE	%PE	N
M^{1-2} area	1.60	0.55	.90	60.17	39.74	73
M_{1-2} area	1.58	0.90	.90	58.39	37.67	73
M^{1-3} length	3.19	−0.57	.91	55.84	35.86	73
M_{1-3} length	3.27	−0.80	.92	53.23	35.24	73
P^4–M^3 area	1.61	0.13	.90	58.96	37.56	73
P_4–M_3 area	1.58	0.39	.91	56.41	36.15	73
P^4–M^3 length	3.21	−0.88	.91	54.14	35.49	73
P_4–M_3 length	3.32	−1.20	.92	51.52	34.56	73
P^4–M^2 area	1.60	0.40	.90	60.06	39.19	73
P^4–M^2 length	3.08	−0.38	.89	62.01	39.93	73
P_4–M_2 area	1.57	0.74	.90	58.09	37.75	73
P_4–M_2 length	3.32	−0.58	.91	55.10	37.19	73
P^3–M^3 area	1.59	0.07	.90	59.35	37.58	73
P_3–M_3 area	1.58	0.32	.90	58.27	37.18	71
PCRU	3.32	−1.50	.92	52.86	34.36	73
PCRL	3.34	−1.59	.90	57.75	37.80	73
UPCTA	1.59	0.03	.89	61.98	38.52	73
LPCTA	1.57	0.31	.90	58.85	37.06	73
Selenodont browsers						
HBL	2.97	−4.55	.98	34.80	23.90	17
M^3 length	3.29	0.80	.93	70.62	44.70	18
M^2 length	3.34	0.73	.94	62.34	39.89	18
M^1 length	3.21	1.07	.96	49.74	35.13	18
P^4 length	3.05	1.60	.91	82.04	52.26	18
P^3 length	3.13	1.40	.86	108.85	63.28	18
P^2 length	3.26	1.31	.86	110.20	66.82	18
M_3 length	3.35	0.35	.93	70.76	47.24	18
M_2 length	3.41	0.72	.94	62.16	38.86	18
M_1 length	3.21	1.12	.94	64.83	46.48	18
P_4 length	3.11	1.30	.90	88.97	57.05	18
P_3 length	3.19	1.31	.84	123.27	80.14	18
P_2 length	2.75	2.06	.81	137.32	75.70	18
M^3 width	2.97	1.25	.89	93.27	53.61	18
M^2 width	3.08	1.08	.90	85.24	53.86	18
M^1 width	3.08	1.20	.91	84.38	57.68	18
P^4 width	2.85	1.62	.91	81.83	54.61	18
P^3 width	2.49	2.14	.88	101.54	66.95	18
P^2 width	2.44	2.32	.86	108.03	69.71	18
M_3 width	2.88	1.81	.89	92.48	58.98	18
M_2 width	2.92	1.76	.89	93.15	58.84	18
M_1 width	2.89	1.89	.92	77.39	50.40	18
P_4 width	2.47	2.48	.87	103.21	69.91	18
P_3 width	2.49	2.60	.90	89.56	56.63	18
P_2 width	2.42	2.90	.84	121.14	86.05	18

Table 16.9. (*cont.*)

Variable	Slope	Intercept	r^2	%SEE	%PE	*N*
M^3 area	1.59	0.98	.92	74.16	45.31	18
M^2 area	1.62	0.89	.93	70.22	44.51	18
M^1 area	1.58	1.11	.94	63.84	44.94	18
P^4 area	1.48	1.60	.91	79.73	53.36	18
M_3 area	1.56	1.11	.92	77.30	50.52	18
M_2 area	1.59	1.25	.92	73.39	47.19	18
M_1 area	1.53	1.51	.93	67.46	46.32	18
P_4 area	1.39	1.94	.89	92.39	62.88	18
M^{1-3} area	1.61	0.19	.93	66.59	42.95	18
M_{1-3} area	1.57	0.50	.93	69.87	45.88	18
M^{1-2} area	1.61	0.50	.94	66.08	43.58	18
M_{1-2} area	1.57	0.88	.93	68.62	45.96	18
M^{1-3} length	3.32	−0.76	.95	55.82	36.61	18
M_{1-3} length	3.38	−0.96	.95	57.58	38.66	18
P^4–M^3 area	1.59	0.12	.93	67.17	43.52	18
P_4–M_3 area	1.54	0.45	.93	71.75	47.15	18
M^4–M^3 length	3.28	−0.99	.95	58.18	38.09	18
P_4–M_3 length	3.34	−1.23	.94	61.52	41.16	18
P^4–M^2 area	1.58	0.39	.93	67.65	44.78	18
P^4–M^2 length	3.23	−0.58	.95	56.68	38.37	18
P_4–M_2 area	1.53	0.78	.92	72.71	49.09	18
P_4–M_2 length	3.29	−0.57	.94	64.98	45.11	18
P^3–M^3 area	1.57	0.07	.93	69.17	43.75	18
P_3–M_3 area	1.53	0.38	.92	73.17	47.47	18
PCRU	3.31	−1.54	.94	65.98	40.83	18
PCRL	3.32	−1.63	.93	70.12	43.80	18
UPCTA	1.56	0.02	.93	70.74	44.60	18
LPCTA	1.52	0.36	.92	74.81	48.22	18
Selenodont nonbrowsers						
HBL	3.32	−5.65	.89	49.65	33.79	53
M^3 length	3.07	0.98	.86	57.57	37.61	55
M^2 length	3.02	1.10	.83	65.34	41.11	55
M^1 length	3.04	1.29	.84	61.65	40.71	55
P^4 length	3.07	1.72	.85	59.77	37.97	55
P^3 length	2.83	1.97	.79	74.32	51.07	55
P^2 length	2.23	2.65	.56	128.10	82.73	52
M_3 length	3.15	0.56	.86	56.29	36.49	55
M_2 length	3.10	1.05	.85	60.07	38.77	55
M_1 length	3.13	1.27	.85	60.63	38.80	55
P_4 length	3.11	1.50	.80	72.38	47.25	55
P_3 length	2.64	2.15	.70	96.31	62.33	53
P_2 length	2.44	2.67	.64	112.82	74.14	50
M^3 width	2.67	1.79	.70	93.25	55.85	55
M^2 width	2.74	1.66	.72	89.00	54.45	55
M^1 width	2.93	1.49	.81	69.44	42.64	55
P^4 width	2.78	1.84	.78	77.07	47.83	55
P^3 width	2.03	2.79	.63	108.51	74.13	55
P^2 width	2.06	2.91	.53	134.40	86.41	52
M_3 width	2.85	2.00	.80	70.94	43.48	55
M_2 width	2.86	1.97	.82	67.27	42.33	55

Table 16.9. (*cont.*)

Variable	Slope	Intercept	r^2	%SEE	%PE	*N*
M_1 width	3.04	1.86	.84	62.41	39.76	55
P_4 width	2.58	2.56	.82	67.76	43.85	55
P_3 width	2.47	2.83	.78	78.79	51.28	53
P_2 width	2.57	3.04	.76	85.54	53.15	50
M^3 area	1.57	1.09	.85	59.74	37.10	55
M^2 area	1.54	1.14	.83	64.51	41.00	55
M^1 area	1.57	1.20	.87	54.09	35.01	55
P^4 area	1.51	1.67	.84	61.99	38.92	55
M_3 area	1.56	1.16	.87	55.21	35.15	55
M_2 area	1.53	1.43	.86	57.99	36.82	55
M_1 area	1.58	1.48	.87	55.50	34.68	55
P_4 area	1.46	1.99	.84	63.60	41.85	55
M^{1-3} area	1.58	0.33	.86	57.05	36.29	55
M_{1-3} area	1.58	0.54	.88	53.03	33.57	55
M^{1-2} area	1.57	0.67	.86	58.54	37.48	55
M_{1-2} area	1.56	0.97	.87	55.57	34.77	55
M^{1-3} length	3.12	−0.46	.86	56.60	36.07	55
M_{1-3} length	3.21	−0.71	.88	52.71	34.24	55
P^4–M^3 area	1.58	0.22	.86	56.44	35.30	55
P_4–M_3 area	1.58	0.41	.88	51.45	32.51	55
P^4–M^3 length	3.15	−0.78	.87	54.00	34.34	55
P_4–M_3 length	3.30	−1.15	.89	49.49	32.22	55
P^4–M^2 area	1.57	0.50	.86	57.44	35.91	55
P^4–M^2 length	3.01	−0.27	.83	64.19	40.48	55
P_4–M_2 area	1.57	0.76	.88	52.77	33.23	55
P_4–M_2 length	3.28	−0.51	.88	52.59	34.02	55
P^3–M^3 area	1.57	0.16	.87	55.74	34.69	55
P_3–M_3 area	1.58	0.35	.88	52.53	33.03	53
PCRU	3.25	−1.35	.89	48.38	32.12	55
PCRL	3.26	−1.42	.88	52.89	34.46	55
UPCTA	1.56	0.13	.86	58.05	35.56	55
LPCTA	1.57	0.34	.88	52.24	32.68	55

Note: Values in mm and g.
Abbreviations: HBL = head–body length; PCRU = upper postcanine tooth-row length; PCRL = lower postcanine tooth-row length; UPCTA = upper postcanine tooth-row area; LPCTA = lower postcanine tooth-row area.
Source: Damuth (this volume).

Table 16.10. *Maximal tooth dimensions, multiple regression equations*

All ungulates
log Mass = 2.01 (log HBL) + 1.20 (log M_{1-3} length) − 3.57
r^2 = .96 %SEE = 38.37 %PE = 26.19 N = 91

Nonselenodonts
log Mass = 2.35 (log HBL) + 0.79 (log M_{1-3} length) − 3.88
r^2 = .99 %SEE = 25.73 %PE = 18.18 N = 19

All selenodonts
log Mass = 1.93 (log HBL) + 1.32 (log M_{1-3} length) − 3.54
r^2 = .94 %SEE = 41.11 %PE = 27.57 N = 70

Selenodont browsers
log Mass = 2.39 (log HBL) + 0.69 (log M_{1-3} length) − 3.90
r^2 = .98 %SEE = 34.16 %PE = 22.25 N = 17

Selenodont nonbrowsers
log Mass = 1.87 (log HBL) + 1.57 (log PCRL) − 4.08
r^2 = .93 %SEE = 38.13 %PE = 27.08 N = 53

Note: values in mm and g.
Abbreviations: HBL = head–body length; PCRU = upper postcanine tooth-row length; PCRL = lower postcanine tooth-row length; UPCTA = upper postcanine tooth-row area; LPCTA = lower postcanine tooth-row area.
Source: Damuth (this volume).

Rodents: craniodental measures

Table 16.11. *Cricetine rodents*

Group	Measure	Slope	Intercept	r^2	%SEE	%PE
Cricetine rodents (33 species)	M_1 length	3.31	.611	.96	21.62	15.58

Note: Values in mm and g.
Source: Martin (this volume).

Macropodoid marsupials: craniodental measures

Table 16.12. *Cranial and occlusal dental measures in 53 species of macropodoids*

Measure	r^2	Intercept	Slope	%SEE	%PE	Dietary differences	Isometry
FLML	.746	1.749	3.562	73.0	45.3	N.E.	N.E.
FLMW	.663	2.356	3.590	88.4	50.4	N.E.	N.E.
FLMA	.737	2.109	1.869	75.0	44.5	N.E.	N.E.
SLML	.802	1.505	3.480	62.6	38.5	N.E.	N.E.
SLMW	.764	2.101	3.598	69.8	41.4	N.E.	N.E.
SLMA	.794	1.813	1.794	64.0	39.1	N.E.	N.E.
TLML	.871	1.295	3.239	47.9	31.3		–
TLMW	.815	2.016	3.700	60.0	36.9		–
TLMA	.868	1.654	1.773	48.6	32.3		–
FOLML	.829	1.226	2.730	57.0	38.7	N.E.	N.E.
FOLMW	.822	1.929	3.300	58.5	39.5	N.E.	N.E.
FOLMA	.852	1.568	1.541	52.1	35.0	N.E.	N.E.
TUML	.855	1.233	3.191	51.4	39.9	I>B*,O; G*>O	–
TUMW	.834	1.681	3.261	56.0	34.4		–
TUMA	.888	1.489	1.695	43.9	34.4	I*>O	–
LMRL	.905	−0.616	3.684	40.0	26.8		–
MZW	.566	0.407	2.647	108	60.4	O,B>I,G*	–
PAW	.919	−0.132	3.456	36.1	25.1		–
MFL	.944	−1.718	3.795	29.4	21.8		–
OCH	.912	−0.893	3.868	38.0	25.2	G*>B	–
PSL	.938	−1.333	3.458	30.9	20.3	O,B>I	–
BCL	.926	−1.204	3.421	34.6	26.6	B>I,O; B>G*	–
TSL	.935	−2.657	3.452	32.1	20.2		–
AJL	.860	−0.807	2.981	50.3	29.7	G*>I,B	–
PJL	.927	−0.538	2.985	34.3	23.0		X
DMA	.868	−0.241	2.800	48.6	32.6		X
WMA	.941	−0.606	3.023	30.3	21.0		X
TJL	.933	−2.296	3.294	32.7	22.2	G*>I	–

Note: Values in cm and kg. Measures defined in Table 13.2 and Figure 13.1. N.E. = not examined.
Source: Janis (this volume).

Index